REDUCTION IN SCIENCE

SYNTHESE LIBRARY

STUDIES IN EPISTEMOLOGY,

LOGIC, METHODOLOGY, AND PHILOSOPHY OF SCIENCE

Managing Editor:

JAAKKO HINTIKKA, *Florida State University, Tallahassee*

Editors:

DONALD DAVIDSON, *University of California, Berkeley*
GABRIEL NUCHELMANS, *University of Leyden*
WESLEY C. SALMON, *University of Pittsburgh*

VOLUME 175

REDUCTION IN SCIENCE

Structure, Examples, Philosophical Problems

Edited by

WOLFGANG BALZER
Universität München

DAVID A. PEARCE
Institut für Philosophie, Freie Universität Berlin

HEINZ-JÜRGEN SCHMIDT
Fachbereich Physik der Universität Osnabrück

D. REIDEL PUBLISHING COMPANY

A MEMBER OF THE KLUWER ACADEMIC PUBLISHERS GROUP

DORDRECHT / BOSTON / LANCASTER

Library of Congress Cataloging in Publication Data

Main entry under title:

Reduction in science.

(Synthese library; v. 175)

Papers presented at the colloquium "Reduktion in der Wissenschaft: Struktur, Beispiele, philosophische Probleme," held in Bielefeld, West Germany, July 18–21, 1983.

Bibliography: p.

Includes index.

1. Science–Theory reduction–Congresses. I. Balzer, W. (Wolfgang) II. Pearce, David A., 1952- III. Schmidt, H.-J. (Heinz-Jürgen), 1948-

Q175.R344 1984 501 84-13389

ISBN 90-277-1811-3

Published by D. Reidel Publishing Company,
P.O. Box 17, 3300 AA Dordrecht, Holland.

Sold and distributed in the U.S.A. and Canada
by Kluwer Academic Publishers,
190 Old Derby Street, Hingham, MA 02043, U.S.A.

In all other countries, sold and distributed
by Kluwer Academic Publishers Group,
P.O. Box 322, 3300 AH Dordrecht, Holland.

Printed in The Netherlands

CONTENTS

PREFACE

The papers in this volume were presented at the colloquium "Reduktion in der Wissenschaft: Struktur, Beispiele, philosophische Probleme", held in Bielefeld, West Germany, July 18-21, 1983. Altogether eighteen talks were delivered at the symposium, and all appear here with the exception of Professor Ehlers' address. In addition, we are pleased to be able to include three papers by invited participants (Kamlah, Ludwig, Scheibe) who were unable to attend the meeting.

The meeting itself brought together a sizeable group of logicians, philosophers and working scientists to discuss and debate the theme of reduction, one that occupies a central place in contemporary philosophy of science. The participants and contributors succeeded in opening up new directions in reduction studies and presenting fresh case studies of reduction from many different areas of scientific practice. Their efforts will greatly enhance our understanding of reduction and, consequently, our grasp of the complex process of scientific change and the unity and growth of scientific knowledge.

We are grateful to the organizers, the Deutsche Forschungsgemeinschaft, who financed the meeting, and to the Ministerium für Wissenschaft und Forschung des Landes Nordrhein-Westfalen, who generously provided additional financial support which enabled us to invite a number of guests from outside West Germany. We should also like to express our thanks to the Directors and Staff of the ZiF (Zentrum für interdisziplinäre Forschung), University of Bielefeld, who kindly offered their expert help and first-rate facilities for hosting four intensive days of lectures and discussions. We are also indebted to Frau A. Schmidt, Frau U. Trentmann, and Frau M. Wick, who assisted the editors by typing a number of the manuscripts which appear below.

THE EDITORS

INTRODUCTION

The theme of reduction bears broadly on the logic, the methodology and the epistemology of scientific change, on the rationality and continuity of scientific progress,and on the unity and growth of scientific knowledge. In short, reduction is established as a central and pressing issue in the philosophy of science, and a topic that exerts an active influence on the actual development of the sciences themselves.

Over the past twenty years or so the study of reduction has undergone a steady and marked development. The classical, empiricist rendering of reduction, associated with the work of Nagel, Hempel, Kemeny and Oppenheim, has given way to an array of varied and sometimes conflicting approaches to the subject. Some have refined and extended the standard, deductive account of reduction; others have branched out in new directions, forged fresh insights and channelled new methods and techniques into the study of relations between scientific theories. Many of the familiar historical examples of reduction have been reanalysed and reevaluated along the way; and a host of new case studies, from all branches of science, have come to light.

The aim of the present volume is to try to do some justice to the current concern for reduction in the philosophy of science and in the special sciences like physics and economics. The volume consists of a collection of new papers reflecting a wide range of different approaches to the explication and analysis of reduction. They tackle a broad variety of the major historical examples of reduction, and they address, either directly or indirectly, many of the key philosophical problems that have come to be associated with reduction and scientific change generally.

It seems clear that much of the impetus for the recent revival in reduction studies can be traced to the prolonged and penetrating criticism that the classical concept of reduction underwent at the beginning of the 1960's. the widespread rejection of the received view of theories, spearheaded by Toulmin, Hanson, Kuhn, Feyerabend and others, carried in its wake a mounting suspicion that the standard, deductive account of intertheoretic reduction was no longer tenable. But although many philosophers were willing to concede consider-

W. Balzer et al. (eds.), Reduction in Science, 1–9.

able structural inadequacies in the received view of theories, few were prepared to entertain a wholesale rejection of the idea that scientific progress is in large measure a continuous and cumulative affair, as some of the critics apparently demanded. Thus, largely in response to this criticism, the concept of reduction was subsequently revised and reformulated, either with a view to consolidating traditional ideas, or with the aim of constructing alternative and more adequate accounts.

Looking back over this period of active development in reduction studies, we can try to single out some of the issues that have come in for special attention and some of the areas where definite signs of philosophical progress are evident.

As far as the analysis of the concept of reduction is concerned, two related themes have emerged that no adequate theory of reduction can afford to ignore. The issues are really two sides of the same coin: one is the problem of conceptual disparity, the other is the question of limits and approximations. Whilst it is fine in principle to say that one scientific theory is reduced to another when its laws have been deduced from it, in practice it seems that very few examples of reduction exhibit a straightforward deductive pattern. Either the languages of the two theories are syntactically different, or perhaps semantically different (which leads to the problem of meaning-variance and possible incommensurability); or the connections established between the theories' laws hold at best only approximately or under some suitable limiting, and often counterfactual, assumptions. In that case it seems one must either modify the claim that a reduction relation obtains, or else try to characterise the relation within an alternative model of reduction.

One way of handling conceptual disparity between theories is of course through the method of translation. Another, and in some sense parallel, approach is to turn away from specifically linguistic features of a scientific theory and focus instead on the theory's models or, more generally, on the mathematical structures that represent fragments of the real or idealised situations that the theory describes. One may then hope to re-express the reduction relation as a structural or model-theoretic link. Both these approaches are explored at length in the present volume.

The problem of approximation has occupied increasing attention in the recent philosophy of science. There are now several well-established methods for describing the imprecise nature of the relation of a theory to its confirming data or its domain of application, or the limiting and approximate character of many intertheoretic relations. Here again the structural or model-theoretic perspective has proved to be highly successful. In this volume both the method of topological uniformites (and generalisations thereof) and the method of nonstandard analysis are proposed as solutions to this problem.

A notable feature of much recent work on reduction is the increased use of sophisticated logical and mathematical concepts, methods and results. It has emerged that many of the key problems connected with reduction and theory change in science cannot be adequately resolved in a simple-minded fashion. One often requires advanced conceptual tools in order to effect a precise reconstruction of the theories concerned and to arrive at a satisfactory solution to the problem at hand. Thus the tools of category theory, abstract logic, topology, differential geometry, nonstandard analysis, and so forth, all find application here to reduction-related problems. One pleasant consequence of the use of such advanced methods is that the level of philosophical discussion surrounding reduction is often raised accordingly.

Turning to the sphere of applications and historical examples, recent work on reduction has made important progress in several key, and formerly recalcitrant, areas of physics. The relations between classical and relativistic theories have now been explored at all major levels: space-time geometry, kinematics, mechanics and electrodynamics. The relations between classical thermodynamics and statistical theories in physics have also been the subject of much attention, as well as the case of classical versus quantum mechanics. Virtually all these examples are treated in depth in the present volume, besides several other reduction cases that are studied here from new perspectives or with new methods. Looking farther afield, we also find here examples from chemistry, economics and sociology that are treated with a degree of rigour that not long ago would have been the exclusive reserve of mathematics and physics.

Of the different basic approaches to the systematic explication and study of reduction, three are particularly well-represented in this volume. First, there is the structuralist view (Balzer, Moulines, Sneed) which has evolved significantly since the first appearance of Sneed's The Logical Structure of Mathematical Physics, in 1971. This account is characterised in general by the use of informally stated set-theoretical predicates - "informal" in the sense that no formal language is employed. On the one hand this allows for precise reconstructions of intertheory relations, on the other hand practically no effort has to be spent on peculiar features of the metatheoretical apparatus, such as properties of the language concerned, the structure of models in general as imposed by the particular metatheory, or foundational aspects of the mathematical theories involved or of set theory itself. This approach therefore permits one to advance directly to make explicit and precise just those features of reduction that are relevant for the example or type of example in question.

A second, formal approach to reduction, originating from the work of G. Ludwig, has influenced a group of people active in the foundations of physics. It is represented here in the papers by Ludwig, Mayr, Schmidt and Werner. Its formal framework is borrowed from axiomatic set theory, especially from Bourbaki's theory of species of structures: thus it takes a somewhat intermediate position between the structuralist account and more traditional, linguistic (or "statement") views of theories. Its typical features are the explicit reference to observational data, its use of an "interpretative hierachy" of theories, and its application of uniform structures to account for empirical imprecision and intertheoretic approximation. Being on the whole rather prudent in its reconstruction of "physical reality", it tends to the view that a large body of physics is conserved during episodes of theory change.

Both of these approaches to reduction have become well-established in recent years, though regrettably the Ludwig school still seems to be little known outside Germany. A third, and more recent, approach which also combines a general account of the structure of theories with a systematic explication of reduction is due to Pearce and Rantala; and also represented here in the paper of Pearce and Tucci. Theirs may be termed a "general model-theoretic" approach. It analyses intertheory relations (reduction, correspondence, etc.) both syntactically and semantically within a rather general model-theoretic framework inspired by abstract logic. The chief point of departure from the traditional use of logic is that theory reconstruction is not effected within a rigid and fixed logical format, but in applications the relevant logic may be allowed to vary according to context. An illustration is afforded by their use of nonstandard analysis to characterise limiting relations between theories. Here, first order logic serves to "justify" the relevant nonstandard entities, and a stronger, infinitary logic defines the requisite proof- and model-theoretic mappings which establish syntactic and semantic connections between the theories. Besides this quite specific and specialised use of logic in reduction studies, logical methods and results - traditional and less so - are put to novel use by a number of other contributors here, for example in the papers by Gaifman, Hoering and Vollmer.

Turning to the overall scope of the volume, we can see it as contributing to four separate but related goals. First, it seeks to clarify the basic concepts of reduction and approximate reduction. The dominant feature is perhaps the use of advanced mathematical methods, and this use is extended to and complemented by the discussion of specific examples. The examples presented may serve as standards against which to check the general explications. The matching of example to explication should lead to a further sharpening and clarification of the general concepts.

Secondly, we hope the volume will also provoke some interest in precisely delineated concepts of reduction among scientists themselves. Though most of the actual case studies here are related to physics, other areas of science should be able to profit from the examples and concepts, which is not to imply that they should take them over uncritically. The examples from chemistry, economics and sociology, included here, will hopefully assist in furthering the study of reduction in those areas and in positively influencing the discussion of progress and theory appraisal in the non-physical sciences generally.

Thirdly, gathering exact studies of reduction from different scientific fields serves another useful purpose. It helps to counterbalance a widespread attitude among philosophers of science of approaching questions of reduction in "a priori" fashion, and merely hinting at examples and applications. The contributions below give us ample evidence that fully-fledged examples of relations between real scientific theories can be precisely and fruitfully reconstructed.

Fourth, but not least, the volume provides material for comparing and evaluating the major formal approaches to reduction, especially the three different groups mentioned above. Analysing the respective merits of each group - the pros and cons of each - may be expected to lead to further progress in the study of intertheory relations.

The various contributions below are not placed according to similarity of approach or outlook, but are loosely arranged according to subject matter. The first and largest group of papers deal with general logical and philosophical aspects of the concept of reduction. Krajewski compares the concepts of reduction and intertheoretic explanation. By means of some informal examples from physics he shows that not all cases of reduction amount to the explanation of the reduced theory by the reducing theory. In conclusion he suggests that an explanation is achieved roughly if the reducing theory is more fundamental and constitutes a unification of the reduced theory with other laws or theories, but not when it is, in his terminology, a"factualization"of the reduced theory. Ludwig presents a general schema of intertheory relations within his own approach to the structure of physical theories. He proposes a novel variant of "restriction with respect to a hypothesis", and makes some remarks on theory nets, pretheories and limiting cases. Hoering discusses the adequacy of some formal explications of reduction. He observes that some purely syntactical and some purely structural criteria of reducibility may admit large classes of unintended reductions. He suggests that such "anomalies" of reduction might be averted if certain further, non-syntactical criteria of adequacy were to be met. In particular, these additional requirements for reduction should refer explicitly to the domains of application of the theories in question.

Ontological features of reduction are dealt with in Moulines' paper. His aim is to enlarge the structuralist account of reduction by including an explicit description of how the "objects" of different theories may be related in reduction contexts. He characterises the main types of reduction, including the homogeneous and heterogeneous cases, with an exact reconstruction of each. The structuralist concept is also further developed in Sneed's paper. He reformulates his original reduction concept in a new metatheoretical setting that employs links and categories. The new framework provides an analysis of various types of intertheory relations, and problems of invariance are also reconstructed within it. Scheibe explores certain progressive and conservative aspects of theory change. Starting with an historical survey of the views of some leading physicists, he goes on to develop a reconstructionist programme which includes a precise concept of "partial approximative theory explanation", and he discusses some problems connected with "empirical explanation". Vollmer considers the concept of reduction and the growth of science from the standpoint of evolutionary epistemology. He surveys the development of entire fields of science alongside the evolution of "real systems" in a hierarchical order - from physical systems to social systems. He argues that whilst the pattern of reduction among scientific theories is bound to be approximative, the evolution of real systems shows that it should be possible to achieve exact reductions, and hence explanations, of complex systems by means of simpler systems.

The second group of papers is explicitly concerned with approximate reduction and the analysis of limiting relations between physical theories. Pearce and Rantala give a general account and defence of their explication of limiting case correspondence using the method of nonstandard analysis. They point out some of the advantages of their approach and illustrate it with a model-theoretic reconstruction of the correspondence of Kepler's laws of planetary motion to Newton's celestial mechanics. Mayr generalises the mathematical concept of Banach manifolds to "pregeodesic contact spaces". He recommends using such concepts, which arise in the axiomatic analysis of general relativistic spacetime, in order to achieve a better grasp of "higher order" approximations between theories. A first attempt in this direction is undertaken by Schmidt, who proposes "tangent embeddings" between structures as a new method for handling approximate reduction relations. The "degree" of approximation may be thereby made precise, and some physical examples, including the case of classical vs. relativistic spacetime, are examined within the framework of the Ludwig approach.

We come next to a series of papers dealing with examples of reduction from outside the narrowly physical sphere. Kamlah's subject is chemistry, and the comparison of the phlogiston theory with the theory of chemical compounds that supplanted it. He proposes a new reconstruction of these theories which reveals their close formal similarity and gives an exact account of the transition from one to the other. He also considers the implications of this study for the current debate over scientific realism. Olson's sociological law concerning collective action relates the decreasing chance of realising a collective good with the increasing size of the group involved. Kuipers' aim in his paper is to reconstruct Olson's law and show how it can be "micro-reduced" to certain assumptions concerning individual behaviour. The pattern of this reduction is shown to be similar to that of the relation of the ideal gas law to the kinetic theory of gases. In economics the names of Sraffa and Wicksell are associated with two conflicting and competing theories of growth, one belonging to the classical and the other to the neoclassical research tradition. However, in their paper Pearce and Tucci show that, from a formal standpoint, the two theories are remarkably similar: using a model-theoretic framework, they are able to define a reduction of a Wicksellian type of theory to the system of

Sraffa; and they discuss some implications of this result for economic methodology. The problem of reduction in economics is also pursued by Hamminga. Beginning with a discussion of the notion of reduction as it usually arises in the philosophy of science, he proposes three new reduction concepts of increasing strength that are tailored to the special structure of economic theories. He draws illustrations from the domain of international trade theories.

A further group of papers is devoted to various aspects of relations between space-time theories. Gaifman's chief concern is with ontological commitment. He argues that ontological differences between scientific frameworks are best elucidated by a linguistic analysis along the lines of: "What are the objectively true-or-false statements of the language ?" His main illustration of the method consists of a detailed comparison of the different ontologies admitted by Newton's "absolute" and Leibniz's "relational" theories of space-time, using a suitable logical reconstruction. He concludes with some general requirements to which translations between scientific languages should conform. In his paper, Balzer proposes a formal reconstruction of a version of classical space-time as well as relativistic space-time along Reichenbachian lines. He establishes a strict relation between these theories of a "reductional kind"; and he goes on to discuss the senses in which it may be considered a genuine reduction. Malhas provides new axiom systems for classical and relativistic space-time for one spatial dimension, based on concepts from affine geometry. His choice of primitives is designed to clarify the comparison of these theories by exhibiting an underlying "observational" structure that is common to each; and he concludes with some general remarks about how an approximate reduction of the classical to the relativistic theory could be performed.

The final group of papers deals with the important issue of how classical and quantum mechanics are logically related. Lahti considers the pair (L,M) of properties and states as a common structure belonging to both classical and quantum theory. In this frame he discusses the role of compatibility, unique decomposability and the projection postulate, and he concludes that the two theories are, respectively, counterfactual ($h=0$) and factual ($h>0$) reductions of an underlying generalised theory. Stachow looks at the problem of reduction between classical and quantum physics at the level of object language, its logic and semantics. In the famework of P. Mittelstaedt's approach to quantum dialog-semantics, he shows how classical and intuitionistic languages can be reduced to an abstract quantum language if certain ontological premises of decidability and objectivity are weakened. The reverse problem, viz. the possible reduction of quantum theory to a hypothetical classical theory of "hidden variables", is taken up by Werner in the paper which rounds off the volume. He analyses the EPR-experiment in terms of statistical dualities and presents a new derivation of Bell's inequalities based on the "principle of directed action".

In arranging the symposium from which these papers are drawn, our original aim was to gather together a limited sample of cases and to provoke discussion and comparison of some of the latest concepts and methods applied in reduction studies. However, the depth and scope of the contributions collected here indicates that this aim was exceeded by an ample margin. Naturally, we cannot pretend that all the important questions have found definitive answers. Many solutions proposed here are tentative; some of the examples treated are exploratory rather than conclusive. But the progress achieved strikes us as welcome and highly encouraging. The signs are that the wealth of ideas and methods developed here points us on the right track for future research and provides a solid basis on which to build. To this extent we feel confident in the claim that the volume as a whole marks something of a turning point and a breakthrough in the systematic study of intertheory relations.

Wolfgang Balzer
David Pearce
Heinz-Jürgen Schmidt

Władysław Krajewski

MAY WE IDENTIFY REDUCTION AND EXPLANATION OF THEORIES?

At the first sight the answer to this question should be positive. We say that a theory T_1, is reduced to a theory T_2 when T_1 may be deduced from T_2. Analogously, we say that T_1 is explained by T_2 when T_1 may be deduced from T_2. Usually some additional assumptions AA are needed - both for reduction and for explanation. Hence, the common pattern is as follows:

$$T_2 \wedge AA \rightarrow T_1 \qquad \text{(I)}$$

Some remarks about AA. When T_1 is a theory of a system and T_2 a theory of its elements, AA must contain the description of the structure of the system (relations among its elements). When T_1 and T_2 use different languages (heterogeneous reduction), AA must contain rules of translation (bridge laws). However, we shall here consider another type of reduction (and explanation): when T_1 is valid only for some extreme and unreal case. Then AA must contain limit conditions (idealizing assumptions). Some factors which are present in all real systems vanish in an ideal system which is the strict model of T_1 (ideal model). We express this assuming that some parameters p_i take on extreme values, usually zero: $p_1=0$. Hence, the pattern is, as follows:

$$T_2 \wedge p_i=0 \rightarrow T_1 \qquad \text{(II)}$$

This is again the common pattern of both reduction and explanation. This does not mean, however, that these two concepts are identical. In order to show the difference, we shall consider four simple examples taken from classical physics.[1] The first two examples are chosen from classical mechanics CM.

As is well known, Galileo's free fall law (GL) may be deduced from Newtonian CM (including the law of general gravitation). In fact, as Duhem and others noticed, the situation is not so simple as usually presented in handbooks of physics. Strictly speaking, CM and GL are incompatible: GL assumes

W. Balzer et al. (eds.), Reduction in Science, 11–15.

the constancy of acceleration $\underline{g}$ during the fall, whereas CM entails the increase of $\underline{g}$ due to the decrease of the distance R between the falling body and the centre of the Earth. Therefore, we say sometimes that only an approximation of GL may be inferred from CM. When, however, we want to obtain a strict inference, we must assume the ideal limit case when the Earth's radius R is infinite, hence the difference ΔR between the distance to the Earth's centre at the beginning and at the end of the fall vanishes: $\Delta R/R=0$. Besides, we assume, of course, that there is no air resistance, i.e. the resistance force $F_r=0$, which is also a counterfactual assumption (idealization). Hence:

$$CM \wedge F_r=0 \wedge \frac{\Delta R}{R}=0 \rightarrow GL \qquad (1)$$

We see that the reasoning follows pattern (II).

Another classical example of reduction is the entailment of Kepler's laws (KL) by CM. The situation is analogous. In fact, the orbit of a planet is never strictly elliptical because it is disturbed by the gravity of other planets (the velocity of planets is disturbed, as well). Therefore, KL are only approximately valid. A strict inference of KL from CM is possible only in an ideal system in which only one planet exists or in which forces exerted by other planets (and all other forces) vanish. If we designate them by F_{ext} (external forces) we may write:

$$CM \wedge F_{ext}=0 \rightarrow KL \qquad (2)$$

The pattern is identical.

We pass to simple examples from XIX century physics. Ideal gas laws, e.g. Boyle-Mariotte's law (BM), may be reduced to van der Waals' law (vW) assuming that $a=b=0$, although in all real gases $a>0$ and $b>0$. Hence

$$vW \wedge a=b=0 \rightarrow BM \qquad (3)$$

The last example. Ohm's (OL) may be reduced to the differential law of current (DL), which takes into account the impact of self-induction and electrical capacity, when we assume that the coefficient of self-induction $L=0$ and the electrical capacity C is infinite. Both conditions are strictly valid only in an "ideal circuit". Hence:

$$DL \wedge L=0 \wedge \frac{1}{C}=0 \rightarrow OL \qquad (4)$$

The pattern is identical, again.

In all four cases the reduced law or theory T_1 is valid only in an ideal system in which $p_i=0$ although in all real systems $p_i>0$. The limit condition $p_i=0$ is always a counterfactual or idealizing assumption. The reducing law or theory T_2 is valid both for the case $p_i=0$ and the case $p_i>0$. We may add that in all four cases T_1 was initially held to be a real (factual) theory. Only after the creating of T_2 the idealizational nature of T_1 was revealed.

Nevertheless, there is a deep difference between the first two and the last two examples. Undoubtedly, CM explains both GL and KL. However, we do not say that vW explains BM or that DL explains OL. We should rather say that in order to explain vW we must first consider BM and then take into account additional factors (volume of molecules and inter-molecular forces). Analogously, we explain DL considering OL and taking into account additional factors (self-induction and electrical capacity). There is no analogous way to proceed from GL or KL to CM.

In all cases T_2 is more general than T_1, however, in different respects. In cases (3) and (4) T_2 is more general than T_1 simply because T_1 is valid for the special case $p_i=0$. They describe the same dependence, e.g. the dependence between the same magnitudes (though T_2 contains additional magnitudes). In cases (1) and (2) the situation is different. We can form a "generalized" GL which takes into account the resistance force and the increase of g during the fall, or "generalized" KL which take into account the disturbance of orbits by external forces. They will contain additional terms, like vW and DL in the last examples, and be more general than GL and KL. However, CM is more general in a deeper sense. It contains more fundamental magnitudes, it describes more fundamental dependences.[2] CM may be applied to various phenomena, like free fall, planetary motion and many others. It is unification of many laws. Analogously BM and OL may be explained by more fundamental theories, the former by statistical mechanics (kinetic theory), the latter by electron theory, using, of course, some additional assumptions. These theories use more general and basic concepts. They have a large family of models.

In cases (3) and (4) we have a reduction of an idealizational law to its factualization: vW is a factualization of BM and DL is a factualization of OL. Factual laws do not explain idealizational laws. Rather, on the contrary, we explain a more complex factual law by a more simple idealizational law and disturbing factors. In cases (1) and (2) we have no factual laws. Of course, we use CM to explain both ideal free fall and real fall, both ideal planetary orbit and real orbit. Nevertheless, CM itself is an idealizational theory: it uses ideal models of material points, inertial systems, etc. The relation between CM and GL (or KL) is not a relation of factualization. CM is a much more general and more fundamental theory than GL and KL, though it is also idealizational one.

We conclude that the explanation of theories cannot be identified with reduction. The reduction of T_1 to T_2 is at the same time an explanation of T_1 by T_2 when T_2 is more fundamental than T_1, when it is a unification of T_1 and other theories (laws). The reduction of T_1 to T_2 is not an explanation of T_1 when T_2 is a factualization of T_1, though in this case T_2 is also more general than T_1. Of course, this is still not a satisfactory criterion. It is desirable to give a strict formal criterion of this difference and to construct a formal definition of explanation.[3)]

NOTES

1) A problem arises about the concept of explanation used in this paper; the problem would disappear if we defined explanation by means of reduction. We shall not employ here a formal definition but use an intuitive concept of explanation instead. The four examples below thus show the intuitive difference between explanation and reduction.

2) Additionally, CM is not a single law but a theory containing several laws.

3) I thank Theo Kuipers for his remarks during the discussion on my paper at the Bielefeld Symposium, July 18, 1983. I thank also Ewa Chmielecka, Małgorzata

Czarnocka, and Witold Strawiński for their remarks during the discussion of my paper at the University of Warsaw, October 7, 1983.

Department of Philosophy
University Warsaw

Günther Ludwig

RESTRICTION AND EMBEDDING

We maintain that complicated relations between physical theories ("*PT*" s for short) can be built up from two basic relations, namely restrictions and embeddings. In this paper we shall describe these two relations more precisely and indicate how they may be applied to analyze complex situations.

1. THE AXIOMATIC BASIS OF A *PT*

In the sequel we shall take *PT*s to be given in the following form, which is described in more detail in Ludwig (1978, 1981, 1983 and 1984): *PT* is composed of a mathematical theory *MT*, correspondence rules[1] (—) and a domain of reality *W*. The correspondence rules are prescriptions how to translate facts into *MT*, which can be detected in the fundamental domain *G* (a part of *W*). It is important that the detection of facts in *G* does not make use of the theory under consideration. That does not mean that we use no theory at all. But we may use only "pretheories" to describe the fundamental domain *G*, that is theories already established before interpreting the theory *PT* under consideration.

Clearly, the fundamental domain *G* has to be restricted to those facts which may be translated into the language of *MT*. However, further restrictions of the fundamental domain *G* are frequently necessary. Such restrictions can be formulated as "normative" axioms in *MT*. All those facts are to be eliminated from the fundamental domain, which, when translated by the correspondence rules, contradict the normative axioms. An example will be given at the end of section 4. The correspondence rules have the following form:

Some facts in *G* are designated with signs, say letters $a_1, a_2 \ldots$. In *MT* some sets are singled out as pictorial sets $E_1, E_2 \ldots E_r$ and some relations $R_1, R_2, \ldots$ as pictorial relations. By virtue of the correspondence rules the facts of *G* are translated into an additional text in *MT* "observational report"), which is of the form

$$(\text{—})_r(1):\ a_1 \in E_{i_1}, a_2 \in E_{i_2}, \ldots$$

W. Balzer et al. (eds.), Reduction in Science, 17–31.

$$(\text{—})_r(2):\ R_{\mu_1}(a_{i_1},a_{k_1},\ldots\alpha_{\mu_1}),R_{\mu_2}(a_{i_2},a_{k_2},\ldots\alpha_{\mu_2})\ldots;$$

$$[\text{not } R_{\nu_1}(a_{u_1},\ldots)];\ldots\ ,$$

where α_μ are real numbers (they may be absent in some of the relations R_μ).

The distinction between $(\text{—})_r(1)$ and $(\text{—})_r(2)$ has no fundamental significance. Instead of $(\text{—})_r(1)$ one may introduce relations $T_i(x)$ which are equivalent to $x \in E_i$. $(\text{—})_r(1)$ then takes the same form

$$T_{i_1}(a_1);\ T_{i_2}(a_2);\ \ldots$$

as $(\text{—})_r(2)$. It is, however, essential that for every sign a_l appearing in $(\text{—})_r(2)$ there is a formula $a_l \in E_{i_l}$ in $(\text{—})_r(1)$.

The pictorial relations R_μ of *PT* can be represented by subsets

$$r_\mu \subset E_{i_1} \times E_{i_2} \times \ldots \times \mathbb{R} \tag{1.1}$$

- that is, by the set of all those $(y_1,y_2,\ldots,\alpha)$ for which $R_\mu(y_1,y_2,\ldots,\alpha)$ is valid. (In this and all analogous formulas the factor $\mathbb{R}$ may be absent.)For the set on the right hand side of (1.1) we shall write $S_\mu(E_1,E_2,\ldots,\mathbb{R})$. Then for $U = (r_1,r_2,\ldots)$ we have

$$U \in S(E_1,\ldots,E_r,\mathbb{R}) \tag{1.2}$$

with

$$S(E_1,\ldots,\mathbb{R}) = PS_1(E_1,\ldots,\mathbb{R}) \times PS_2(E_1,\ldots,\mathbb{R})\ldots \tag{1.3}$$

(P denotes "power set of").*MTA* will denote the mathematical theory *MT*, complemented by the observational report $(\text{—})_r(1)$ and $(\text{—})_r(2)$. We say, that the theory *PT* is not in contradiction with experience if we have no contradiction in *MTA*.

Every mathematical theory as a part of a *PT* has the form called by Bourbaki (1968) a "theory of the species of structure Σ" which we will denote by MT_Σ. *MT* contains only set theory together with the theories of real and complex numbers, $\mathbb{R}$ and $\mathbb{C}$. (Bourbaki considers also arbitrary theories stronger than the theory of sets). The additional ingredient in MT_Σ is the structure Σ which is given by:

(1) a collection of letters $x_1,\dots,x_n$ called principal base sets,
(2) the terms $\mathbb{R}$ and (sometimes also) $\mathbb{C}$ as auxiliary base sets,
(3) a typification $s \in S(X_1,\dots,x_n,\mathbb{R},\mathbb{C})$ where S is an echelon construction scheme, and s is called the structural term,
(4) a transportable formula $P(x_1,\dots,x_n,s)$ called the axiom of the species of structure Σ.

We assume that the pictorial sets E_i are intrinsic terms (cf. Bourbaki 1968) $E_i(x_1,\dots,x_n,s)$.

In order to develop the concept of an "axiomatic basis" we now consider two theories *PT'* and *PT* which are related as follows:

The mathematical theories $MT_{\Sigma'}$ and MT_Σ have the form as described above, only the specifications of $MT_{\Sigma'}$ are given by primed letters. Let $U,E_1,\dots,E_n$ be (intrinsic) terms in $MT_{\Sigma'}$ such that

"$U \in S(E_1,\dots,E_n,\mathbb{R})$ and $P(E_1,\dots,E_n,U)$"

is a theorem in $MT_{\Sigma'}$. In this case U will be called a derived structure of species Σ in $MT_{\Sigma'}$ (cf. Bourbaki (1968), IV §1.6). We shall say, that U is a representation of Σ in Σ' if, in addition, every Σ-structure s is isomorphic to some derived structure U, or more explicitely, if

$$\begin{aligned}&(\exists x'_1)\dots(\exists x'_{n'})(\exists s')(\exists f_1)\dots(\exists f_n)\{s'\in S'(x'_1,\dots,x'_{n'},\dots)\\ &\text{and } P'(x'_1,\dots,x'_{n'},s') \text{ and all } f_i \text{ are bijective}\\ &\text{mappings } f_i: x_i \to E_i(x'_1,\dots,x'_{n'},s') \text{ such that}\\ &\langle f_1,\dots,f_n,1\rangle^{S'} s = U(x'_1,\dots,x'_{n'},s')\}\end{aligned} \tag{1.5}$$

is a theorem in MT_Σ.

For the relation between *PT'* and *PT* to be defined we shall require two things:

(i) The pictorial sets E_i of *PT'* and the term U, which comprises the pictorial relations (cf. (1.1) to (1.3)), define a representation of Σ in Σ'.

(ii) If in *PT'* the observational report $(-)_r(1),(2)$ is of the form given above, in *PT* it will be of the form

$(-)_r(1)$: $a_1 \in x_{i_1}$, $a_2 \in x_{i_2}, \ldots$

$(-)_r(2)$: $(a_{i_1}, a_{k_1}, \ldots, \alpha_{\mu_1}) \in s_{\mu_1}$,

$(a_{i_2}, a_{k_2}, \ldots, \alpha_{\mu_2}) \in s_{\mu_2}, \ldots,$

where s_μ are the components of $s = (s_1, s_2, \ldots,)$.

The two theories *PT* and *PT'* are then physically equivalent. We shall say that MT_Σ is an axiomatic basis (of the first degree) of *PT'* and that *PT* is an axiomatic basis of *PT'*. In the axiomatic basis the pictorial sets are identical to the principal base sets $x_1, \ldots, x_n$ of Σ and the pictorial relations are identical to the components of the structural term s of Σ.

In this sketch of the concept of an axiomatic basis we have omitted the complications caused by taking into account the imprecision of the correspondence rules. These only give rise to mild alterations.

It is not difficult to solve the problem of finding an axiomatic basis *PT* for a given theory *PT'*: One only has to take the statement (1.5) as the axiom $P(x_1, \ldots, x_n, s)$ of Σ. There is considerable physical and philosophical interest in finding an axiom of greater physical content than simply (1.5). But we need not go into details (cf. Ludwig (1981) and the last chapter of Ludwig (1984-85)), because the form of $P(\ldots)$ is irrelevant for explaining the concept of restriction and embedding.

2. RESTRICTIONS

We start with a theory PT_1 and wish to construct another theory *PT* which makes less restrictive assertions about reality. The theory PT_1 is assumed to be given in the form of an axiomatic basis (of the first degree). Let the species of structures Σ_1 of the mathematical part MT_{Σ_1} of PT_1 be characterized by the base sets (which are the pictorial sets) $x_1, \ldots, x_n$ and the structural term s (the components s of s being the pictorial relations). Further, consider in MT_{Σ_1} intrinsic terms $E_1, \ldots, E_g$ of the following type:

(α) E_ν is a subset of a product set $x_{\nu_1} \times x_{\nu_2} \times \ldots x_{\nu_p}$, or

(β) E_ν is the range of a mapping f from a set F of type (α) into an echelon set over $x_1,\ldots,x_n$

Clearly, (α) may be considered as a special case of (β) if f is taken to be the identity map.

As the mathematical part of the new theory *PT* we simply take $MT = MT_\Sigma$, the same as for PT_1. (Of course, this need not render *PT* an axiomatic basis.) As pictorial sets for *PT* we single out the sets $E_1,\ldots,E_g$. To obtain pictorial relations we seek intrinsic terms $u_\mu(x_1,\ldots,x_n,s)$, $\mu = 1\ldots m$, which have an appropriate typification over the pictorial sets:

$$u_\mu(x_1,\ldots,x_n,s) \subset E_{\mu_1} \times E_{\mu_2} \times \ldots \times \mathbb{R}. \qquad (2.1)$$

These subsets represent the new pictorial relations as above in (1.1).

The last step of defining *PT* consists of transforming the correspondence rules of PT_1 into the new ones of *PT*, without changing the fundamental domain G_1. Thus we have to translate the observational report on a special situation occuring in G_1, $(\text{---})_r(1)$, $(\text{---})_r(2)$, into a new text employing the new pictorial sets and relations. In order to give the most general definition we have to allow for the case that the observational report does not refer to a real situation but is of a hypothetical kind. In this event it is called a <u>hypothesis</u> H_1 (of the first kind, see my (1978)). The form of H_1 is the same as of a "real" observational report, say:

$$a_1 \in x_{\nu_1}, \; a_2 \in x_{\nu_2}, \ldots \qquad (2.2)$$

and

$$(a_{i_1}, a_{i_2}, \ldots) \in s_{\mu_1}; \ldots \qquad (2.3)$$

The mathematical theory MT_{Σ_1} extended by the letters a_i and the "axiom" H_1 will be denoted by $MT_{\Sigma_1} H_1$. We thus have to define a theory $MT_{\Sigma_1} H$ from $MT_{\Sigma_1} H_1$. Let E_ν be a term of type (α). If

$$(a_{i_1}, \ldots, a_{i_p}) \in E_\nu \qquad (2.4)$$

is a theorem in $MT_{\Sigma_1} H_1$, (that is, if $H_1 \Rightarrow$ (2.4) is a theorem in MT_{Σ_1}), we introduce a new sign b_1 as an abbreviation for the p-tuple $(a_{i_1},\ldots,a_{i_p})$:

$$(a_{i_1},\ldots,a_{i_p}) = b_1. \tag{2.5}$$

(2.4) and (2.5) imply

$$b_1 \in E_\nu. \tag{2.6}$$

b_1 is called a new pictorial element of the pictorial term E_ν^1. Now let E_ν be of type (β), say $E_\nu^\nu = f[F]$, $F \subset x_{\nu_1} \times x_{\nu_2} \times \ldots x_{\nu_p}$. We are looking for sets of p-tuples $(a_{i_1},\ldots,a_{i_p}^{2})$ which are in F

$$(a_{i_1},\ldots,a_{i_p}) \in F \tag{2.7}$$

and are mapped under f onto the same element b_k:

$$f(a_{i_1},\ldots,a_{i_p}) = f(a_{j_1},\ldots,a_{j_p}) = \ldots = b_k. \tag{2.8}$$

In this event we introduce the sign b_k as a new pictorial element of the pictorial term E_ν:

$$b_k \in E_\nu. \tag{2.9}$$

The new hypothesis H will consist of statements of the form (2.6), (2.9) and of all sentences of the form

$$(b_{i_1}, b_{i_2}, \ldots) \in u_\mu \tag{2.10}$$

which are theorems in $MT_{\Sigma_1} H_1$. This completes the definition of the new correspondence rules of PT. PT will be called a <u>restriction</u> of PT_1, in symbols: $PT_1 \rightarrow PT$. It is plausible that PT_1 is a more comprehensive theory than PT (for a proof, see my (1978)).

As an example, take for PT_1 the quantum theory of one electron with spin. Let x_1 be the corresponding set of ensembles (states), x_2 the set of effects (general yes-no-measurements, see my (1983)), and $\mu(w,f)$ the probability of the effect $f \in x_2$ in the ensemble $w \in x_1$. x_1 is represented by the set K of selfadjoint operators W in a Hilbertspace H with $W \geq 0$ and $\mathrm{tr}W = 1$, and x_2 by the set L of selfadjoint operators F with $0 \leq F \leq 1$; such that $\mu(W,F) = tr(WF)$. This theory shall be restricted to a theory *PT* without spin. So we introduce the set $E_2 \subset x_2$ of all those effects, which are not influenced by the spin. If H is written as a tensor product space $H = H_b \times \mathfrak{h}_s$, where H_b is the orbital space and $\mathfrak{h}_s$ the two-dimensional spin space, these effects are all operators of the form $F = F_b \times 1$. Hence E_2 is of type (α). The corresponding equivalence relation

$W_1 \sim W_2$ iff $tr(W_1-W_2)F = 0$ for all $F \in E_2$ on x_1 gives rise to a set of equivalence classes E_1 and the canonical mapping $f\colon x_1 \to E_1$. E_1 is therefore a term of type (β).

As a new pictorial relation we define a probability function $\tilde{\mu}\colon E_1 \times E_2 \to \mathbb{R}$ by

$$\tilde{\mu}(v,f) = \mu(w,f)$$

for $f \in E_2$ and $w \in v \in E_1$.

It may be left to the reader to define the correspondence rules for *PT* from those of PT_1. Note that *PT* is not yet the usual theory of an electron without spin described by the Hilbertspace H_b and the corresponding sets of operators K_b and L_b, since the "ensembles" of *PT* are equivalence classes of ensembles of PT_1. The transition to the usual description is achieved by means of an embedding to be explained in section 3.

Sometimes it will be necessary to generalize the concept of restriction in the following way. We consider a special hypothesis H_s in MT_{Σ_1}, i.e. a set of sentences of the form $(\overset{s}{—})_r(1)$, $(\overset{s}{—})_r(2)$ as described above. We now admit also new pictorial sets E_i and pictorial relations u_μ which are only definable in the stronger theory $MT_{\Sigma_1}\ H_s$. If the restricted theory *PT* obtained in this way is applied one always has to assume a realisation of H_s in G_1, that is a collection of facts in the fundamental domain which can be identified with H_s. In the case just described we shall call *PT* a <u>restriction</u> of PT_1 <u>relative to H_s</u>, in symbols:

$$PT_1 \xrightarrow{H_s} PT.$$

Restrictions relative to a hypothesis H_s are widespread in physics. Frequently H_s describes a situation in a laboratory relative to which the theory PT_1 will be restricted. A fixed H_s can be realised by a variety of possible facts in G_1, for example by many different situations "of the same kind" in different laboratories. As an example let us take for PT_1 special relativity and for H_s the description of a particular inertial frame. Then we could restrict the set of all inertia frames to those with a velocity less than $10^{-6}c$ (c: velocity of light) relative to the frame of H_s. The resulting restricted theory PT, however, is not yet Galilean relativity!

3. EMBEDDING

Another important procedure to obtain from PT a less comprehensive theory PT_2 is an "embedding". The mathematical part MT_Σ of PT is not supposed to be an axiomatic basis, but rather to be given in the form of section 1, i.e. by certain intrinsic terms $E_1,\dots,E_r$ for pictorial sets and a term $U = (r_1,r_2,\dots)$ representing pictorial relations and thus satisfying

$$r_\mu \subset E_{i_1} \times E_{i_2} \times \dots \times \mathbb{R} = S_\mu(E_1,E_2,\dots\mathbb{R}) \tag{3.1}$$

Now consider a second species of structures Σ_2 with principal base sets $y_1,\dots,y_p$ $(p \le r)$, typification $t \in T(y_1,\dots,y_p,\mathbb{R})$ and axiom $P_2(y_1,\dots,y_p,t)$. We suppose that T has the special form

$$\begin{aligned} &T = PT_1 \times PT_2 \times \dots \\ &\text{and } T_\mu(y_1,\dots,y_p,\mathbb{R}) = y_{\mu_1} \times y_{\mu_2} \times \dots \times \mathbb{R}. \end{aligned} \tag{3.2}$$

Hence

$$t = (t_1,t_2,\dots) \text{ and } t_\mu \subset y_{\mu_1} \times y_{\mu_2} \times \dots \times \mathbb{R}\ . \tag{3.3}$$

Let $\complement$ denote the set-theoretic complement, such that the negation of $x \in r_\mu$ may be written as $x \in \complement r_\mu$, analogously for t_μ. Let further $i \to \mu_i$ be a mapping of indices. Then we shall assume that the following theorem ("embedding theorem") can be proven in MT_Σ:

$$\begin{aligned} &(\exists y_1)(\exists y_2)\dots(\exists t_1)(\exists t_2)\dots(\exists f_1)(\exists f_2)\dots \\ &[t_\mu \subset T_\mu(y_1,\dots,y_p,\mathbb{R}) \text{ for all } \mu, \\ &\ P_2(y_1,\dots,y_p,t), \\ &f_i\colon E_i \to y_{\mu_i} \text{ for all } i, \\ &\langle f_1,\dots,f_r,1\rangle^{S_\mu} r_\mu \subset t_\mu, \text{ and } \langle f_1,\dots,f_r,1\rangle^{S_\mu} \complement r_\mu \subset \complement t_\mu]. \end{aligned} \tag{3.4}$$

In many cases we have simply $p = r$ and $i \to \mu_i$ is a bijection. The construction of PT_2 is performed as follows: MT_{Σ_2} is the mathematical part of PT_2. The fundamental domain G_2 is taken as G endowed with the same signs as in PT. To obtain the correspondence rules of PT_2 we replace every sentence of the form $a \in E_i$ in $(—)_r(1)$ of PT by $a \in y_{\mu_i}$ and every sentence of the form

$$(a_{i_1}, a_{i_2}, \ldots) \in r_\mu, \text{ resp. } \notin r_\mu$$

$$\text{by } (a_{i_1}, a_{i_2}, \ldots) \in t_\mu, \text{ resp. } \notin t_\mu.$$

This completes the definition of what we call an <u>embedding</u> of PT into PT_2, in symbols $PT \rightsquigarrow PT_2$. It turns out that in this event MT_{Σ_2} will be an axiomatic basis of PT_2. For the proof that PT is more comprehensive than PT_2 the reader is again referred to my (1978). Note that the case where PT_2 is an axiomatic basis of PT is a special instance of embedding.

In order to give an example let us return to the quantum theory of electrons "without spin" PT that we obtained in section 2 by restricting the quantum theory of electrons "with spin" PT_1. Let the species of structures Σ_2 be defined by base sets y_1, y_2, a structural term μ_2: $y_1 \times y_2 \to [0,1] \subset \mathbb{R}$ and an axiom of the following kind:

> There exist a Hilbertspace H_b and bijections
> α: $y_1 \to K_b$, β: $y_2 \to L_b$ (cf. section 2) such that $\mu_2(w,g) = \text{tr } \alpha(w)\ \beta(g)$.

Then the embedding theorem reduces to the statement: there exist bijections f_1: $E_1 \to y_1$, f_2: $E_2 \to y_2$ with $\tilde{\mu}(v,f) = \mu_2(f_1(v), f_2(f))$. To prove this one defines $(\beta \circ f_2)(F_b \times 1) = F_b$ and $\alpha \circ f_1 \circ f$ as the adjoint map by $\text{tr}\ ((\alpha \circ f_1 \circ f)(W)F_b) = \text{tr}(W(F_b \times 1))$ for all $F_b \in L_b$.

In this way we arrive at PT_2 as the usual theory of electrons without spin.

On the other hand, it is not possible to embed the restriction PT of special relativity (the last example of section 2) into Galilean relativity PT_2 in the sense defined above, but only in some "imprecise" manner. This is the typical feature for most cases of interest. We cannot go into details here [for the extension of restrictions and embeddings to the case of imprecise correspondence rules see may (1978)] but only give a sketchy exposition of "imprecise embeddings". In this case the idealized pictorial relations r_μ and t_μ are blurred by means of "imprecision sets" (defined via uniform structures, see my (1978, 1980)) yielding terms $\tilde{r}_\mu$ and $\tilde{t}_\mu$. Then one tries to prove the following weaker version of the embedding theorem:

$$\begin{aligned}
&(\exists y_1)(\exists y_2)\ldots(\exists t_1)(\exists t_2)\ldots(\exists f_1)(\exists f_2)\ldots\\
&[t_\mu \subset T_\mu(y_1,\ldots,y_p,\mathbb{R}) \text{ for all } \mu,\\
&P_2(y_1,\ldots,y_p,t),\\
&f_i : E_i \to y_{\mu_i} \text{ for all } i,\\
&\langle f_1,\ldots,f_r,1\rangle^{S_\mu} r_\mu \subset \tilde{t}_\mu, \text{ and}\\
&\langle f_1,\ldots,f_r,1\rangle^{S_\mu} C\tilde{r}_\mu \subset Ct_\mu].
\end{aligned} \tag{3.5}$$

The theory PT in the last example of section 2 obtained by restriction of special relativity can be imprecisely embedded into the theory PT_2 of Galilean relativity.

This example shows that the sets $f_i[E_1]$ will in general be proper subsets of y_{μ_i}. Thus not all elements of y_{μ_i} need appear as pictorial elements of facts in the fundamental domain G_2 of PT_2. In the above example only such inertial frames will appear in G_2 which have a velocity smaller than $10^{-6}c$ relative to a given inertial frame. This extra condition restricting the fundamental domain remained unknown until the advent of special relativity. This is typical: in the historical development of a physical theory PT_2 one often starts with too large a domain and has then to learn by experience that only under certain restricting conditions the theory can be applied successfully. Later when PT_2 is understood as an embedding of a more comprehensive theory PT, the embedding procedure yields a systematic derivation of those constraints which were formerly encountered only as empirical limitations if not completely unknown.

After the embedding $PT \rightsquigarrow PT_2$ has been achieved, it will often be more adequate to reinterpret the restricting conditions as normative axioms added to MT_{Σ_2} (cf. section 1). In our example this would mean that restricting to small relative velocities of inertial frames appears as a normative axiom in Galilean relativity.

Another example of imprecise embedding in statistical theories is given in Werner (1982).

4. NETWORKS OF THEORIES

The concept of theory-nets was introduced by Balzer and Sneed (1977, 1978). They emphasized that many "physical theories" (in the intuitive sense) like classical mechanics should be considered as a network of PT's.

Combining the two processes described in sections 2 and 3 we arrive at the following connection between three theories:

$$PT_1 \longrightarrow PT \rightsquigarrow PT_2 \tag{4.1}$$

(4.1) represents a transition from the axiomatic basis PT_1 to the axiomatic basis PT_2, where PT_2 is less comprehensive than PT_1. Since we encounter this special case very often in physics, we shall abbreviate (4.1) by

$$PT_1 \longrightarrow\!\!\!\rightsquigarrow PT_2\ . \tag{4.2}$$

Since in some cases the second step in (4.1) will only be a transition from PT to its axiomatic basis PT_2, we can say that (4.2) represents the simplest case of a transition from a theory PT_1 to a less comprehensive PT_2, when both theories are to be given in the form of an axiomatic basis.

We claim that all physical theories can be ordered in a network of relations of the form (4.2). Thus by a network we mean a set of theories PT_i and relations of the form (4.2) such that one can reach every theory PT_k from any other PT_i by going along arrows $\rightsquigarrow$ (but possibly disregarding their directions). If one can reach PT_k from PT_i by going only in the direction of the arrows, we say that PT_k is less comprehensive than PT_i.

When reconstructing a physical theory it is advisable not to begin with the most comprehensive theory, but to pass step by step from one theory PT_i to the next more comprehensive theory PT_{i+1} according to the scheme

$$PT_1 \leftsquigarrow PT_2 \leftsquigarrow PT \leftsquigarrow \ldots \leftsquigarrow PT_n \tag{4.3}$$

In [3] and [6] quantum mechanics is presented in the form

of such a chain. However, the relations between teories can be much more complicated than (4.3). For instance we can have the following situation:

$$\begin{array}{ccccc} & & PT_4 & & \\ & \swarrow & & \searrow & \\ PT_3 & & & & PT_2 \\ & \searrow & & \swarrow & \\ & & PT_1 & & \end{array} \qquad (4.4)$$

For an example of such a mesh, take as PT_4 the quantum mechanics of electrons, as PT_3 the description of electrons as classical point masses, as PT_2 the description of cathode rays as classical waves (which does not describe individual electrons!) and as PT_1 the analogue of "geometrical optics" for such waves.

Often in the development of physics a situation like (4.4) occurs, in which PT_4 is missing. As the above example shows, this lack is acutely felt by the physicists who then try to invent a new theory PT_4 completing the diagram (4.4). The network linking the various physical theories is very complicated and has so far been explicitly elaborated only in very few places. Nevertheless physicists find their way in this network like spiders in their webs.

5. PRETHEORIES

The relation between a theory PT and a pretheory PT_p is a special case of the situation

$$PT_p \longrightarrow PT'_p \rightsquigarrow PT_1 \longleftarrow PT \qquad (5.1)$$

$PT \rightarrow PT_1$ is a special restriction in which some of the basic pictorial sets and some of the pictorial relations are omitted. The remaining relations in PT_1 are "explained" by PT_p through the restriction $PT_p \rightarrow PT'_p$ and the embedding $PT'_p \rightsquigarrow PT_1$. PT_1 is less comprehensive than PT_p. But one has to avoid the mistake of judging PT, too, as less comprehensive than PT_p, although the embedding $PT'_p \rightsquigarrow PT_1$ entails that also between PT'_p and PT a mathematical theorem like (3.4) holds. But only the physical theory PT_1 is defined by this embedding and not PT! Even if all pictorial sets and relations of PT are explained by $PT_p \rightsquigarrow PT_1$ the theory PT may be more comprehensive than PT_1, since it can comprise more axioms.

There may be more than one pretheory to a given theory PT to explain all the pictorial sets and relations. For instance instead of (5.1) we can have a scheme

$$\begin{array}{lllll} PT_{p_1} & \rightsquigarrow & PT_1 & \nwarrow & \\ & & & & PT \\ PT_{p_2} & \rightsquigarrow & PT_2 & \swarrow & \end{array} \qquad (5.2)$$

It might become necessary to weaken the concept of restriction by considering hypotheses (instead of the existence of mappings f as in β of section 2) to account fully for the relations between theories and their pretheories. But so long as the necessity for such modifications is not demonstrated by a concrete example, we shall not purpose this possibility further.

The following example for the relation (5.1) is given in [3]. In this case the pretheory PT_p is the theory of preparation and registration procedures (Ludwig (1983), II to III, 1). As PT we take the theory with the basic sets K and L and the pictorial relation $\mu \subset K \times L \times [0,1]$. (Ludwig (1983) III, 3 and III, 5). The interpretation of K and L as sets of states and effects and of μ as a a probability function is then "explained" by the pretheory PT_p.

The form of relation (5.1) suggests the search for a theory more comprehensive than both PT and PT_p. In the above example this would demand the invention of a theory more comprehensive than the usual quantum mechanics on the one hand, and the macroscopic description of the preparation and registration procedures on the other. After the discovery of quantum mechanics many physicists believed (contrary to the opinion of N. Bohr) that quantum mechanics itself already was this more comprehensive theory. The author shares Bohr's point of view and has presented an approach to a theory comprising quantum mechanics as well as theory of preparation and registration in Ludwig (1984).

6. LIMIT PROCESSES

The technique of limit processes to obtain one theory from another is an elegant way to overcome some of the considerable mathematical difficulties related to imprecise restrictions and embeddings. (For an example of such difficulties see Werner (1982)). For instance one may consider the limit $c \to \infty$ to study the transition from the special theory of relativity to the Galilean theory of relativity. The use of imprecision sets is, however, indispensable when one wants to decide the question whether a limit is a good approximation to the real situation. Severe mistakes can result from ignoring this question. We shall only sketch an example:

Consider the classical point mechanics of N point masses of mass m in a volume V. The interaction is given by a potential v(r), where r is the distance between two of the point masses. With $M = Nm$ and $V(r) = N^2 v(r)$ we consider the limit $N \to \infty$, $m \to 0$, $v(r) \to 0$, keeping M,V and V(r) fixed. In this limit we obtain a kinetic equation of the Vlasov type (see for instance, Balescu (1963)). But is this equation a good approximation to a real situation with given N,V, and v(r)? It is a good approximation only if the range r_o of the interaction v(r) is large compared to the typical distance $(V/N)^{\frac{1}{3}}$ between the particles. If $r_o \ll (V/N)^{\frac{1}{3}}$ the Vlasov equation is no approximation at all. Instead, the Boltzmann equation provides a much better approximate description of the system. Only for $r_o \gg (V/N)^{\frac{1}{3}}$ can we expect an imprecise (but not too imprecise) embedding of the Vlasov equation into the classical mechanics of N point particles to exist.

Günther Ludwig
Fachbereich Physik der Philipps-Universität
Renthof 7
3550 Marburg/Lahn
Fed. Rep. of Germany

NOTE

1. In my (1983) the wording "mapping principles" has been used for the same concept.

ACKNOWLEDGEMENT

I would like to thank H.-J. Schmidt and R. Werner for many suggestions improving the exposition of my article and D. Pearce for final linguistic corrections.

REFERENCES

Balescu, R.: 1963, '<u>Statistical Mechanics of Charged Particles</u>', Interscience Publ., London,I. 2.

Balzer, W. and Sneed, J.D.: 1977, 1978, 'Generalized Net Structures of Empirical Theories I and II', <u>Studia</u>

Logica 36 (3), 195-212; and Studia Logica 37 (2), 168-194.
Bourbaki, N.: 1968, 'Theory of Sets', Addison-Wesley .
Ludwig, G.: 1978, 'Die Grundstrukturen einer physikalischen Theorie', Springer .
Ludwig, G.: 1981, 'Axiomatische Basis einer physikalischen Theorie und theoretische Begriffe' Zeitschrift für allgemeine Wissenschaftstheorie XII/1, Seite 55-74 .
Ludwig, G.: 1983, 'Foundation of Quantum Mechanics' vol 1, Springer .
Ludwig, G.: 1980, 'Imprecision in Physics';in A. Hartkämper and H.-J. Schmidt (eds.),Structure and Approximation in Physical Theories Plenum Press.
Ludwig, G.: 1984-1985, 'An Axiomatic Basis for Quantum Mechanics', 2 volumes, Springer.
Werner, R.: 1982, 'The Concept of Embeddings in Statistical Mechanics',Doctoral dissertation, Marburg.

W.Hoering

ANOMALIES OF REDUCTION [1]

Rarely, if ever, does one find in the literature actual proofs of irreducibility of one theory to another. So for all we know, most theories could be reducible to each other, a situation which would deprive positive reducibility results of much of their interest.

Of course it would be hard to substantiate claims about "most of the theories" and "reducibility" in general. We shall show however that different kinds of reduction relations, which have been proposed by Eberle, and Sneed (Stegmüller, Mayr) admit of large classes of unintended interpretations - for different reasons.

There is Eberle's (1975) purely syntactic reduction: relative interpretability. Well known results of Feferman, Kreisel, Orey, and most recently, Lindström demonstrate, on the one hand, that reducibility in this sense depends very critically on the way in which arithmetic is formalized in the respective theories; on the other hand, true first order arithmetic can reduce all recursively enumerable theories.

Then there is the set-theoretically defined reduction concept of Adams, as modified by Sneed (1971), Stegmüller (1973) and Mayr (1976) in similar but different ways, which takes reduction to be a relation between the (syntactically unanalyzed) sets of (partial) models of the respective theories. It has been argued that this relation is so weak that not even very elementary conditions of adequacy are maintained (Pearce 1982). This conception goes little further than making claims about the relative cardinalities of the model sets, which can, without any change of physical interpretation, easily be controlled, e.g. by just altering the representation of the reals in the models.

After some destructive work of the kind just indicated, we shall, with much less assurance, I'm afraid, proceed to make some constructive proposals. To this end we briefly review several good reductions, real or imagined, and attempt to abstract some common features from them. We propose to distinguish clearly between analogy which can hold between theories treating totally different subjects

W. Balzer et al. (eds.), Reduction in Science, 33–50.

and reduction which is only possible for theories with a shared area of application. We shall argue for considering a three place relation as basic for reduction (the third place being reserved for the area of experience with respect to which theory one is deemed to be better, more adequate, than theory two), and for a continuity condition even for "exact" reductions. We say a word about the empirical meaning of approximative reductions and finally we put forward some conditions, which are designed to rule out analogies and otherwise "crazy" correspondences. They require, to put it roughly, that the two theories can be described as having overlapping areas of application and at least partly deal with the same entities. Conditions of this kind can, by their very nature not be expressed in a purely syntactical form. This may be one of the reasons why they've been somewhat neglected hitherto.

RELATIVE INTERPRETABILITY

R.A. Eberle was the first to propose relative interpretability as reconstruction of reduction (1971). This idea was taken up by Vollmer in 1975. It had been known well enough from the theory of undecidability, Tarski et al. 1953.

If T and T' are two first order theories, T is relatively interpretable in T', if we can first relativize T to T'' (by changing all quantifiers in formulas of T into quantifiers relativized to a new predicate P) and then find a consistent common extension T''' of T'' and T' by adjoining to T' definitions for all predicates of T'' (in terms of predicates of T') s.t. T''' yields all valid sentences of T.

The relative interpretability kind of reduction has built into its very definition the definability of the basic vocabulary of T in terms of the vocabulary of T' as well as the validity of the 'laws' of T within the 'new' theory T'. So the basic conditions of adequacy which one would like to hold are fulfilled.

The one example which is treated by Eberle in some detail is the reduction of Galilean Mechanics (=law of free fall in the form "for all bodies g(h) = const.") to Newtonian mechanics (=gravitation law in the form "for all bodies $g(h) = M/(R + h)^2$)". To achieve reduction (= relative interpretation) of G by N we must relativize the bound variable of the law in G to a new predicate, say P, and find an interpretation of P in G which makes the law of G true in N. Eberle choses P such, that in effect it says: "if x is a body, then h = 0".

How does this method fare in more complicated theories, where one might want to relativize with respect to different predicates in different contexts? The most natural way out, in these cases, seems to relax relative interpretability to "multi relative interpretability", which allows just what is needed, i.e. relativization of some "types" of variables - as seen from the reducing theory - to different predicates P_i and no relativization of others. This seems especially adequate, if one wants to be able to deal with theories, whose purely mathematical part has been made explicit.

Alternatively one could also fit the multi relativization case into the formal straight jacket of normal relative interpretability by the simple device of packing those individual P_is and their respective conditions of application into the definition of one "master relativization predicate" P, when the relative interpretation is actually written down.

But all this would not help in a case like the attempt to reduce the Copernican theory of planetary motion to Newton's, where the conditions to which one would have to relativize (zero mass for planets), are outside the range of application of the reducing theory. Perhaps it was for this reason that Vollmer 1977, when discussing this example, did not attempt to deduce the Copernican equations of motion from the relativized Newtonian ones, but restricted himself to the discussion of Copernicus' third law. The appropriate method, however, in such a case seems to be to compare models of the respective theories with respect to a suitable topology, as is done in Scheibe 1973 or Mayr 1981.

At least as important as the limitation of applicability just mentioned are results, which in effect say that you can control the relative interpretability of two (consistent, recursively enumerable) theories containing (first order) arithmetic, by merely chosing a suitable axiomatization of arithmetic in one of them.

Let us suppose then that T is in the following always a consistent, recursively enumerable theory, Pe designates Peano (first order) arithmetic, Con(t) is the standard sentence expressing the consistency of T as given by the recursive enumeration t of its axioms, T + S is the theory obtained from T by adjoining the sentence S . We write "rel.int." for "is relatively interpretable in".

1957 Feferman proved:

I.	T	rel.int.	Pe + Con(t)	for all T
II.	T + Con(t)	not rel.int.	T	for all T $\supseteq$ Pe

Moreover the interpretation claimed to exist in I. can be chosen to be faithful if T is Σ_1^0-sound (Feferman, Kreisel, Orey 1960).

That calls for some explanation of terminology. A (relative) interpretation of T by T' (which we think of as represented by a function f mapping T into T') is said to be faithful, if $T' \vdash f(\varphi) \rightarrow T \vdash \varphi$ for every sentence φ of T. In that case (since via f some T'-sentences can be said to express T-truths), T' does not express more T-truths than T itself; but indeed, it expresses all of them, since we have a (relative) interpretation. Lindström 1983 deduces some necessary and sufficient conditions for faithful interpretations, which are too technical to be given here.

Actually I suspect that Eberle meant faithful (relative) interpretation, when he spoke of (relative) interpretation. But that could be disputed.

SNEED'S REDUCTION

Sneed's conception of reduction exhibits two striking traits. It is (almost) language independent, and therefore simple. Only models count, even if later it may be a problem which models to count. Second, since the distinction between theoretic and non theoretic functions is brought into play, it does become quite complicated. Six reduction concepts emerge with rather lengthy definitions. For these we have to refer you to the literature. The standard reference is Stegmüller 1973/76; a feeling for the motivations can be gathered from Sneed 1971. Mayr 1976 proposes a variant definition. Rantala 1978/80 insists on modeltheoretic foundations in a stricter sense. Pearce 1982 complies and discusses questions of adequacy.

Especially worth mentioning is Balzer/Sneed 1977/78, a brisk, authortative account of structuralist theory, which states and proves many useful results, correcting on the way some minor flaws of Sneed's original presentation. A puzzling point here is the reluctance to state sufficient and necessary conditions for theory elements. Only the latter are given. This feature is, as one immediately checks, carried over from Sneed 1971 D 38, and Stegmüller 1973 D 11. So it must be intended - probably as an act of caution. But unfortunately this omission renders discussions of the adequacy of strong reduction almost impossible: you can never be sure that what you put into this relation will actually be recognized as theory element. And indeed the authors trip into their own trap, when they, trying to prove the independence of their intertheoretical relations, act as if one could use certain simple structures as examples of theory elements. On their own definition one can't. Before the authors make up their mind to strengthen D 4 into some kind of equivalence, they also have no chance of disproving the conjecture attributed by them to D.Mayr and W.Hoering[2]. To prove it will also hardly be possible - as we shall see soon.

We shall mainly be interested in questions of applicability and adequacy. Strong claims have been made for Sneeds conception: By dint of the aforementioned feature of almost total language independence it will be able to bridge the gap between rather different languages. The transition from one theoretical vocabulary to a totally different one, which is the hallmark of scientific revolutions, would no longer lack rationality, if one could establish a rational bridge across it: reducibility of basic concepts and laws of the old theory to those of the new one. This in very broad terms is the proposal made in the last chapter of Sneed 1971. Stegmüller's 1973 (p.249) idea is somewhat more global and daring: to use Sneedian reduction of whole theories for that purpose, explicitly disregarding translatability requirements.

We suspect that Sneedian reduction is too unspecific to be useful. Let us first give two very general reasons, then two simple theorems, one on core- the other on theory-reduction; a third general reason will emerge, when we turn to approximate reduction.

Reason 1. All of Sneeds reducibility definitions are expressed, or expressible at least, in the language of set theory. They have the following form: There exists a function f (the reduction function) and a function r (the restriction function), s.t. $\underline{A}$; where $\underline{A}$ is a term of set theory, which contains the components of the theory elements (of the two theories) involved as constants, and expresses the conditions of inclusion and elementhood imposed by the reduction which is claimed to hold. VfVr $\underline{A}$ will not in general suffice to derive all the consequences of reducibility, because it does not express all we know, but

$$\mathrm{VfVr}(\underline{A}\ \&\ \underline{B}) + \mathrm{ZF}$$

will, where $\underline{B}$ is a term expressing in the same language, the requirements which Sneed imposes upon relations between components of theory elements $\tau[Mp] = Mpp$, $M \supseteq Mp$. etc.), and ZF is a standard axiom system of set theory, e.g. Zermelo-Fraenkel. Claims of derivability (of the laws of T in T') or definability (of the concepts of T in T') which figure in the usual adequacy conditions for reducibility are not part of this axiom system, they are not even expressible in its language, so they cannot be among its consequences. Therefore we cannot expect these adequacy conditions to hold.

This may not look like a strict proof, but it seems to be a plausible argument.

We can go a little further and ask ourselves, what the existence of a function, which is normally taken to map its range D1 onto its domain D2, means in set theoretical terms. The answer is simple and well known:

$$\mathrm{Vf}\ (\ f[D1] = D2 \quad \rightarrow \quad \mathrm{card}(D1) \geq \mathrm{card}(D2)\ )$$

Thus in set theory function existence statements can be translated into claims concerning cardinalities - in our case into claims concerning certain sets which are closely connected to or identical with components of theory elements. It may, but need not be the case that cardinality claims which correspond to reducibility claims are more perspicuous than the original formulations. This brings us close to

Reason 2. Sneed reducibilities can be equivalently transformed into statements about the cardinalities of certain sets. But it seems clear that this is not the kind of question which scientists investigate, when they try to find out, whether one theory explains the laws or defines the concepts of another. We don't count (partial, potential) models in order to settle questions of reducibility. Thus intuitive reducibility must be something very different from Sneed's concept.

We wish to illustrate the above method of transformation into cardinality claims for a relatively simple case. We take Pearce's 1982 (p. 311) core reduction concept, which has close ties, with Balzer/Sneed's $\overline{RD}$ and Stegmüller's RED, but neglects constraints.

It is easy to see in this case, that
K' is reducible to K
(there is a function $\mathfrak{F}$ reducing K' to K) iff

(1) card(Mp) $\geq$ card(Mp')

(2) card(Mpp) $\geq$ card(Mpp')

(3) a partitioning condition holds:

there is a mapping $\underline{m}$ of Mp-parts, $t \in$ MpT, onto the set of Mp'-parts, Mp'T, s.t.

$$\bigwedge t \in \text{MpT} \; (\text{card}(t) \geq \text{card}(\underline{m}(t)))$$

$$\text{Def.: } t \in \text{MpT} \leftrightarrow \bigvee y \in \text{Mpp} \bigwedge x \in \text{Mp} \; (x \in t \leftrightarrow e(y))$$

e is the "inverse" of r (cf. Stegmüller 1973 p. 126.)
(1) is a consequence of the requirement that Pearce's be a "function on K, K'", (1) follows from P.'s (i), (3) from P.'s (ii) and (iii). On the other hand, if (1) - (3) hold, a function can be pieced together, which satisfies (i) - (iii) and thus reduces K to K'.

Similar, but less easy to obtain, is the following theorem, which Mayr 1976 in effect proves (p.286), but does not formulate explicitly. The proof in this case relies essentially on the presence of constraints.

$$\text{RED}(\rho, T, T') \leftrightarrow M = \rho^{-1}[M' \cap D2(\rho)] \;\&\; J = \rho^{*-1}[J' \cap D2(\rho^{*})]$$

$$\& \; \text{rd}(\rho, \text{Mp}, \text{Mp}') \;\&\; \text{card}(\text{Mpp}) \leq \text{card}(R(D2(\rho)))$$

$$\& \; \rho \text{ is Mpp-unambiguous}$$

Here D2 and D1 denote the domain and range of ρ and R a projection function (cf. Stegmüller 1973 p. 126). The condition of Mpp-unambiguity guarantees that we can define from ρ in an unambiguous way as a kind of r-projection.[5] It corresponds to our partitioning condition.

The theorems show two things. Theory reducibility, $\bigvee \rho$ RED(ρ,T,T'), is very unspecific, it can almost entirely be rephrased in terms of cardinality conditions. Secondly, it is not quite as trivial as one might think.

It is instructive to compare this kind of result

with Pearce's valiant effort to squeeze some syntactic juice, out of the, as we saw, rather dry fruit of core reduction. He does have to add some model-theoretic dressing first, in order to be able to use Gaifman's or Feferman's results. And what he gets is a not effectively, but model-theoretically determined kind of translatability, which, as he remarks himself, does not yield "a dictionary" of definitions of concepts of K' by those of K - and thus falls short of the expectations usually associated with that term.

These remarks are just meant to put some translatability claims into perspective. The general approach however, followed by Pearce and Rantala, of applying the full fledged machinery of modern logic to philosophy of science, in order to obtain rigid and non trivial results, deserves recognition and full support (cf. our REFERENCES to their work).

DIRECTIONS FOR IMPROVING REDUCTION

One possible reaction to the results just presented,would be to maintain that the conditions defining these reduction concepts were just intended as necessary conditions, not as sufficient ones, for "good" reductions. But then these results show how precious little we have in our hands, if we just have proved a theory to be reducible to another, in one of the formal senses mentioned. That, which makes "good" and "interesting" reductions good and interesting, must lie elsewhere. But where?

We shall in this section collect some rules, ideas, conditions, and distinctions, which one might regard as essential ingredients of a future unified treatment of reduction.

Find and respect conditions of adequacy

Let us look at an example.
If we claim that one day we shall be able to reduce biology to physics, what do we mean? - That we shall then be able to describe plants and animals in physical terms, in such a way that their properties, i.e. the laws of biology, follow from those descriptions plus the laws of physics. This means

1. We must add something to the mere laws of physics, construct models (of plants etc.), which are admitted but not prescribed by those laws, in order to get a basis for deriving the laws of biology.

2. Deriving the biological laws means also representing

them - especially their concepts - within the new framework, i.e. physics as applied to those models. This is a requirement of (at least approximate) definability.

But perhaps we don't want to claim this reduction for certain higher animals, such as homo sapiens.

Quite generally it seems useful to set up the concept of reduction in such a way that the two theories concerned, are only evaluated, compared, with respect to a limited area of experience, with respect to a limited range of data, or even with respect to an arbitrary condition, expressible in either of the two languages concerned. Especially we want to be able to claim reduction, already if T' is reproducing the empirical successes of T, and to disregard the empirical failures if we want to. If we explicitly stake our claims concerning the data correctly represented by both theories, we relieve the reducing theory from the burden of being taken to represent the "true" state of affairs without exception. Therefore we propose:

3. Reduction is to be conceived as a 3-place relation: T' reduces T with respect to condition C.

A related proposal was made e.g. by Kemeny and Oppenheim 1956. A similar, not quite as hypothetical story could be told about the reduction of chemistry to physics, cf. Bunge 1982.

Remarks 1. and 2. suggest that we should return to the two well known conditions of adequacy: <u>derivability</u> (of laws) and <u>definability</u> (of concepts). These conditions are also explicitly acknowledged by Stegmüller 1973 (VIII,9 pp. 145/6), implicitly, however, discarded by the approach chosen for dealing with scientific revolutions (IX,7c, p.249), which relies on the feature of language independence in an essential way. Pearce 1982 notes that a fulfillment of the conditions, although desirable, is everything but guaranteed in the structuralist setting. He goes on to give sufficient conditions for a weak form of translatability.

If we decide to insist on the above adequacy conditions, aren't we giving away a chance for jumping (reducing) across revolutions in a lightfooted and very useful manner?

We have been suggesting, that these leaps would be just too easy and too far. - But how does the <u>science</u>, in whose development revolutions most obviously occured, deal with them? How are revolutions treated in physics? How does one explain old laws by new laws, which flatly contradict them, or, more characteristically, make no contact with them, since they use entirely new concepts? How does one handle e.g. the transitions from Keplerian to Newtonian theory of planetary motion, from Faraday's to Maxwell's conception of electrical phenomena, from Clausius' to Maxwell's, Boltzmann's and Gibbs' thermodynamics?

This is done first, by limiting the area of comparison (remark 3. above), secondly by approximations, and very often also by a blend of these methods.

This answer to our questions is what a physicist would read off the development of his science. This kind of answer has been the subject of extended investigations by Ludwig, Moulines, E.Scheibe (e.g.1973, 81), D.Mayr 1981 and others. Of course it deviates from the structuralist account mentioned.[4]

But there is also a deviating answer in the vein of a reformed structuralism, so to speak: in the paper already refered to, Pearce proposes to use certain higher logic systems to cover at least normal science developments in a way which respects (his concept of) translatability. However this kind of machinery appears to be unfit for dealing with approximation processes and thus for the task of reconstructing some important actual developments.[3] Perhaps if physicists one day will get a better education in logic, they will employ reductions which have to be reconstructed using the results of Gaifman and Feferman.

Let us pause for a second and reflect on the two answers. They seem to illustrate a metamethodological decision which time and again confronts the philosopher of science: should he try to reconstruct science or to restructure other people's reconstructions - which may be already pretty far removed from science?

A further condition of adequacy

There seems to be no sharp division between laws and constraints. Some laws could equivalently be conceived as constraints. If such a choice is available, it should, one would think, not affect reducibility.

It seems evident, that the known structuralist reductions do not have the desired property.

Examples for this ambivalent status include e.g. constraints in thermodynamics, as treated by Moulines,[2] and the vector additivity of forces, which is implicit in Sneed's formulation of Newton's second law. Conceiving the latter as constraint would also be in better agreement with experimental procedures, where one tends to first investigate "pure cases" - systems where only one kind of force is present -

before going on to more complex ones.

In thermodynamics, incidentally, we encounter constraints which look sensible, but are not "transitive"![6]

Conditions of adequacy involving truth and identity claims

True reduction

We have up to now only been concerned with the formal side of reduction, with conditions which can be ascertained independently of the actual state of the world - with "formal reduction".

Adequacy conditions for DN-explanation of contingent facts, as put forward by Hempel and Oppenheim, include the stipulation that laws and facts used be true. We need analogous conditions, when we talk about explaining laws by reducing; even if it's non-trivial to explain truth for laws involving theoretical terms. Therefore:

The reducing theory and the definitions used for reduction must be true.

How can definitions be said to be true or false? This is no problem in the empirical setting considered: They must turn out true by the rules of interpretation of the two theories concerned, and, taken together with (the axioms of) the reducing theory, must allow to deduce the part of the reduced theory considered to be true.

Let us, for the sake of illustration, go back to our physics/biology example. Here, among other things, we may require a description in physical terms of a certain plant. The features of this description will in principle be accessible to experimental testing and so can turn out to be false. Furthermore the laws governing, say, the growth of the plant should be deducible from the description (plus laws of physics).

A blind spot of structuralist terminology

In structuralist jargon we talk about the set of intended interpretations. But this term does not quite mean what it suggests. And for that we have no term left.

We quote from Balzer/Sneed 1977 p.199:

"<u>D5</u>. If $\langle \underline{K}, \underline{I} \rangle$ is a theory-element then that $\underline{I} \in \underline{A}(\underline{K})$ is the <u>claim</u> of $\langle \underline{K}, \underline{I} \rangle$.

The idea here again is simple. Mpp is all possible non-theoretical descriptions of some body of phenomena; $\underline{I}$

consists of phenomena that actually occur. The core K narrows down Pot(Mpp) to A(K) - it restricts the range of possibilities. It claims to do this in a way that narrows down onto what is actually to occur."

So here at first I is taken to be a set of structures ($\subseteq$ Mpp). But then it is supposed to be a set of descriptions of phenomena. This sometimes convenient ambiguity is already present in Sneed 1971, beginning of Ch.II, where the claim Q $\in$ S can be filled either by a physical system or by a mathematical structure Q.

In a benevolent reading one could speak of phenomena and physical systems interpreted as mathematical structures. Perhaps that is, what Sneed means, when he speaks of the possibility that I or its members be given by an "intensional description" (1971, p.274 ff.).

But certainly there is no unique interpretation of phenomena as mathematical structures - not even with respect to a given theory. Think of the solar system interpreted as possible partial model of Newton's theory. It can be taken as sun plus planets, or ... plus natural satelites, or ... plus all satelites - etc.

And, more importantly, there are cases of physical systems, exhibiting the same non-theoretic structure, some of them intended interpretations of the theory in question - some not. Think for instance of light-points moving like particles colliding elastically.

We don´t want unintended interpretations of this kind to rule out the existence of reduction functions. Therefore D2(ii) of Pearce 1982 which follows Sneed and stipulates that, whether a potential model belongs to the domain of a reduction function, depends only on its non-theoretical part, should presumably be relativized to intended applications.

Perhaps the oddities just mentioned do arise out of the extensional use of intensional entities.

In the next section we want to talk about physical and biological systems, however not about those systems interpreted as mathematical structures, etc.

Reduction, analogy, similarity

We want to set off reduction from analogy and similarity which are related, but are out to work in different ways and carry different presuppositions.

Let us, for the purpose of orientation, go back to our

physics/biology example. Here we may want to explain the reproduction of strain A of virus V by its description in physical terms and laws of physics. This would be part of the supposed reduction of biology to physics.

If we were to try to explain the reproduction of strain A referring to the reproduction of strain B, we would be using a similarity argument that presupposes certain laws of uniformity across the strains of virus.

If we tried to explain details in the mentioned reproduction process, say, by reference to acoustical wave propagation, we would be invoking an argument of analogy that would have to be founded on transference principles for laws between the two disciplines involved.

Formally, the way we arrive at all these relations is the same: we adjoin definitions to a reducing, similarity or analogy providing basis theory which, together with the axioms of the theory, are intended to yield the laws of a target theory.

We have explained them, if, with reduction, the definitions are true, in the two other cases, if the similarity and analogy principles, expressed by the definitions, are valid.

So, if we know enough, we can use all three relations for explanation. But the explanations of the second and third kind will generally not be considered to be as reliable as those by reduction. The reason is that analogy has to bridge the gap between distinct areas of experience and presupposes the validity of laws of nature of a rather unusual form. Similarity stays within one area of experience, but switches systems; it needs invariance laws, i.e. second order laws for support.

If reduction is understood as connecting identical systems described in different languages, then a true basis theory and true definitions (cf. last but one section) have to yield a true reduced theory. The notion of validity, by contrast, which can be associated with the coordinating definitions of analogy and similarity, has no meaning independently of truth in basis and target theory.

In order to be able to distinguish reduction from the other relations, we propose the following condition of adequacy for reduction:

Only identical (physical, biological, etc.) systems are to be put into correspondence; more exactly: every description in the language of the reduced theory of a set of systems becomes a description of the same set of systems, when translated into the language of the reducing theory.

This is not a condition which can be rendered in syntactical terms. But nevertheless it seems to be important. It is not fulfilled for approximate reduction.

Another adequacy condition, which we just want to mention in passing at this point, concerns the identity of quantities. We want to exclude the possibility that a reduction translation connects isomorphic but physically distinguishable quantities (or properties). This is a possibility only in almost pathologically symmetrical theories and covered in principle by our condition on the "truth" of definitions.

Continuity for exact reduction

Unintended correspondences intentionally constructed in the framework of structuralist reduction typically exhibit strong discontinuities. So it is clear that enforcing a condition of continuity would help to weed out quite a few unintended reductions - but certainly not all. We shall not attempt to follow up on this hint here.

Approximation

In the literature approximative reduction has been treated from a theoretical point of view involving the topological ideas of uniformity and completion (much like the completion of rationals to the reals). This line of development is connected with the names of G. Ludwig, C.U. Moulines and D. Mayr (1982).

On the other hand there were case studies (on Copernicus/Kepler, relativity and quantum mechanics) by E. Scheibe and his school (note especially Scheibe 1973/83).

Underdetermination of reduction, why it does not bother us much more

Regarding our central question of underdetermination of reduction it is apparent at once that the situation is not improved by introducing approximation on top of an exact reduction relation.

It creates an additional degree of freedom, whether taken in the sense of inexact correspondence, correspondence within a certain range of inexactness, or exact correspondence in the limit. In the latter case it is the choice of uniformity and limiting process, which brings about the additional latitude. Once this choice has been made, however, the rest is determined - or not quite determined - as with the corresponding "normal" exact reduction.

If there is as great an underdetermination of reduction as has been maintained, why does it not become more apparent? The question is relevant already for exact reduction. Some of the answers are valid for both. 1. With many theories we know rather well, which models of the two theories we want to regard as representing the same physical system and therefore as elegible to be put into correspondence. 2. (for approximate reduction) The topology can be very nearly determined by conventional ideas on "closeness" of curves and therefore often is not really a matter of choice. 3. There is another very general reason for the fact that very "exotic" reduction relations very rarely turn up in real life (real science) situations. It is our innate reluctance and often downright unability to deal with very complicated definitions and concepts. So we are saved by our practical limitations.

How to count models, or
How thin do we want to slice nature?

We wish to make evident that the model counting mechanism built into structuralist reduction can be easily confused by physically irrelevant changes in the mathematical machinery of one two theories concerned. That applies also to the most recent formulation given in Balzer, Moulines, Sneed 1983. The general idea is simple. A finite number of measurents cannot force upon us a decision to introduce models of any infinite power. Therefore we can make do with a finite number of finite models. After all nature seems to be atomistically constructed. We could also use the Löwenheim Skolem theorem to reduce the power of models (or even the power of the set of models) to, say, countable - after switching, if necessary, to a suitable embedding of (the meta-theory of) our theory in set theory.

Points of conflict

Even if we can argue for the individual adequacy conditions, we cannot expect them to form a consistent whole, when put together. Where then are the likely points of conflict? It seems that we'll have to give in and weaken the requirements of definability and derivability, evident as they are, in some cases, so that they become consistent with approximative reduction and the undefinabilities associated with theoretical functions.

CONCLUSION

In our discussions of formal properties of reduction concepts we tried not to completely lose sight of theory comparison as practiced by the scientists.

We have stressed the importance of some well known conditions of adequacy: definability (of concepts) and deducibility (of laws). We have pointed towards some adequacy conditions apparently neglected before: continuity for exact reductions and a certain invariance requirement (with respect to the placement among laws or constraints) for axioms of theories formulated in a structuralist setting.

These conditions, which in principle can be described on the level of formal syntax and semantics, will have to be weakened in order to be made consistent. But then they are not strong enough to exclude anomalies. They deal only with one coordinate, as it were, of the problem. The other, "inhaltliche" component, concerns questions of identity of physical systems and quantities, and truth claims made for theories. It is much harder to tackle, but equally important.

Universität Tübingen

NOTES

1. This paper grew out of work connected with DFG project Ho 620 in the mid seventies. I wish to thank the group of philosophers of science then assembled in Munich for discussions, especially Dieter Mayr and Joseph. D. Sneed. It was largely the "Schwerpunkt Wissenschaftstheorie" of the Deutsche Forschungsgemeinschaft which made this fortunate temporary constellation possible.
2. cf. Balzer/Sneed 1977, p.206
3. There is a treatment of approximation using non standard analysis: Pearce and Rantala, A logical study of the correspondence relation, J. Philos. Logic, 1983.
4. A shift in Stegmüller's position is indicated in his <u>The structuralist view of theories</u>, New York 1979.

5. To avoid unnecessary confusions please note differing priming conventions in Mayr and Pearce. Pearce reduces K' to K (following Sneed 1971, and Balzer/Sneed 1977/78), Mayr reduces T to T' etc. (following Stegmüller 1973). Mayr's definition of Mpp-unambiguousness (Mayr 1976, p. 281) amounts to this: Assume $rd(\rho, Mp, Mp')$, i.e. one-many, $D1(\rho) = Mp$, $D2(\rho) \subseteq Mp'$. Let E(x) denote the set of those possible models (in Mp) which result from x through addition of theoretical functions: $r[E(x)] = x$. Then ρ is called Mpp-unambiguous, iff for all $x' \in r'(y')$, $y' \in D2(\rho)$ there is an $x \in Mpp$, such that the inclusion $\rho^{-1}[E'(x')] \subseteq E(x)$ holds.
For other related results we refer to D.Mayr's DFG-report of 1975, "Ein Beitrag zur Untersuchung des Sneed-Stegmüllerschen Reduktionsbegriffs" (reference [M] of Mayr 1976), which can be obtained from the author.
6. Balzer/Sneed 1977, p.196, DOc. require, that constraints be "transitive". This is equivalent with the possibility of formulating them as purely universal sentences. Some of Moulines' constraints, however, which connect thermodynamic systems, have mixed quantifiers.
7. C.U. Moulines, Diss. Munich, 1975.
8. cf. Erkenntnis, 10, 1976, p.201-227. Approximate application of empirical theories: a general explication, and G.Ludwig, Deutung des Begriffs 'physikalische Theorie' und Grundlegung ..., Berlin 1970.
9. Englische Ausgabe: W.Stegmüller, The Structure and dynamics of theories, New York 1976
10. e.g. J.Nitsch, J.Pfarr, E.-W. Stachow (ed), Grundlagenprobleme der modernen Physik, Mannheim 1981.

REFERENCES

W.Balzer, J.D.Sneed, 1977/78 "Generalized net structures of empirical theories I" and "II", Studia Logica 36/37, pp.195-211 and 176-194.

W.Balzer, C.U.Moulines, J.D.Sneed, 1983, The Structure of empirical science: local and global, 7th International Congress of Logic methodology and philosophy of science, Salzburg 1983, mimeographed.

M.Bunge, "Is chemistry a branch of physics?", Z. f. allg. Wissenschaftstheorie 1982, pp. 209-223.

R.A.Eberle, 1971,"Replacing one theory by another under preservation of a given feature", Philosophy of Science 38, pp. 486-501.

S.Feferman, 1957, "Formal consistency proofs and interpretability of theories", J.S.L. 22, p. 107.

S.Feferman, 1960, "Arithmetization of metamathematics in a general setting", Fund.Math. 49, pp. 35-92.

S.Feferman, G.Kreisel, S. Orey, 1960, "1-consistency and faithful interpretations", Arch.f.math.Logik u.Grundlagenf. 6, pp. 52-63.

P.Lindström, 1983, On faithful interpretability, Logic Colloquium 1983, Aachen, mimeographed.

J.G.Kemeny, P.Oppenheim, 1956, "On reduction", Philosophical Studies 7, pp. 6-19.

S.Orey, 1961,"Relative Interpretations", Z.f.math. Logik u. Grundlagen d. Math. 7, pp. 146-153.

D.Mayr, 1976, "Investigations of the concept of reduction I" Erkenntnis 10, pp. 275-294.

D.Mayr,1981, "Investigations of the concept of reduction II" Erkenntnis 16, pp. 109-129.

D.Pearce, 1981, "Is there any theoretical justification for a non-statement view of theories?", Synthese 46, pp. 1-39.

D.Pearce, 1982, "The structuralist concept of reduction" Erkenntnis 18, pp. 307-333.

V.Rantala, 1978, "The old and the new logic of metascience", Synthese 39, pp. 233-247.

V.Rantala, 1980, "On the logical basis of the structuralist philosophy of science", Erkenntnis 15, pp.269-286.

E.Scheibe, 1973, "Die Erklärung der Keplerschen Gesetze durch Newtons Gravitationsgesetz", in E. Scheibe, G.Süßmann (ed.), Einheit und Vielheit,Festschrift f. C.F. v. Weizsäcker, pp. 98-118.

E.Scheibe, 1981, "Eine Fallstudie zur Grenzfallbeziehung in der Quantenmechanik", in J.Nitsch, J.Pfarr, E.-W.Stachow (ed.), Grundlagenprobleme der modernen Physik, Mannheim, pp. 258-269.

E.Scheibe, 1983, "Two successor relations between theories" Z. f. allg. Wissenschaftstheorie 14, pp. 68-80.

J.D.Sneed, 1971, The logical structure of mathematical physics,Dordrecht 1971.

W.Stegmüller, 1973, Wissenschaftstheorie, Bd. II,2

A.Tarski, A.Mostowski, R.M.Robinson, 1953, Undecidable Theories, Amsterdam.

G.Vollmer, 1977, "Theoriendynamik und Ablösung einer Theorie durch eine bessere: Simulation statt Erklärung", in G.Patzig, E.Scheibe, K.Wieland (ed.), Logik, Ethik, Theorie der Geisteswissenschaften, XI. Deutscher Kongress für Philosophie, Göttingen 1975, Hamburg 1977, pp.493-499.

C. Ulises Moulines

ONTOLOGICAL REDUCTION IN THE NATURAL SCIENCES (1)

In The Structure of Science, Ernest Nagel distinguishes two formal necessary (though not sufficient) conditions of reduction which he characterizes as the requirement of connectability and the requirement of derivability. The first says that, in order to reduce a theory T to another theory T', some "coordinating definitions" or "bridge laws", which have the form of conditionals, have to be stated such that they connect all basic predicates of the reduced theory to some basic predicates of the reducing theory. The second requirement is that the laws of the reduced theory have to be deducible from the laws of the reducing theory plus the connecting statement plus, perhaps, some singular statements about initial conditions.

This account of reduction has been much criticized in the last twenty years, but it has also found continuing support. Some critics have argued from general epistemological and semantical positions, apparently grounded on the history of science. This is notably the case with meaning-variance theorists like Kuhn, Feyerabend, and his followers. They have tried to undermine the concept of reduction altogether. Others have argued from a more moderate point of view. Without abandoning the unexplicated concept of reduction itself, they have tried to show that particular instances of would-be reductions in present-day science don't fit the standard account. A case for this has been made, for example, by Spector [1978] for the physical sciences and by Kimbrough [1979] for genetics. Some counter-attacks on the criticisms of the meaning-variance theorists have come from proponents of somewhat modified versions of the standard view of reduction, like Krajewski

W. Balzer et al. (eds.), Reduction in Science, 51–70.

[1978] and Yoshida [1977], who have tried to show that the criticisms can be circumvented by appropriate modifications or that the alleged counterexamples are not really such.

Whatever the relative merits of all these attacks and counterattacks on reduction, they all have centered around the idea that what is at stake is the possibility or impossibility of stating term-by-term connections or identifications between each theory's predicates, and of deducing one's statements from the other one's statements.

As is already well-known, at least to the present audience, the structuralist view of science proposes a concept of reduction that does not make use of term-by-term connections nor of deduction of statements. Essentially, its conditions keep the spirit though not the letter of the standard account, and at the same time point at some features of reduction that make the criticisms to which the standard account has been subjected to understandable and to some extent correct. By means of the globalistic approach to reduction characteristic of the structuralist view, one can easily see that it is actually wrong, as the adherents of the standard view do, to expect:

1) single predicate-to-predicate conditional statements or identifications;
2) deduction of one theory's statements from the other theory's statement in a straightforward manner.

On the other hand, it would be equally wrong to think that, because of the negation of 1) and 2), there is no logical relationship at all between theories that are said to constitute a case of reduction. There is indeed an ascertainable relationship between them, though not a relation given by identification of meanings or by direct deducibility of statements; it is a relation between whole structures established through a global link.

Since these features of the structuralist notion of reduction have been abundantly discussed in the relevant literature on the subject, (2) I suppose there is no need now to dwell upon them. Let me only summarize how the concept of reduction appears in the present-day formalism employed by the structuralist program. (For the present purposes, I shall leave aside the theoretical-non-theoretical distinction, constraints, and special laws. For abbreviation's sake, I shall also use "theory" as synonymous with "theory-element".)

Let T be a theory consisting of a class M_p of potential models, a subclass M of actual models, and a set I of intended applications. In the present context we shall assume that $I \subseteq M_p$ and that the empirical claim of T is that $I \subseteq M$. (Though these are gross over-simplifications within the structuralist metatheory, they are harmless for the present discussion and we could easily correct them if needed.) Correspondingly, let T' consist of M'_p, M', and I'. Further, if we have a relation $R \subseteq A \times A'$ between two classes whatsoever, then $\bar{R} \subseteq \mathcal{P}(A) \times \mathcal{P}(A')$ will be the relation induced by R on the next set-theoretic level.

With these notions in mind, we may define the most recent version of the structuralist notion of reduction as follows.

If $T = \langle M_p, M, I \rangle$ and $T = \langle M'_p, M', I' \rangle$, then:

(R) T is reducible to T' by means of ϱ iff:

There are M'_r, I'_o so that:

(1) $M'_r \subseteq M'_p$ and $M'_r \cap M' \neq \emptyset$

(2) $\emptyset \neq I'_o \subseteq I' \cap M'_r$

(3) $\varrho : M'_r \longmapsto M_p$ and ϱ is many-one

(4) $\bar{\varrho}\,(M'_r \cap M') \subseteq M$

(5) $\bar{\varrho}\,(I'_o) \subseteq I$

It is obvious that this scheme of reduction does not require semantic predicate-by-predicate connections nor deducibility of statements - thougł it does not prohibit them either. On the other hand, conditions (R)-(4) and -(7) roughly correspond to the "spirit", though not at all to the "letter", of Nagel's requirement of connectability, whereas condition (R)-(5) could be seen as a parallel to Nagel's requirement of derivability.

By introducing appropriate topological uniformities defined on the classes M_p and M'_p respectively, scheme (R) can be expanded in a natural way to cover the different cases of <u>approximative</u> reduction, which are so important for actual science. How this can be done I have shown in my 1981 . The distinction between reduction and approximative reduction is by no means trivial, and in many contexts of discussion it is important to keep the two concepts apart, since different general consequences may be drawn from one and the other case. However, in the present context the distinction is not essential, so that I shall use "reduction" and "approximative reduction" as synonyms in what follows.

This notion of reduction has proved to be very fruitful. On the one hand, it has allowed to shed a new light on the general controversies over reduction, by identifying the right as well as the wrong aspects of the standard view, and at the same time allowing for a higher degree of conceptual differentiation. On the other hand it has led to very detailed reconstructions of particular "real-life" examples of reductions like those of rigid body mechanics to Newtonian particle mechanics, (Sneed [1971]), classical collision mechanics to Newton's particle mechanics (Balzer/ Mühlhölzer [1982]), and Kepler's planetary theory to Newtonian particle mechanics (Moulines [1980], [1981]). There are good prospects that scheme (R) (or one of its approximative versions) will be applicable to other interesting examples of purported reductions as well.

However, the present structuralist concept of reduction still does not tell all there is to tell about reduction. There is at least one further aspect of reduction that is overlooked by scheme (R). This is what I would like to call "the ontological aspect". I wish to argue that, for a complete picture of a reductive relationship between two theories, one has to take into account some sort of relation between the respective domains. Otherwise, when confronted with a particular example of a reductive pair, we would feel that all we have is an *ad hoc* mathematical relationship between two sets of structures, perhaps by chance having the mathematical properties we require of reduction but not really telling something about "the world". We could have a reductive relationship between two theories that are completely alien to each other. The possibility that we find a formally appropriate ρ just by chance or by constructing it in an *ad hoc* way cannot be ruled out in general. Would we then say that we have reduced one theory to the other? I think we would feel such a reduction is not "serious".

A standard reply to this sort of objection is that all you can do in formal philosophy of science when dealing with any kind of subject is to try to build up a general scheme (like (R)) that on the one hand has as much information content as possible and on the other hand fits the intended instances; but you can never rule out someone coming up with a contrived non-intended example that fits the scheme though nobody would like to have it inside. (Of course, the same is true of the axiomatic approach in the sciences themselves.) This reply is correct. But it does not rule out the possibility of enriching our previous concept through appropriate modifications so as to keep outside it as many of the unwanted cases as possible. If this is possible, it should be done.

In the present case of reduction, I think it can be done. I think many cases of counter-intuitive reductions would come out from the fact that ρ is defined too globally. (The globalistic methodology of the structuralist program may have some disadvantages sometimes.) It does not specify any particular kind of link between the respective base sets upon which the relations and functions of both theories are constructed. And we should expect of a "real" reduction that it connects some of the respective base sets at least - though not necessarily by means of "coordinating definitions" or "bridge laws", as Nagel would have it. The base sets of one theory stand for a specific way of carving up reality (most likely to be described in rather intuitive, presystematic, "operational" terms) - a specific way of telling what "the furniture of the world" is; and we should expect of a reduction relation that it tells us how two such mobiliaries are related to each other.

Global links between different theories of the kind structuralism is now considering are normally composed of some partial links that connect some terms of each theory in specific ways. (Again, these are intuitively analogous to Nagel's term-to-term connections, though their actual form is quite different.) What I suggest for the present case is that, among all other requirements put on a reductive link , it be viewed as composed of some partial links between some of the base sets of each theory. That this is a reasonable requirement on reduction appears to be something that not even critics of Nagel's approach would like to deny. They even appear to come closer to Nagel's views than I think it safe to do. For example, Kimbrough proposes, after all his criticisms, a "weak sort of reduction [where] individuals described in the language of one science are identified with individuals described in the language of another science" (Kimbrough [1979] , p. 403, my italics). Even an apparent iconoclast on reduction like Feyerabend, in the very few cases when he becomes a bit clearer than usual, proves to be more orthodox than it seems to be the case. It is amusing to quote him at some length:

> "Consider two theories, T' and T, *which are both empirically adequate inside domain D'*. In this case the demand may arise to explain T' on the basis of T, i.e. to derive T' from T and suitable conditions *(for D'). Assuming T and T' to be in quantitative agreement inside D'*, such derivation will still be impossible if T' is part of a theoretical context whose "rules of usage" involve laws inconsistent with T. It is my contention that the conditions just enumerated apply to many pairs of theories which have been used as instances of ... reduction" (Feyerabend, [1962], p. 75, my italics).

This is, of course, Feyerabend's famous thesis of incommensurability. But the interesting thing to note here is that, in order to assert the thesis, a radical meaning-variance theorist like Feyerabend has to assume that both theories apply to *the same domain D'*. (3) After all, then, incommensurable theories are not all that unrelated: There is a common domain D' to which we can refer to (in whose language? - we may ask; and: how do we know it is the same D'? - But these are questions for Feyerabend to worry about, not for me.)

It is clear that Kimbrough, Feyerabend, Kuhn, and other critics of the standard view on reduction (and, of course, the proponents of the standard view as well) can conceive of a reductive relationship, for all incommensurabilities and the like, only when there is a common domain of individuals to be treated by both theories - with other words, only when the basic ontology is the same (though the relations and functions might differ both extensionally and intensionally). However, we need not go that far in approaching the standard view as Feyerabend does. We can allow even for ontologically different domains D and D', such that neither D = D', nor $D \subset D'$

nor $D' \subset D$. All we need for having reduction is <u>some sort</u> of structural relationship between D and D', so that from some specific way of conceiv--ing the world we can pass over, in a cannonical manner, to some other specific way of conceiving it. ("The world" is here, of course, only a <u>façon de parler</u>.)

To sum up: When a reduction of T to T' takes place, not only it might be the case that the laws of T are not deducible from the laws of T' and that the relation predicates that appear in T are extensionally and intensionally different from those in T', but it might even be the case that the basic ontologies of T and T' differ radically. All we need for reduction, besides scheme (R), is that there be some ascertainable structural relationship between the base sets of T and those of T'. Since the rest of relations and functions are constructed out of the base sets, any relation-ship between the latter will indirectly induce some kind of relationship between the former, which will be a part of the reductive link too. The fundamental problem, then, is to find out a possible way of formalizing the relationship between the base sets in general. Our conceptual framework will help us in this task.

Suppose we want to reduce $T = \langle M_p, M, I \rangle$ to $T' = \langle M'_p, M', I' \rangle$. Any potential model x^p of T will have the general form:

$$x = \langle D_1, \ldots, D_n, A_1, \ldots, A_m, r_1, \ldots, r_p \rangle ,$$

where the D_i are the "real" base sets (= sets of empirical entities like particles, molecules, fields, genes, commodities, persons, etc.), the A_i are auxiliary base sets (mainly mathe-matical spaces like $\mathbb{N}$ or $\mathbb{R}$), and the r_i are relations or functions <u>typified</u> (4) by the base sets; that is, for each r_i there will be a typification τ_i allowing to state:

$$r_i \in \tau_i(D_{i_1}, \ldots, D_{i_h}, A_{j_1}, \ldots A_{j_k})$$

where $\{D_{i_1}, \ldots, D_{i_h}\} \subseteq \{D_1, \ldots, D_n\}$ and

$$\{A_{j_1}, \ldots, A_{j_k}\} \subseteq \{A_1, \ldots, A_m\}$$

Analogously, we shall have that each potential model x' of T' is of the form

$$x' = \langle D'_1, \ldots, D'_{n'}, A'_1, \ldots, A'_{m'}, r'_1, \ldots, r'_{p'} \rangle$$

where for each r'_i there is a τ_i with

$$r'_i \in \tau'_i(D'_{i_1}, \ldots, D'_{i_{h'}}, A'_{j_1}, \ldots, A'_{j_{k'}})$$

It is convenient to have the following notational device at hand. If x is a potential model constituted by n+m+p entities, then $\pi_i x$, for $1 \leq i \leq n+m+p$, is the i-th component of x. (Such π_i are usually called "projection functions".)

Now, consider a link $\rho \subseteq M'_p \times M_p$ between T' and T. We shall say that ρ is an <u>ontological reductive link</u> if, besides fulfilling a scheme of form (R), ρ is partly composed of a relationship between some of the D_i and some of the D'_i and, perhaps, some of the A'_i as well. (It need not be <u>all</u> of them.) The A_i are of no concern to us here, since they come from mathematics and are not specific of empirical theories.

What kinds of relationship may have the D_i with the D'_i? There are basically two different kinds of ontological reduction that show up immediately when considering the matter in this way. They correspond to what Spector 1978 has called, in more intuitive terms, "domain preserving reduction" and "domain eliminating reduction"(5) Spector thinks that domain eliminating reduction is just what is normally called "micro-reduction";

but there is no reason for thinking this: micro-reduction is at most a special case of domain eliminating reduction; there are examples of the second which are not micro-reductions, as Yoshida has convincingly argued. Anyway, I prefer to state Spector's distinction in somewhat different terms, that seem to me to be intuitively more adequate. I shall call the first kind "(ontological) homogeneous reduction" and the second kind "(ontological) heterogeneous reduction". Homogeneous reduction is a sort of partial or total identity relation on the ontological level: It is total when some base set of the reduced theory is identified with some base set of the reducing theory. It is partial when a base set of the reduced theory is identified with a proper subset of a base set of the reducing theory. (In the following, unless otherwise stated, when I speak of "base sets" I always mean non-auxiliary, i.e. empirical base sets.)

An ontological reductive link is said to be heterogeneous if at least one base set of the reduced theory is related to the corresponding base set(s) of the reducing theory in a way that does not imply identification of elements. Note that, by definition, a global reductive link between two theories can be (ontologically) homogeneous and heterogeneous at the same time. That is, some base sets of the reduced theory may be partially or totally identified with some base sets of the reducing theory; while others not so. This is a case of partly homogeneous and partly heterogeneous reduction. Let us call this a case of "mixed (ontological) reduction". At the opposite ends of it we may have "completely homogeneous reductions", where all base sets of the reduced theory are identified with some base sets of the reducing theory, and "completely heterogeneous reductions", where none is so. As far as I can see, there may be examples of all these kinds of reduction in real science, though it is likely that mixed reduction is the most frequent case.

Before we further analyze the different kinds of ontological reduction in formal terms let us consider how specific examples of purported reductions appearing in existing expositions of natural science may fit into the categories just distinguished. For this purpose, I introduce the following abbreviations:

CCM = classical collision mechanics
NPM = Newtonian particle mechanics
RBM = rigid body mechanics
KP = Kepler's planetary theory
SR = special relativity
GR = general relativity
GO = geometrical optics
WO = wave optics
CED = classical electrodynamics
QM = quantum mechanics
STG = simple thermodynamics of gases
KG = kinetic gas theory
MeG = Mendelian genetics
MoB = molecular biology

The following examples of reductions have been reconstructed with all detail within the structuralist frame (the first theory mentioned is the reduced, the second the reducing one):

CCM - NPM
RBM - NPM
KP - NPM

Further, NPM - SR is in a working process.
Of these, CCM - NPM and KP - NPM are cases of completely homogeneous reduction, since all base sets of CCM and KP (namely, in each case, the set of particles P, the set of space points S, and the set of instants T) is identified either with a base set or with a subset of a base set of NPM. RBM - NPM is a case of mixed reduction: Whereas S and T are identical in both theories, the base set of a model of RBM, which consists of one or several rigid <u>bodies</u> is not identical with any set P of a model of NPM (6). The pair NPM - SR is also a case of mixed reduction, but for different reasons: The set P of particles is here the same

(or a subset) in both theories; but instead of the Newtonian separate structures S and T, in SR we have a single Minkowski space.

The next examples of would-be reductions are considered as hypothetical cases. They are frequently mentioned in the literature, though to my knowledge none of them has been reconstructed so far with a sufficient degree of completeness. It is therefore difficult to make any definite assertion about their true nature. Nevertheless, it seems plausible to assume that none of them is a completely homogeneous reduction. They are either mixed reductions or completely heterogeneous reductions. They are the following pairs (roughly by chronological order):

GO - WO
STG - KG
WO - CED
SR - GR
NPM - QM
MeG - MoB

By listing these examples and the previous ones I have tried to make plausible that the distinction between different sorts of ontological reduction is both interesting for a general discussion of reduction and useful for analyzing specific examples found in the literature. Let us see now how the different types can formally be defined. Homogeneous reduction is easy to define.

Def. 1: If T is reducible to T' by means of ρ, then: ρ is a homogeneous reductive link of T to T' iff:
For all $x \in M_p$, $x' \in M_p'$, if $\rho(x') = x$,
then there are $i, j \in \mathbb{N}$
so that $\pi_i x$ is a base set of x, $\pi_j x'$ is a base set of x',
and $\pi_i x \subseteq \pi_j x'$.

Generally speaking, in the case of heterogeneous reduction the individuals of the corresponding base sets will not be in an identity relationship; however, it is plausible to think that, if we are going to have a "real" reduction at the ontological level, the corresponding sets of physical individuals will be in a relation, so to speak, of "next-to-identity". This means that in this case there will be a biunivoque correspondence between the base sets; in other words the ontological reductive link will be a one-one function. (Notice that this does not imply in any way that the global reductive link be itself a bijection - in general, when there are T-theoretical functions involved, it will not be one-one but many-one.) Perhaps this somewhat strong requirement of biunivoque correspondence could be weakened by requiring just a functional correspondence in one or the other direction. But, first, the examples I am familiar with seem to suggest the presence of a bijection; and, second, from a more general point of view it seems to me that less than a bijection would appear as a rather unsatisfactory kind of ontological reduction. (We could not use anymore typical phrases such as that, by the purported reduction, one can show that "this kind of thing actually is nothing but this other kind of thing".) At any rate, whoever is not convinced that heterogeneous reduction should be construed as a bijection may easily modify the scheme to follow in the sense of weakening it. However, I think that at least some kind of functional correspondence must be kept if we are going to speak of reduction at all. Before trying to formalize heterogeneous reduction, we have to consider different possible subtypes of this kind.

The simpler cases of heterogeneous reduction are those where a base set D_i of the reduced theory is biunivoquely related to just one base set D'_j of the reducing theory. Intuitively, this means that the ontological units of the reduced theory are regarded as entities composed of just one sort of ontological units of the reducing theory. More exactly, this may happen in two ways. In the simplest case, we consider that each element of the reduced base set corresponds to a <u>set</u> of elements of the reducing base set. (Note that we say that it "corresponds" - not that it is identical with.) This means that the global reductive link ρ is composed, among other things, of a specific one-one function ω of the form:

$$\omega: D_i \longmapsto \mathcal{P}(D'_j),$$

or, to put it in a way that will prove to be enlightening, ω is such that

$$\bar{\omega}(D_i) \in \mathcal{P}\mathcal{P}(D'_j).$$

We find an example of this sort in the reductive link RBM - NPM: The domain of bodies corresponds to a class of sets of particles, i.e. to each body, a set of particles biunivoquely corresponds.

A slightly more complicated case is the one where the elements of the reduced domain correspond to a <u>sequence</u> of elements of the reducing domain. In this case, there is an integer n such that ρ is composed, among other things, of a specific one-one function ω of the form:

$$\omega: D_i \longmapsto D'_j \times \overset{n)}{\ldots} \times D'_j$$

i.e. $\bar{\omega}(D_i) \in \mathcal{P}(D'_j \times \overset{n)}{\ldots} \times D'_j)$

An example of this sort seems to be (at least part of) the reductive link MeG - MoB (see Kimbrough [1979]): If D_i is the set of genes of an organism and D_j a certain set of organic molecules, then to each gene a sequence of organic molecules is supposed to correspond biunivoquely.

In general, however, we cannot expect that all cases of ontological reduction will belong to one of the previous types, namely correspondence of one base set to another single base set. It is very likely that, in more complicated cases of reduction, the reduced base set will be put in correspondence with <u>several</u> base sets of the reducing theory. Since there still are so many instances of reduction pairs that have not been fully reconstructed, especially as far as the ontological aspects are concerned, I cannot offer any clear-cut example of this more complex kind. Nevertheless, it is likely that, on closer inspection, we would find cases of this sort among the examples of the putative list above. For example, it is plausible to view the ontological aspect of STG - KG in this way: If my own reconstruction of simple thermodynamics (see Moulines [1975]) is adequate, then the only base set that really comes into play in STG is a set of states (see also Giles [1965] and Falk and Jung [1959] for further support of this view of thermodynamics). On the other hand, it is clear that, if we try to reduce a phenomenological gas to a kinetic system, to any state of the gas not only a set of particles (= molecules) corresponds, but also the space-points and instants localizing each particle. Whatever further entities we might consider necessary for a proper reconstruction of KG, it seems reasonable to require that a set of molecules, a spatial region and a time interval be among its base sets.

In a case of this sort, the reductive link would be composed of a one-one function of the sort

$$\omega : D_i \longmapsto D'_{j_1} \times \overset{n)}{\ldots} \times D'_{j_n} \quad \text{for some } n \in \mathbb{N}$$

with $\{D'_{j_1}, \ldots, D'_{j_n}\} \subseteq \{D'_1, \ldots, D'_{n'}\}$.

Some of the various D'_j may be identical. We could also imagine that instead of some D'_{j_i}, their power-sets appear in the correspondence relation. In other words, we could have the case that to the reduced base set a combination of Cartesian products and power sets of some reducing base sets corresponds. That is, to the reduced base set an <u>echelon set</u> over some reducing base sets would correspond. We may write it in this way:

There is a τ_i such that

$$\omega : D_i \longmapsto \tau_i(D'_{j_1}, \ldots, D'_{j_n}) \text{ where } \omega \text{ is one-one;}$$

consequently, $\bar{\omega}(D_i) \in \mathcal{P}(\tau_i(D'_{j_1}, \ldots, D'_{j_n}))$

which is again of the form, for a typification τ'_i derived from τ_i:

$$\bar{\omega}(D_i) \in \tau'_i\,(D'_{j_1}, \ldots, D'_{j_n})$$

We could generalize the former scheme to cases where not only some of the empirical reducing D'_j come into play to reduce a given D_i but also some <u>auxiliary</u> base sets of the reducing theory. This means that the correspondence would look like this:

$$\bar{\omega}(D_i) \in \tau_i(D'_{j_1}, \ldots, D'_{j_n}, A'_{k_1}, \ldots, A'_{k_m})$$

A still more complicated situation could well arise: A reduced base set not only corresponding to several base sets of the reducing theory, but also to several relations and/or functions defined over the reducing base sets. For example:

$$\omega : D_i \longmapsto D'_{j_1} \times \overset{n)}{\ldots} \times D'_{j_n} \times A'_{k_1} \times \overset{m)}{\ldots} \times A'_{k_m} \times R'_{k_1} \times \overset{p)}{\ldots} \times R'_{k_p}$$

for some $n,m,p \in \mathbb{N}$, such that, for each R'_{k_i} there are

$$C'_1, \ldots, C'_r \in \{D'_1, \ldots, D'_{n'}, A'_1, \ldots, A'_{m'}\} \text{ with}$$

$$R'_{k_i} \subseteq C'_1 \times \ldots \times C'_r$$

Formally, however, a case like the last one can be "reduced" to the previous one, that is, we need not mention the relations $R'_{k_1},\ldots,R'_{k_m}$ explicitly. To see this, remember that the relations R'_{k_i} may be defined as elements of the echelon sets over their respective domains. Thus, after all, this most complicated case may be expressed in exactly the same way as before, viz.

$$\bar{\omega}(D_i) \in \tau_i(D'_{j_1},\ldots,D'_{j_n},A'_{k_1},\ldots,A'_{k_m})$$

Now, this last scheme also applies to the simpler, "degenerated" cases, where D_i corresponds to a power set or to a Cartesian product, since these are also echelon sets. Thus, the last formula expresses the general scheme for any heterogeneous ontological reduction of any degree of complexity. Let us summarize this result in the following definition.

<u>Def. 2:</u> If T is reducible to T' by means of ρ, then ρ is a heterogeneous reductive link of T to T' iff:
For all $x \in M_p, x' \in M'_p$, if $\rho(x') = x$, then there are $j_1,\ldots,j_n \in \mathbb{N}$ a one-one function ω, and a typification τ_i so that $\pi_i x$ is a base set of x, $\pi_{j_1} x',\ldots,\pi_{j_n} x'$ are (auxiliary and/or non-auxiliary) base sets of x', and

$$\bar{\omega}(\pi_i x) \in \tau_i(\pi_{j_1} x',\ldots,\pi_{j_n} x')$$

A conclusion we may intuitively draw from this general scheme is that, in heterogeneous reduction, the reduced base set will appear as a <u>relational structure</u> typified over some of the base sets of the reducing theory: The amorphous basic entities of the reduced theory become structured through reduction.

This is, of course, something we should have expected intuitively anyway, but perhaps it is not completely idle to have our vague intuition formally developed. In this way, also, a confirmation or disconfirmation of this general metatheoretical hypothesis about reduction by means of case studies will become more feasible.

Universität Bielefeld

NOTES

1. Some of my friends in the Reduction Colloquium held in Bielefeld, notably D. Mayr, H.J. Schmidt, and R. Werner, drew my attention upon the fact that the concept of reduction offered in this paper implicitly shows striking similarities with some intertheoretical relations (named with different terms) introduced by G. Ludwig and that it would be worthwhile further to analyze the close connection of the present structuralist concept of reduction with Ludwig's counterpart. I think it likely that they are essentially right; however, since I have had no time to pursue this question further on, I leave the paper as it is now, with the good hope that I shall be able to make a better use of their remarks in a future investigation. I also would like to thank D. Pearce for his helpful remarks concerning a first draft of this paper, which led to some amendments of the original text.
2. For the standard structuralist account of the matter, see Stegmüller [1976] and [1979]; for some other, more or less divergent interpretations of the structuralist notion of reduction or critical discussions of it, see Kuhn [1976]; D. Mayr [1976]; Feyerabend [1977]; Pearce [1982].
3. Very similar passages could be found in Kuhn's discussions of incommensurability in his [1970] and afterwards.
4. In the sense of Bourbaki or in a similar sense. To typify a set means to give a procedure τ_i to construct it from some array of previously given sets by means of repeated application of

the set-theoretic operations of power-set and Cartesian product. For example, a relation R between the sets A and B is said to be typified by the following constructing procedure: first build the Cartesian product of A and B, then take its power set; R will be an element thereof, i.e. $R \in \mathcal{P}(A \times B)$, which will be the particular form the typification $\tau(A,B)$ takes in this case. The set obtained by the typification τ, in this case $\mathcal{P}(A \times B)$, is called "the echelon set" (over A and B). For more details and applications to the reconstruction of physical science, see e.g. Ludwig [1978].

5. Actually, Spector prefers to speak of "theory replacement" instead of "reduction"; but Spector's prejudices against the word "reduction" are of no concern to us here.
6. A (rigid) body is not identical with a particle: A body is something that has a volume; a particle does not - or, as physicists say, a particle is something that has "a negligible volume compared to the size of the system considered"; but "neligible volumes" are no volumes.

REFERENCES

Balzer, W. and F. Mühlhölzer, [1982]: 'Klassische Stoßmechanik', Zeitschrift für allgemeine Wissenschaftstheorie.

Falk, G. und H. Jung, [1959]: 'Axiomatik der Thermodynamik', in S. Flügge (ed.), Handbuch der Physik, vol. III/2.

Feyerabend, P.K., 1962: 'Explanation, Reduction, and Empiricism', in H. Feigl and G. Maxwell (eds.), Minnesota Studies in the Philosophy of Science, vol. III, Minneapolis.

Feyerabend, P.K., [1977]: 'Changing Patterns of Reconstruction', British Journal for the Philosophy of Science, vol. 28.

Giles, R., [1964]: Mathematical Foundations of Thermodynamics, New York.

Kimbrough, S.O., [1979]: 'On the Reduction of Genetics to Molecular Biology', Philosophy of Science, 46.

Krajewski, W., [1977]: Correspondence Principle and Growth of Science, Dordrecht.
Kuhn, T.S., [1970]: The Structure of Scientific Revolutions, 2nd ed. Chicago.
Kuhn, T.S., [1976]: 'Theory-Change as Structure-Change: Comments on the Sneed Formalism', Erkenntnis, vol. 10
Ludwig, G, [1978]: Die Grundstrukturen einer physikalischen Theorie, Heidelberg.
Mayr, D., [1976]: 'Investigations of the Concept of Reduction, I', Erkenntnis, vol. 10.
Moulines, C.U., [1975]: 'A Logical Reconstruction of Simple Equilibrium Thermodynamics', Erkenntnis, 9/I.
Moulines, C.U., [1980]: 'Intertheoretic Approximation: The Kepler-Newton Case', Synthese, 45.
Moulines, C.U., [1981]: ' A General Scheme for Intertheoretic Approximation', in A. Hartkämper and H.J. Schmidt (eds.), Structure and Approximation in Physical Theory, New York.
Nagel, E, [1961]: The Structure of Science, New York.
Pearce, D., [1982]: 'Logical Properties of the Structuralist Concept of Reduction', Erkenntnis, vol. 18.
Sneed, J.D., [1971]: The Logical Structure of Mathematical Physics, Dordrecht.
Spector, M., [1978]: Concepts of Reduction in Physical Science, Philadelphia.
Stegmüller, W., [1976]: The Structure and Dynamics of Theories, New York.
Stegmüller, W., [1979]: The Structuralist View of Theories, Heidelberg.
Yoshida, R., [1977]: Reduction in the Physical Sciences, Halifax.

E.Scheibe

EXPLANATION OF THEORIES AND THE PROBLEM OF PROGRESS IN PHYSICS

I would like to begin my exposition of theory explanation and the problem of progress with some remarks on "progress". To talk about "progress" has become somewhat delicate nowadays, since the common consciousness at present is subject to a critical if not hostile tendency towards science.

The political, social and economic utilization of the results of research in science, especially physics, has put mankind in a awkward position. One may rightly doubt whether those promises of benefit from scientific progress, which have been constantly repeated since Bacon and Descartes, will ever be unambigously fulfilled. However, I shall not dwell upon the actual mixed feelings caused by the practical consequences of research, but rather on the problem of theoretical progress of science, in particular of physics. As is well known this question has become a much debated issue in the philosophy of physics during the last two decades. The view was put forward that there is no intrinsic evaluation of a transition like that from prerelativistic to relativistic physics or classical to quantum mechanics showing the transition to constitute progress in any straightforward sense. It was argued that so called meaning changes and conceptual incommensurabilities involved in the process would defy all attempts to present the new theories as simply adding to our previous knowledge rather than being, as was claimed, total replacements of the earlier views, that as a matter of principal could be appraised only by their own new standards. Put in broader terms, as has actually been done, the development of the natural sciences and of physics in particular was assimilated to the various other changes of our cultural

W. Balzer et al. (eds.), Reduction in Science, 71–94.

achievements essentially depending on external factors,bearing the stamp of historical relativity, perverted by the idiosynchrasies of their producers and at any rate of a highly questionable rationality.[1)]

Again, I would not like to become directly involved in this debate,which eclipses the formation of a concept of progress adapted to modern physics.It appears to me more satisfying to devote this contribution -as well as my research proper- to a positive program,and to give a glimpse of its basic ideas and its present state of elaboration.This program is in the tradition of logical empiricism,being reconstructionist in method and acknowledging the objective of unity of science. Reconstructionism means to exhaust the conceptual and logical means at hand in order rationally to reconstruct a discipline,like physics, and its history -adequately and as far as is feasible.For the relevant principles of reconstructionism I refer to my (1984),whereas this contribution,as a case study,should exemplify a rational reconstruction more concisely.On the other hand,creating the unity of physics is,of course,not primarily the business of philosophy of science,but of physics itself.This business of establishing the unity cannot be done in a trice, we are still faced with a variety of individual physical theories entangled at first sight.Even in those cases where,according to general belief, unifications have been advanced,it may remain a matter 'of principle',without reflection in the scientific practice.But this very situation provides the secondary task for the reconstructionist minded analysis of science to make complicated circumstances as transparent as possible.

The motif of unity of physics in particular connects the two concepts of theory explanation (or reduction) and of theoretic progress.Explaining a theory by another one always reduces the variety of physical theories in favour of a greater unity of physics: the explained theory in principle is absorbed in a better,more extensive and superior theory,although it may survive for practical purposes.Indeed,the systematic reductions,which we can accomplish nowadays,often reflect theoretical progress in the history of physics.A glance at present day physics is always a glance into the past -like looking into the starry sky.Snell's law of refraction and Ohm's law are older than Maxwell's equations,which in turn are older than quantum electrodynamics.But all these are mentioned in the systematic context of present day physics.Moreover,Maxwell's equations were not simply joined by the two aforementioned laws,nor did quantum electrodynamics simply supervene Maxwell's theory.But these pieces of physics stand in a historical relation of predecessor and successor <u>and</u> in a systematic relation of subordination or reduction.This is at least part of my thesis that the historical steps duly correspond to theoretical progress.

I am conscious that this idea of reconstructing theoretical progress as an intertheoretical relation,viz.explanation or reduction,clashes with the aforementioned trend to regard this kind of analysis as irrelevant if not hopeless.Before entering into my systematic considerations it will therefore be instructive to see how the <u>physicists themselves</u> understand the development of their discipline.There exists something like a 'standard' view' of physicists with respect to this problem, ignored by philosophy of science,which deserves citing in more detail.As far as I see,it goes back to L.Boltzmann,who says[2)]:

"The layman may have the idea that to the existing basic notions and basic causes of the phenomena gradually new notions and causes are added and that in this way our knowledge of nature undergoes a continuous development.This view,however,is erronous,and the development of theoretical physics has always been one by leaps.In many cases it took decades or even more than a century to articulate fully a theory such that a clear picture of a certain class of phenomena was accomplished.But finally new phenomena became known which were incompatible with the theory;invain was the attempt to assimilate the former to the latter. A struggle began between the adherents of the theory and the advocates of an entirely new conception until, **eventually, the latter was** generally accepted."

Obviously,this squares with the account of the development of physics given by Th.Kuhn in nuce: A physical theory or discipline first passes through a time of normality and consolidation. Problems of adapting the theory are a prelude to the crisis,which leads to a rather dramatic break-through of a new theory.On the other hand,the rupture between the old theory and the new one is mitigated by Boltzmann,who continues:

"Formerly one used to say that the old view has been recognized as false.This sounds as if the new ideas were absolutely true and,on the other hand,the old (being false) had been entirely useless.Nowadays,to avoid confusion in this respect,one is content to say: The new way of ideas is a better,a more complete and a more adequate description of the facts.Thereby it is clearly expressed 1) that the earlier theory, too,had been useful because it gave an,if only partially,true picture of the facts,and 2) that the possibility is not excluded that the new theory in turn will be superseded by a more suitable one."

In the light of the first part of this story the second part tells us therefore the following: Although there has been some leap in the transition to the new theory,something from the old theory is conserved.The new one is not completely true, but the old one not completely false either.The transition is a leap,but not a salto mortale.

This Boltzmannian dialectic -as I would like to call it- sounds like prophesy,being formulated in 1895.It was to determine the way physicists understand the development of their discipline also in our century,as I shall show by further citations.A closer inspection,however,reveals some individual variations in estimating the problems of bridging the gap between innovation and tradition.For instance,in 1922,Walter Nernst in an article on "The Domain of Validity of the General Laws of Nature"[3)] referred to the passage just cited and again emphasized that

> "the modification of general laws of nature by no means entirely overthrows the earlier laws; rather the latter are modified only for <u>more or less extreme</u> cases..."

However, made wiser by the lesson of relativity theory, Nernst goes on to point out some difficulties that may occur in the actual proof that, for instance, Newton's gravitational theory is contained in Einstein's in the sense that all empirical successes of Newton's theory are also successes of Einstein's. Anticipating an argument of Feyerabend he says:

> "The modifications that have to be made with respect to the earlier theory are so small that according to the present state of research they can be neglected except for the computation of the orbit of Mercury. But as a matter of principal every computation that astronomers have performed so far must be changed. And it is this principal aspect of the problem, not the numerical amount of the correction, that is our point. To avoid any misunderstanding: the works of Galileo and Newton are "as glorious as on that first day", but they have not brought us the final laws of the motions of the celestial bodies, and nobody would claim this for the theory of relativity..."

But Nernst not only perceives problems in the domain, where computations have already been performed. After mentioning quantum theory which was in those days in statu nascendi, he continues:

> "One might think that...the laws of nature... are valid with absolute precision in certain domains and that the matter could be settled very easily by pointing out the limits within which they remain valid. For all practical applications this is true enough, and it was for this reason that we could ascribe eternal values to the discoveries of Galileo, Newton, ...etc. From a strictly logical point of view, however, the matter appears much more disastrous. If a general law of nature becomes significantly inaccurate beyond certain limits, then the curse of this imprecision comes to roost on every application of the law even with-

in these limits, though the magnitudes of the errors are below the threshold of measurement for the time being."

This rather detailed account of the situation may be followed by the evaluation of Einstein and Infeld (1938) where the conservative component of Boltzmann's scheme again prevails. Starting from an older example, namely the introduction of the field concept into classical physics, which had been available to Boltzmann too, the authors -by the way also following Boltzmann's style- point out:[4)]

"A new reality was created, a new concept for which there was no place in the mechanical description. Slowly and by a struggle the field concept established for itself a leading place in physics and has remained one of the basic physical concepts."

On the other hand the authors finish with the strong generalization:

"But it would be unjust to consider that the new field view freed science from the errors of the old theory of electric fluids or that the new theory destroys the achievements of the old. The new theory shows the merits as well as the limitations of the old theory and allows us to regain our old concepts from a higher level. This is true not only for the theories of electric fluids and field, but for all changes in physical theories, however revolutionary they may seem."

Again a reverse accentuation is given to Boltzmann's dialectic in a recent comment with which I shall close these citations. It is given by Bondi (1975) and again refers to the replacement of Newton's theory of gravitation by Einstein's. He points out:[5)]

"It had been thought that whatever in the world might be difficult, might be complex, might be hard to understand, at least Newton's theory of gravitation was good and solid, tested well over a hundred thousand times. And when such a theory falls victim to the increasing precision of observation and calculation, one certainly feels

> that one can never again rest assured.This is the stuff of progress.You cannot therefore speak of progress as progress in a particular direction,as a progress in which knowledge becomes more and more certain and more and more all-embracing.At times we make discoveries that sharply reduce the knowledge that we have, and it is discoversies of this kind that are indeed the seminal point in science.It is they that are the real roots of progress and lead to the jumps in understanding."

But in spite of these jumps also Bondi turns to the second part of Boltzmann´s story,saying:

> "It is,of course,important to remember that when a theory has passed a very large number of tests,like Newton´s theory,and is then disproved -and we can certainly speak of its disproof now- you would not say that everything that was tested before -all those forecasts- were wrong.They were right,and you know therefore that although the theory qua general theory is no longer tenable,yet it is something that described a significant volume of experience quite well."

Besides the first,progressive part we are again faced with a second,conservative part of theory change constitutive for the notion of progress in a proper sense.

I hope that these citations (another selection can be found in my 1982,291-309) make sufficientily clear what physicists think about progress in science.We may discriminate between two aspects,a progressive and a conservative one.In this paper I want to emphasize the latter.It expresses that a new theory must not be less efficient than its predecessor.It has to explain all those phenomena which the old theory already explained.At least in this respect the theory has to be absorbed,explained or reduced by its successor.Otherwise we would not speak of progress in the usual sense.In order properly to exceed the old theory the new one must have empirical successes which were absent in the old one. Whereas this is obvious and left between the lines of our texts, it is stressed here that this kind of progress is not quite a straightforward matter, since the new successes as a rule will be failures of the old theory, falsifying it or anyhow menacing its right to exist. Thus the break in the succession of theories, repeatedly conjured, might consist in a contradiction or an even more intricate relation between the theories forcing us definitely to abandon the old theory or some parts of it when passing to a new one. From this point of view it is not simply a matter of cumulative progress.

In the following systematic part I want to survey possible concepts of progress among theories. In spite of the situation just described I will start with a concept of explanation according to which the respective explanatum, i.e. what really is explained by means of the explanation, is identical with the respective explanandum, i.e. what originally was to be explained. This simple pattern we meet in the orthodox concept of a deductive-nomological (DN-) explanation of Hempel and Oppeneim.[6] According to their original conception, which has its roots in philosophical tradition, in such an explanation a contingent proposition E, e.g. the occurrence of a certain event, is explained by means of some statement T of lawful dignity. This is achieved with the help of some further contingent statement C which, together with T as premise of the explanation, logically implies proposition E:

$$T \wedge C \vdash E$$

The point of the explanation therefore is that in the domain of law T the contingent proposition E is recognized with logical force as inescapable once the additional contingent state of affairs as expressed by C is realized. In pragmatic respects, e.g. responding to the need of an explanation of E, it goes without saying that the explanation will be regarded as successful only if just this E, and not something else, is actually explained.

The requirements mentioned up to now may be regarded as a priori conditions of adequacy for the concept of DN-explanation.But this concept in addition must conform to further empirical conditions.One of these is that proposition E actually subsists or at least that we have good reasons to believe it subsists.Another one is the corresponding condition for C.Further,there is the condition that law T is well confirmed,and finally,the condition that all those reasons to believe have been acquired independently of each other,if possible.If the explanation is offered to us for the sudden death of Mr.X that Mr.X was poisoned with arsenic then in order to have a proper,and not an ad hoc,explanation we have to know independently of this case that poisoning with arsenic regularly leads to death.

The concept of a DN-explanation thus recalled has to satisfy further requirements,and the conditions just mentioned need further specification. All this cannot be taken into account here,as well as certain intrinsic difficulties which arose for this concept of explanation.For,as usual in the course of the introduction of new concepts in the continuation of older ones,there are other and more important things to worry about at first.My first concern now is with the logical frame in which all further reconstructions are to take place.With respect to the recapitulation of the Hempel-Oppenheim concept of explanation just given,I have left the frame open.This is common practice;if not classical first-order logic is taken as such a frame.Here I cannot remain with the first alternative,but also I cannot concur with the last one.For I want my general account to obtain a certain degree of definiteness,whereas on the other hand the customary use of standard logic will not be sufficient for my examples, which are to be taken from present day physics.

In my opinion a sufficient degree of definiteness,and optimal possibilities of application, can be best achieved today by using set theory as a logical frame.For one thing in this frame the reconstruction of theoretical physics is possible, insofar as theoretical physics makes use of modern mathematics. Everybody knows today that this is so to a considerable degree. But, on the other hand, the hypothesis that in formal respects modern mathematics completely covers the conceptual frame of physics, and that all phsical questions in the narrower sense are questions of interpretation has been given too little consideration and exploitation. In order to indicate at least what possibilities exist and become clearer by means of a set-theoretical reconstruction, I want to elucidate briefly the theory-concept to be used in the following.[7)]

This theory-concept preserves the feature that, seen from a formal point of view, a theory is a statement, although what it speaks about may have some status strangely oscillating between reality and possibility. The statement is about some finite system of sets, and it says **that this** system is a structure of a certain type. Furthermore, it expresses something particular, although still general, by means of some axioms, and in this way it fixes the species of the structure under consideration. The first part of such a statement, called typification, says of some sets s of the system that they are "echelons" over the other, proper base-sets X of the system.[8)] Here, an echelon over the set X is the result of iterated application of forming cartesian products and power sets. Because of this construction the typified sets s of a structure may function as predicates of arbitrary types, in particular of arbitrarily high order and arity. In this way the expressive power of theories falling under our theory-concept mentioned before is guaranteed. This power is not restricted by the second component of a theory, the theory's axiom proper which has only to satisfy the requirement of being invariant under transitions to isomorphic structures. In shorthand notation the form of a theory thus is as follows:

$$s \in \sigma(X) \wedge \alpha(X;s) \qquad [\equiv \Sigma(X;s)]$$

where $\sigma(X)$ is an echelon over the sets X.

Species of structures, and therefore theories in my sense, are abundant in modern mathematics and -as already said- partly even in textbooks of theoretical physics. In mathematics, on the one hand algebraic species of structures like groups, rings, vector spaces etc., and on the other hand topological spaces and the species of structures based on them are fundamental. But I want to illustrate the idea of a theory with an example which has a close relation to physics for a much longer time already: Euclidean geometry. Here, X is space, regarded as a set of points, and s may be introduced -as one of several possibilities- as a distance function on X. That is, s would obtain the typification

$$s \in \mathrm{Pot}(X^2 \times \mathbb{R})$$

where $\mathbb{R}$ is the set of real numbers. Again as one of several other possibilities of axiomatization there is the especially simple -though somewhat brutal- one of expressing the proper content of Euclidean geometry in one sentence, requiring that there exists a global coordinate system with respect to which the distance function s becomes the Euclidean distance of $\mathbb{R}^3$. This would be the other part, the axiom proper of our theory. The geometrical example makes clear that -as already announced- a physical structure has the ontological status of real possibilities: the points of space and their distances are never realized in total but they <u>can</u> be realized in principle in each case to the extent presumed by the theory. Also, the example of geometry gives grounds for hoping that not only some of its concomitants but also a physical theory <u>in toto</u> may be reconstructed as a species of structures. And just this is part of the abovementioned program of reconstruction, as performed today by some theoretical physicists like Ludwig, Truesdell, Noll, Sachs and others, but especially by the first.[7)]

Eventhough the examples of species of structures just mentioned show that not each such species can be used for the reconstruction of a physical theory itself, and though there would be much to say about the possibility of further restricting the theory-concept in this way, I have to remain content with these general remarks in order to return now to my proper subject of explanation and progress. Looking back at DN-explanations as our point of orientation, I first want to take up its deductive part and now replace this part by a deduction of the form

$$\Sigma(X;s) \wedge C^{[\Sigma]}(X,s,Y,t) \vdash \Theta(Y;t) \qquad (I)$$

Here, the symbol for deduction already belonging to the logical frame is meant to denote set-theoretic deduction, i.e. so that the theorems of set theory may always be tacitly used as premises. For the rest let Σ and Θ be theories in the sense just explained, and let $C^{[\Sigma]}$ extend Σ to a theory in which the Y occur as new base-sets, proper or typified through the X, and in which the t occur as typified through the X.

With respect to our aims of application I will always assume that Σ stands for the new theory, and I will assume in the beginning that Θ denotes the old theory which is to be explained by Σ. In this sense (I) yields a total explanation of the original explanandum from which, however, we will have to move later on in the direction of more modest possibilities. For the moment the real problem consists of the relation between the third component, i.e. the additional requirements $C^{[\Sigma]}$, and the theory Σ. The problem is that by means of Σ we can explain nearly everything we want, as long as the C-requirements are not further restricted. If here we allow for too much one would say immediately that the explanation is no longer performed by Σ. Therefore admissible C-extensions must not substantially change this theory. But when does this happen, and when not? Through experience in dealing with theories we have developed a certain feeling for this frontier, but at the moment our question cannot be answered precisely. A relatively clear case of a C-extension which is generally admissible is the case in which the structure <Y;t> is defined out of <X;s> by means of $C^{[\Sigma]}$, i.e. the case in which $C^{[\Sigma]}$ has the form

$$Y=P(X;s)\wedge t=q(X;s)$$

where P and q are so-called intrinsic terms for Σ which in particular require that all Y become sets typified by the X. But definitions alone won't do, as becomes clear immediately if we look at how, for instance, classical mechanics is developed. Essentially this is done by means of specializations, and specializations are extensions but not extensions by definitions. On the other hand we cannot allow for arbitrary extensions. For classical mechanics is an extension of Euclidean geometry, and one does not want to say that for instance the law of conservation of momentum is explained by geometry. A minimal condition of adequacy of a C-extension in my present opinion is that no proper base-sets are added. If this condition is also to be sufficient, the introduction of new concepts into a theory will have to be regulated by means of prescriptions more severe than the ones used today. From the point of view of physics the introduction of an electromagnetic field into a Lorentz-manifold, for instance, is the introduction of a new, ontologically independent entity. But technically this is done with a tensor field and without adding a new, proper base-set.

Ultimately the problem which extensions of a theory are admissible for reasons of explanation is of course a problem of the hierarchy of theories which at the orthodox level of DN-explanations has been discussed -though in somewhat veiled fashion- as the problem of demarcation of lawful and contingent statements, and which has not been solved there either. In leaving for the moment the problem of whether an extension is admissible to our respective intuitions, I now want to point to the paradigm of an <u>exact deductive</u> explanation of a theory -as I want to call case (I)- as given by the explanation of Kepler's three laws by means of Newton's theory of gravitation, <u>if</u> the latter is already described in its central field approximation, that is as a theory of the motion of some body idealized as a mass-point in the gravitational field of a big mass, <u>and if</u>

Kepler's laws (as explanandum) are understood as referring to a distinct set of bodies, for instance to the six planets known in Kepler's times. The C-extension of Newton's theory then would say essentially that the energy of each of the bodies involved (in the central field of the sun) is negative.

It has to be noted, however, that according to a slightly different view of Kepler's (and correspondingly of Newton's) laws this case can be subsumed only under some more general type of explanation. If Kepler's theory is understood so that <u>every</u> (finite) set of bodies in the gravitational field of the sun moves according to Kepler's laws, then this statement would be false from the point of view of Newton's theory, for this theory allows for the possibility of hyperbolic and parabolic paths (in case of non-negative energy). This is exactly the point in which the (simplified) Newtonian theory corrects Kepler's. Here we have the phenomenon we already met in the introductory citations, namely that a new theory <u>cuts down</u> the <u>domain of validity</u> of its predecessor. We have to note then that, strictly speaking, the new theory does <u>not</u> allow for a deduction of the original theory but only of some restriction of the latter.

This class of explanations by restriction of the domain of validity belongs to a much bigger class of <u>partial</u> explanations of a theory. I briefly want to introduce the latter for the exact deductive case, although only its approximative counterpart plays an important role in applications. In the present context of exact explanations, partial explanations are those for which not the original theory, the explanandum proper, can be deduced in the sense of (I), but only some consequence of an admissible extension of it. If Θ_o denotes the explanandum and Θ the explanatum in (I), then these two are related as follows

$$\Theta_o(Y_o;t_o) \wedge C^{[\Theta_o]}(Y_o,t_o,Y,t) \vdash \Theta(Y;t) \qquad (A)$$

and Θ would be totally explained by Σ while Θ_o would be explained by Σ only partially. Again, it is essential that $C^{[\Theta_o]}$ in (A) is an admissible extension of Θ_o because otherwise one could hardly speak of a partial explanation of Θ_o, let alone of an explanation of Θ_o. Well known exact, but only partial, explanations of classical statistical mechanics (Θ_o) are provided by quantum mechanics (Σ), for instance by means of the first theorem of Ehrenfest (Θ). To see this one proves the *classical* connection for the mean values of the distribution of position and momentum from quantum mechanics, that is, the relation which can *also* be derived for those mean values from classical statistical mechanics. Both theories therefore, under respective additional assumptions, have a common consequence for the same thing -as required by (I) and (A).

Explanations by restriction of the domain of validity are special cases of partial explanations. For them the *form* of the explanandum and of the explanatum is the same, i.e. we have $\Theta \equiv \Theta_o$. But in the explanatum a statement of this *one* form is made for some domain of validity $\langle Y;t \rangle$ *restricted* with respect to $\langle Y_o;t_o \rangle$. Therefore in this case (A) takes the form

$$\Theta_o(Y_o;t_o) \wedge Y=P(Y_o;t_o) \wedge t=t_o | Y \vdash \Theta_o(Y;t) \quad (I')$$

and (I) the form

$$\Sigma(Y_o;t_o) \wedge Y=P(Y_o;t_o) \wedge t=t_o | Y \vdash \Theta_o(Y;t) \quad (A')$$

where in addition the requirement of restriction

$$\Sigma(Y_o;t_1) \vdash P(Y_o;t_o) \subseteq Y_o$$

is satisfied and the stroke | denotes an automatic restriction the definition of which I have to suppress here. In this way, in the above example of Kepler-Newton, a set of bodies in the gravitational field of the sun which is arbitrary at first, and for which originally Kepler's laws Θ_o were claimed to hold, is restricted to the set $P(Y_o;t_o)$ of those among them the total energy of which is negative. And for *this* set Kepler's laws then follow from those of Newton (in central field approximation).

The forced partial explanations give a first clear idea of those aspects of theoretical progress which were touched on in the introducing citations. Exact explanations with restriction of the domain of validity are just those mentioned by Nernst, where he said

"One might think that...the laws of nature... are valid with absolute precision in certain domains and that the matter could be settled very easily by pointing out the limits within which they remain."

But Nernst is quite correct in concluding that "the matter appears much more disastrous". For the concept of exact deductive explanation considered above has no real application within the scope of evolution and progress of theories, and is here of purely heuristic value. It is better suited to understanding how a new theory develops in exploiting its as yet unknown consequences, although insuperable mathematical difficulties may force one to use approximations. This idea which is well known to every physicist leads to the much more useful concept of approximative deductive explanation. Strictly speaking even the Kepler/Newton case belongs here, since a system of gravitating bodies almost never obeys the Kepler laws exactly.[9] But under certain contingent assumptions it behaves approximatively like a Kepler system. It is said to contain the Kepler solution as an approximate limiting case, and the problem arises how to comprehend generally and precisely this kind of explanation.[10]

Following an idea of Ludwig[7] this problem can be solved by means of a species of structures called by mathematicians uniform spaces. In some cases, therefore, approximation may be achieved by means of a (topological) metric. In a uniform space a family of sets of pairs of points, so-called entourages is singled out. Two points of a pair belonging to an entourage U are, so to speak, indistinguishable up to U. These limits will have to suffice if we are now going to sketch how to treat theories and theory relations topologically. We assume that a uniform structure is given on an echelon set over the structure the theory is talking about. Now we may coarsen the theory

$$\Theta(Y;\ldots t_*\ldots)$$

by means of an entourage U obtaining a statement $\Theta_U(Y;\ldots t_*\ldots)$ given by

$$\Theta_U(Y;\ldots t_*\ldots) \equiv \exists\tau \,.\, \Theta(Y;\ldots\tau\ldots)\ \tau \underset{U}{\approx} t_* \qquad (B)$$

This statement -by the way logically weaker than Θ - thus says that the previous statement is not fulfilled by t_*, but -ceteris paribus- by some which is U-close to t_*. Using this coarsening we may replace the exact deductive explanation (I) by an approximative deduction

$$\Sigma(X;s)\wedge C_\varepsilon^{[\Sigma]}(X,s,Y,..t_*..) \vdash \Theta_U(Y;..t_*..) \qquad (II)$$

ε being some parameter of approximation.

Let us suppose for the moment that the above Θ is our explanandum and that thus far a total explanation is achieved. However, it is plain that by the very feature of approximation we wander far from our primitive idea of explaining a given theory. Indeed, one would assume an ε to exist for every entourage U such that (II) holds, but this would generally not approximate any definite statement on the right hand side. For if U becomes smaller and smaller those structures <Y;t> satisfying the premise will become rarer and rarer and none may survive the limit process. Let us demonstrate this with a simple example which can be straightforwardly reconstructed within the present frame. In this example Σ asserts that an arbitrary set of gases satisfies the van der Waals equation

$$\left(p+\frac{a}{v^2}\right)(v-b)=RT$$

with given constants a, b and R. If we restrict this set to those elements which fulfil the C-conditions

$$\frac{a}{pv^2} < \varepsilon_1 \wedge \frac{b}{v} < \varepsilon_2,$$

we may conclude that there exist values p', v' and T', which are for suitable ε_i arbitrary close to the given values and satisfy the ideal gas law

$$p' \cdot v' = R \cdot T'$$

In this sense the latter can be approximately explained by the van der Waals equation. But it is plain, too, that both equations have no common solution and hence there is no limit statement.

Such examples of approximate explanations involving only simple empirical laws are abundant in physics. The mentioned Kepler/Newton case employing rigorously Newton's law of gravitation belongs to the more complicated cases, which, nevertheless, I confidently expect to be reconstructable according to my proposal (II).[10) In all these cases we have in particular a contradiction between the involved theories which forces us -strictly speaking- to reject the overthrown theory. But still there are more uncomfortable occasions where in close analogy to the type of exact explanations we have to consider <u>partial</u> explanations in the approximative case, too. Before giving the general formulation, I shall premise the following simple example. In any textbook of quantum mechanics, the reader is taught the energy spectrum of the hydrogen atom first in the context of the problem of an electron within a classical Coulomb field. Some textbooks later on improve on this theory by treating the atom correctly as a 2-particle-system thus replacing the Coulomb field by a co-moving proton. Again the energy levels are calculated and then it turns out that they have changed, although the difference is smaller the higher the mass ratio proton/electron. If any case deserves to be reconstructed in the present frame, it is this one. But textbooks in no way provide an approximative explanation of the full 1-particle-theory from the 2-particle-theory of the H-atom, but only explain the approximation of the <u>energy spectrum</u> on the grounds of the mentioned mass ratio.

I do not want to claim that in this example a total explanation is impossible[11), but rather I want to generalize what is actually done in this case towards a concept of partial approximative explanation. As in the case of an exact, partial explanation only a consequence of the extended theory is deduced, but now only approximatively. Thus we have to consider a combination of formulas (II), (A) and (B), where a conclusion of (A) is to be directly approximated according to (II) and (B). In the example of the H-atom one requires besides the 2-particle treatment that the state function not only solves the Schrödinger equation but is also an eigenfunction of the inner energy operator. Then it follows from the condition on the mass ratio, that there is an eigenfunction of the 1-particle energy operator with an eigenvalue close to the given 1-particle energy. This approximation is a weak one in so far as the approximation is performed only with respect to the energy but not with respect to the pertaining evolution of the state of the electron, let alone to an arbitrary evolution.

With the concept of a partial approximative explanation at hand one could attack the really intricate cases of theory comparison, viz. special and general relativity or the relation of quantum theory to certain parts of non-relativistic classical physics. This enterprise, however, would be the subject of another paper and here I shall confine myself to the following remarks.[12] Total, or even approximate explanations of greater parts of classical physics have not yet been achieved, say explanations of Newton's theory of gravitation from Einstein's or general particle mechanics from the corresponding quantum mechanics. And there are good reasons to doubt that they could ever be achieved. But it might be plausible, that partial approximate explanations are possible and a great deal of existing material could be reconstructed in terms of the concepts proposed here. I take it to be plausible for the following reason. In principle, the difficulty arises that quantum and classical mechanics, for instance, rely on basic concepts far apart from each other, or, in my terms, that the physical structures they postulate are very different. But restricting the problem to one of partial explanation reduces the difference because one is free to choose suitable <u>parts</u> of the structures for comparison, or even identification, and there are still sufficiently many such parts.

In my concluding remarks I want at least to mention a new issue, new in a still more extended sense, without being able to dwell upon it. Deductive explanations alone, let it be exact or approximate, total or partial, will be inefficient if certain empirical conditions are not imposed. This may become clear at once if we consider partial explanations which are symmetric on both sides, if not in all specializations at least in the general formulation. However, a condition of progress requires a properly asymmetric relation, which plainly comes into play in the real state of affairs or -empiristically purified- in the experience actually made. The aforementioned comparison of two energy spectra of the H-atom says nothing about which theory is empirically better as long as the measured spectral lines are left out.

A similar feature has been mentioned with respect to the orthodox DN-explanation of a contingent fact: besides logical conditions we need conditions of empirical confirmation to characterize such explanations. Also in theory explanation we shall have to require such conditions. But now something new is added: the explanandum is itself a theory, which -as opposed to a contingent fact- we may sometimes successfully use for empirical explanations, i.e. deriving observed phenomena, whereas sometimes the same theory fails in doing so. As indicated above, the new theory will be empirically superior to the old if it can reproduce the latter's empirical successes and turn its failures into own successes. These remarks already indicate that it is not only a matter of finding additional empirical conditions for an otherwise deductive concept of explanation, but an autonomous, empirical and ab initio asymmetric concept of progress is looming up. For if those conditions are fulfilled why should we then trouble over additional deductive relations so far discussed?

Indeed, there is a type of empirical theory explanation independent of deductive explanation, or better: the desideratum for such a type exists. For it is not possible at present to introduce this concept in as similarly precise a way as the other one treated in this paper. The reason is that with this concept we enter into the realm of inductive systematizations which does not posses such an accepted basis as the logical or rather set-theoretical tool of deduction. With respect to a single empirical explanation and its reproduction by the superior theory the question arises not only which concept of such an explanation one has to build. One also has to ask whether every form of empirical confirmation whatsoever, together with each of its factual instances, has to be reproduced by the better theory. This problem of intertheoretic relations can hardly be tackled at present, since the situation has not even been clarified for single cases of sufficient interest.

I am not saying that this contingent state of affairs in research is the only reason for me to take refuge in the better known realm of deductive relations. There are pertinent links between the two domains showing that empirical explanations cannot obviate deductive explanations. On the one hand we have already touched on the problem of empirical explanations when dealing with partial deductive explanations: it has not been excluded in formula (A) that Θ refers to a contingent fact. If thus Θ_o is explained by Σ via a set of such contingent Θ's representing our experience, we have accounted for at least the rudiments of one kind of empirical explanation, exact or approximate. In orthodox terms it would be founded on the concept of hypothetic-deductive confirmation.

On the other hand there are more general connections of fundamental significance. Physicists do not ever and anew try to establish extensive deductive relations between theories, say deriving Newton's gravitational theory from Einstein's, as a goal in itself. Rather they hope to achieve at one blow what Nernst postulated in the Newton/Einstein case, namely that "every computation that astronomers have performed so far must be changed." One hopes that a deduction, an at most approximative one in this example, would concisely show how to change the calculations in every instance, and would generally guarantee the feasibility of such changes. This would constitute the importance of deductive explanation. But the question is which condition would in fact establish such a guarantee. Besides the problem of formulating concepts of empirical explanations we thus encounter another problem, hardly tackled up to the present, to find sufficient conditions for applying deductive explanations. I shall finally represent this problem by the formula

$$\mathrm{Ded}(\Sigma, \Theta, C) \wedge \mathrm{Apl}(\Sigma, \Theta, C, \beta) \Rightarrow \mathrm{Emp}(\Sigma, \Theta, \beta)$$

where Ded is of the form (I) or (II), β renders some experience, Emp stands for empirical explanation and Apl is the desired condition guaranteeing empirical explanation by deduction. Almost nothing is known at the general level[13] about the relations expressed by this formula, and nothing is more apt than this situation for demonstrating that our philosophy of science is only as advanced as physics was in the days of Galileo. At the same time there is however no reason to despair of the attempt to explicate an adequate concept of progress, as some current trends do.

E. Scheibe
Philosophisches Seminar der Universität
Marsiliusplatz 1
D-6900 Heidelberg 1

NOTES

1) Especially by P. Feyerabend (1975).
2) L. Boltzmann (1979), p.95 and p.60.
3) W. Nernst (1922). Here: p.489 and pp.491.
4) A. Einstein and L. Infeld (1938), p.158.
5) H. Bondi (1975), p.3.
6) C.G. Hempel and P. Oppenheim (1948). See also C.G. Hempel (1965) and (1977).
7) Compare G. Ludwig (1978) and E. Scheibe (1979).
8) For a system of sets $Y_1, \ldots, Y_n$ we here always write a vector Y as a shorthand.
9) E. Scheibe (1973a).
10) See note 9) and also E. Scheibe (1973b), C.U. Moulines (198o) and A. Hartkämper and H.-J. Schmidt (1981).

11) E.Scheibe (1981).
12) Compare A.Ashtekar (1980) and J.Ehlers (1981) for the situation in quantum mechanics and relativity theory.
13) See E.Scheibe (1982) and E.Scheibe (1983).

REFERENCES

A.Ashtekar, 1980, "On the Relation Between Classical and Quantum Observables",Commun.Math.Phys. 71, 59-64

L.Boltzmann, 1979, Populäre Schriften,Braunschweig,original edition: Leipzig 1905

H.Bondi, 1975, "What is Progress in Science?",in: R.Harré (ed.): Problems of Scientific Revolution, Oxford

J.Ehlers, 1981, "Über den Newtonschen Grenzwert der Einsteinschen Gravitationstheorie",in: J.Nitsch et al. (eds.): Grundlagenprobleme der modernen Physik,Mannheim

A.Einstein and J.Infeld, 1938, The Evolution of Physics, New York, in German: Die Evolution der Physik,Wien (1950),Hamburg (1956).

P.Feyerabend, 1975, Against Method,London,German translation: Wider den Methodenzwang, Frankfurt (1976).

A.Hartkämper and H.-J.Schmidt, 1981, (eds.), Structure and Approximation in Physical Theories, New York

C.G.Hempel and P.Oppenheim, 1948, "Studies in the Logic of Explanation", Phil.Sci. 15, 135-175

C.G.Hempel, 1965, Aspects of Scientific Explanation and other Essays in the Philosophy of Science, New York, German translation: Aspekte wissenschaftlicher Erklärung, Berlin (1977)

G.Ludwig, 1978, Die Grundstrukturen einer physikalischen Theorie, Berlin-Heidelberg-New York

C.U.Moulines, 1980, "Intertheoretic Approximation: The Kepler-Newton Case", Synthese 45, 387-412

W.Nernst, 1922, "Zum Gültigkeitsbereich der Naturgesetze", Naturwissenschaften 10, 489-495

E.Scheibe, 1973a, "Die Erklärung der Keplerschen Gesetze durch Newtons Gravitationsgesetz",in: E.Scheibe and G.Süßmann (eds.),Einheit und Vielheit.Festschrift für C.F.von Weizsäcker,Göttingen, 28-118

E.Scheibe, 1973b, "The Approximative Explanation and the Development of Physics",in: P.Suppes et al. (eds.), Logic,Methodology and Philosophy of Science IV, Amsterdam, 930-942

E.Scheibe, 1979, "On the Structure of Physical Theories",in: I.Niiniluoto and R.Toumela (eds.), The Logic and Epistemology of Scientific Change, Amsterdam, 205-224

E.Scheibe, 1981, "Eine Fallstudie zur Grenzfallbeziehung in der Quantenmechanik",in: J.Nitsch et al. (eds.), Grundlagenprobleme der modernen Physik, Mannheim

E.Scheibe, 1982, "Zum Theorienvergleich in der Physik",in: K.M.Meyer-Abich (ed.),Physik,Philosophie und Politik,Festschrift für C.F.von Weizsäcker, München, 291-309

E.Scheibe, 1983, "Two Types of Successor Relations between Theories", Zeitschrift für allgemeine Wissenschaftstheorie 14, 68-80

E.Scheibe, 1984, "Zur Rehabilitation des Rekonstruktionismus",in: H.Schnädelbach (ed.), Rationalität, Frankfurt

Joseph D. Sneed

REDUCTION, INTERPRETATION AND INVARIANCE

0. INTRODUCTION

0.0

This paper sketches a way of describing reduction relations among theories as a special case of a more general view of the logical structure of related empirical theories. The key concepts in this view are "model element" (Sec. 1) and "intertheoretical link" (Sec. 2). Model elements provide a formal characterization of the conceptual apparatus of individual theories. Intertheoretical links provide a unified treatment of particular, well-known intertheoretical relations such as reduction, specialization and theoritization and include, as well, other types of relations among empirical theories. The global structure of empirical science is represented as a net of linked theories (Sec. 4). Central to the understanding of empirical science are interpreting links. These links provide a kind of "empirical semantics" for the mathematical apparatus associated with individual theories. Interpreting links are characterized and distinguished from reducing links (Sec. 3). The concept of an interpreting links provides us with a formal characterization of the distinction between theoretical and non-theoretical concepts, relative to a given theory and the net in which it lives (Sec. 5), as well as a formal characterization of the intended applications of a theory in a net (Sec. 7). The role of invariance principles in relation to interpreting links is described (Sec. 6). Finally, these ideas are exploited to provide an account of the way interpreting and reducing links work together (Sec. 8).

1. MODEL ELEMENTS

1.0

Theories may be represented as an intricate array, web or net of linked theory elements. Indeed, one may envision the whole of empirical science - not just single theories - as

W. Balzer et al. (eds.), Reduction in Science, 95–129.

a "net" of linked theory elements. Intuitively, a theory element, in isolation from the context in which it occurs, consists of some "concepts" - call them K - that are used to say something about some array of things, the intended applications for the concepts - call them I. Thus a theory element is a certain kind of ordered pair

$$T = \langle K, I(K) \rangle.$$

1.1

On our view of empirical theories, the "conceptual apparatus" K of the theory element consists of certain categories (Mac Lane (1977)) of set-theoretic structures. A category $\|\chi\|$ is a an ordered triple of classes

$$\|\chi\| = \langle |\chi|, \chi, \chi_c \rangle,$$

where $|\chi|$ is the class of "objects", χ the class of morphisms and χ_c the class of morphism compositions. Intuitively, $|\chi|$ is a class of set-theoretic structures or models associated with the theory. In all categories associated with empirical theories it appears that $|\chi|$ may always be regarded as a "species of structures" in the sense of Bourbaki ((1968). χ is a function:

$$\chi\colon |\chi| \times |\chi| \rightarrow |SET|.$$ [1]

For x, y $\in |\chi|$, $\chi(x,y)$ is the set of morphisms "from x to y". Intuitively, morphisms are things like structure preserving transformations from the domain of x into the domain of y, e.g. group homomorphisms. In categories whose class of objects is a species of structures the morphisms will be structure preserving mappings relating the basic sets in different structures. In categories associated with empirical theories χ_c will be nothing more than composition of functions. Each object $x \in |\chi|$ has a unique "identity morphism" $\iota_x \in \chi(x,x)$. Among these morphisms which may be in $\chi(x,y)$ are "isomorphisms". A morphism $\mu \in \chi(x,y)$ is an isomorphism iff there is a $\mu^{-1} \in \chi(y,x)$ so that μ and μ^{-1} composed both ways yield the identity morphisms. We denote the set of isomorphisms by '$I\chi$' so that

$$I\chi(x,y) \subseteq \chi(x,y)$$

and when $I\chi(x,y) \neq \Lambda$, we say 'x and y are χ-isomorphic' which we abbreviate by '$\chi^{=}(x,y)$'. In the categories associated with empirical theories the members of $I\chi(x,y)$ are intuitively thinks like "scale transformations" and "coordinate transformations" that relate different models of the same theory. They are essential to describing the "invariance properties" of the theory's claims. Very roughly, we want to view the conceptual apparatus of mature empirical theories as categories of set-theoretic structures - classes of models for the theory, together with morphisms which relate "empirically equivalent" structural features of these models. Intuitively, the models for a theory element are to be identified with the theory's empirical laws.

1.2

The conceptual core K of a theory element consists of two categories of structures which we call "potential models" and "models". Potential models are just the kinds of structures that one might intelligibly claim to be models for a theory. They determine the formal properties of the theory element's conceptual apparatus without imposing any additional laws. We may make this distinction somewhat more precise by defining a "model element". A model element is the kind of structure that could serve as the formal core K of a theory element T. It is an ordered pair

$$K = \langle \| M_p \| , \| M \| \rangle$$

in which $\| M \|$ is a "full sub-category" of $\| M_p \|$

$$\| M \| \subseteq_f \| M_p \|$$

i.e. $|M| \subseteq |M_p|$; for objects in $\|M\|$ all M_p-morphisms are also M-morphisms and identity morphisms remain the same. $\| M_p \|$ is the category of potential models for theory element T and $\| M \|$ is the category of models for T. We also require that the laws of T be "invariant under M_p-isomorphisms" in the sense that, for all $x,y \in |M_p|$,

A) if $M_p^{=}(x,y)$, $x \in |M|$ iff $y \in |M|$.

Intuitively, the concept of "empirically equivalent" structures is completely captured by the conceptual apparatus layed out in the potential models and all laws stated using this conceptual apparatus are compatible with this concept of empirical equivalence.

1.3

We consider theory elements

$$T = \langle K, I(K) \rangle$$

where

$$K = \langle \| M_p \| , \| M \| \rangle$$

is a model element and I(K) is the range of intended applications of K. We might think of the "claim" of T as that

$$I(K) \subseteq |M| ,$$

but this is not quite correct. In Sec. 7 the claim of T will be discussed in connection with the characterization of I(K).

2. MODEL ELEMENT LINKS

2.0

Globally, scientific theories and empirical science as a whole are a collection of "local" thories - to a first approximation, what we call 'model elements' - connected by what we will call 'intertheoretical links'. We may consider a concept of intertheoretical link that is general enough to describe a model element's "external links" with other model elements as well as "internal links among the model element's own potential models. Generally, intertheoretical links serve to carry information about the values of relations and functions from the applications of one theory to those of another or across different applications of the same theory. External links appear to be a necessary condition for a theory element core to be associated with an empirical theory. Among other things, these links provide a kind of "empirical semantics" for a model element that make it more than "just a piece of mathematics".

2.1

We may define a general concept of "intertheoretical link" between model elements and say what it is for an intertheoretical link to link specific components of the potential models of the linked model elements. In general, intertheoretical links may link any finite number of theories or any finite number of potential models of the same theory. We shall not consider the general case here. We begin by simply trying to characterize the purely formal properties of a binary intertheoretical link between model element cores K' and K. In order to formalize the most general notion of a binary intertheoretical link, we take a step back and consider "relators" between categories $\|\chi\|$ and $\|\Psi\|$. We may think of a relator between $\|\chi\|$ and $\|\Psi\|$ - a $\|\chi\|$, $\|\Psi\|$ relator - as a category $\|R\|$ that is a sub-category of the category $\|\chi\| \times \|\Psi\|$, i.e.

$$\|R\| \subseteq \|\chi\| \times \|\Psi\| .$$

Thus

$$|R| \subseteq |\chi| \times |\Psi| ;$$

for $\langle x,y\rangle, \langle x',y'\rangle \in |R|$,

$$R(\langle x,y\rangle,\langle x',y'\rangle) \subseteq \chi(x,x') \times \Psi(x,x');$$

$$\iota_{\|R\| \langle x,y\rangle} = \iota_{\|\chi\| \times \|\Psi\| \langle x,y\rangle} = \langle \iota_x, \iota_y\rangle .$$

and morphism composition in $\|R\|$ is the same as in $\|\chi\| \times \|\Psi\|$. Intuitively, $|R|$ pairs objects of $\|\chi\|$ with objects of $\|\Psi\|$ and, for pairs of pairs that are in $|R|$, R pairs their morphisms. Further, the identity morphism for $\langle x,y\rangle$ is just the pair of identity morphisms $\langle \iota_x, \iota_y\rangle$. A relator is a generalization of the usual category theoretic concept of a functor. Relators are required to preserve isomorphism in the same way that functors do in that χ- isomorphisms are paired only with Ψ-isomorphisms and conversely. But note that, for $\langle x,y\rangle$ and $\langle x',y'\rangle \in |R|$, there is no guarantee that all, or indeed any, isomorphisms in $\chi \times \Psi(\langle x,y\rangle,\langle x',y'\rangle)$ will be in $R((\langle x,y\rangle,\langle x',y'\rangle)$, since $\|R\|$ is generally not a full sub-category.

2.2

Intuitively, we may begin by noting that a link between K' and K is a restriction of the potential models of both theories. That is, it is a sub-set of $|M_p'| \times |M_p|$. But we also note that a link may have associated with it some restriction of the morphisms of both $\| M_p' \|$ and $\| M_p \|$. Intuitively, the morphisms associated with the links will have to do with transformations of just those components of the structures in $|M_p'|$ and $|M_p|$ whose values are correlated by the link. This suggests that we might regard a binary link between the model element cores K' and K as an $\| M_p' \|$, $\| M_p \|$ relator $\| \lambda \|$ in which

$$|\lambda| \subseteq |M_p'| \times |M_p|$$

characterized the mutual restriction of the potential models and

$$\lambda(\langle x',x\rangle,\langle y',y\rangle)) \subseteq M_p'(x',y') \times M_p(x,y)$$

characterized the pairs of M_p' and M_p morphisms associated with transformations of the linked components. For example, suppose $\| \lambda \|$ is the moment of inertia, mass function link between rigid body mechanics and particle mechanics. Suppose $\langle x',x\rangle$ and $\langle y',y\rangle$ both consist of a rigid body system and a particle system with correspondending values for the moment of inertia and mass functions and differ only in that these quantities are measured in different units. Then $\lambda(\langle x',x\rangle,\langle y',y\rangle))$ would contain the "corresponding" pair of transformations of the moment of inertia function and the mass function required to transform x' to x and y' to y and $I\lambda(\langle x',x\rangle,\langle y',y\rangle)$ would contain such pairs of isomorphisms. This example suggests that links should have certain properties that relators, in general, do not have. Generally, a $\| \lambda \|$, $\| \Psi \|$ relator need not contain any $\chi \times \Psi$-isomorphisms at all since it is not a full sub-category of $\chi \times \Psi$. But in case of links between empirical theories the isomorphisms are of crucial importance. We might require that K', K links contain all $M_p' \times M_p$ - isomorphisms. That is, for $\langle x',x\rangle,\langle y',y\rangle \in |\lambda|$

$$I[M_p' \times M_p](\langle x',x\rangle,\langle y',y\rangle) \subseteq \lambda(\langle x',x\rangle,\langle y',y\rangle).$$

This would, however, be too strong. Intuitively, we do not want all $M_p' \times M_p$-isomorphisms, but only those that are associated with the components in the potential models that are correlated by the link. The weaker requirement:

$$\text{If } I[M_p' \times M_p](\langle x',x\rangle,\langle y',y\rangle) \neq \Lambda \text{ then}$$
$$I[M_p' \times M_p](\langle x',x\rangle,\langle y',y\rangle) \cap \lambda(\langle x',x\rangle,\langle y',y\rangle) \neq \Lambda$$

will suffice. Intuitively, if $\langle x',x\rangle$ and $\langle y',y\rangle$ are λ-linked, x' is M_p'-isomorphic to y' and x is M_p-isomorphic to y, then there is some λ-linked pair of isomorphisms "connecting" x' with x and y' with y. λ-isomorphic pairs will typically have many $M_p' \times M_p$-isomorphisms that are not λ-isomorphisms. Intuitively, these correspond to transformations of components of the potential models other than those that are correlated by λ.

2.3

It also seems intuitively clear that, whenever x' and x are λ-linked and y' is empirically equivalent to x', there should be some y that is empirically equivalent to x that is λ-linked with y'. Roughly, links between model element cores should be "M_p' and M_p isomorphism invariant" in the same way that the laws in single model element cores are. More precisely, a link $\|\lambda\|$ should be isomorphism invariant in the following sense.

If $\langle x',x\rangle \in |\lambda|$ and $y' \in |M_p'|$ is M_p'-isomorphic to x', then there is some $y \in |M_p|$ which is M_p-isomorphic to x so that $\langle y',y\rangle \in |\lambda|$.

This requirement has the effect that links may be regarded as linking isomorphism equivalence classes in $|M_p'|$ and $|M_p|$.

2.4

To talk about specific kinds of links, it is convenient to identify the corresponding functions (or generally non-auxilliary components) in K' and K that are linked. To identify these functions we will denote them by the places they occupy in the structures in the species of potential models using the following notation. If $|\Sigma|$ is a class of structures with m-components and $i_1,\ldots,i_n \in \{1,\ldots,m\}$ with $i_n \leq i_{n+1}$ and $X \subseteq |\Sigma|$ then

'$|X: i_1,\ldots,i_n|$'

denotes the class of all sequences of components appearing in the places $i_1,\ldots,i_n$ in some structure in the class X. In particular, for $x \in |\Sigma|$,

$$'|\{x\}: i_1,\ldots,i_n|'$$

denotes the "value" of the components $i_1,\ldots,i_n$ in the structure x. We also need to be more explicit about what the components in the objects of the category of potential models are doing. We say K is a 'k-l-n model element core' when the first k-components are the "base sets" of the structures, the next l are ;auxiliary base sets" having to do with auxiliary mathematical structures like the real numbers and the last n components are "non-basic components" like relations and functions over the basic sets. (See Bourbaki (1968) Ch. IV). Thus when K' and K are respectively k'-l'-n' and k-l-n model element cores and $\langle i_1,\ldots,i_r\rangle, \langle J_1,\ldots,J_s\rangle$ are sequences of non-auxilliary "component positions" in members of $|M_p'|$ and $|M_p|$ respectively, we may define a $\langle i_1,\ldots,i_r\rangle$-K', $\langle J_1,\ldots,J_s\rangle$-K link to be a K', K link $\|\lambda\|$ in which the "values" of the components $\langle i_1,\ldots,i_r\rangle$ in the structures in $|M_p'|$ are correlated with the "values" of the components $\langle J_1,\ldots,J_s\rangle$ in the structures of $|M_p|$ and the values of no other components. What this means formally is that, for all the correlated components $h \in \{1,\ldots,r\}$ and $g \in \{1,\ldots,s\}$, some values that these components take in M_p' and M_p do not appear in the structures that are related by $|\lambda|$. That is

$$|D_1(|\lambda|): h| \neq |M_p': h|$$
$$|D_2(|\lambda|): g| \neq M_p : g|.^2$$

Further, for all uncorrelated components $\{a_1,\ldots,a_t\}$ and $\{b_1,\ldots,b_u\}$, all values that these components take in M_p' and M_p appear in the structures that are related by $|\lambda|$. That is

$$|D_1(|\lambda|): a_1,\ldots,a_t| \times |D_2(|\lambda|): b_1,\ldots,b_u| =$$
$$|M_p': a_1,\ldots,a_t| \times |M_p: b_1,\ldots,b_u|.$$

We further require that the M_p'-morphisms linked by $\|\lambda\|$ with M_p-morphisms shall be only those that exist between structures in which the uncorrelated components $\{a_1,\ldots,a_t\}$ and $\{b_1,\ldots,b_u\}$ have the same values. That is, if

$\lambda(\langle x',x\rangle,\langle y',y\rangle \neq \Lambda$

$$|\{x'\}: a_1,\ldots,a_t| = |\{x\}: a_1,\ldots,a_t|$$
$$|\{y'\}: b_1,\ldots,b_u| = |\{y\}: b_1,\ldots,b_u|.$$

Intuitively, $\|\lambda\|$ links only morphisms that "transform" the components it correlates. We summarize these ideas in the following definition.

D1 For all categories $\|\lambda\|$, $K,K' \in |SET|$, $k,k',l,l',n,n' \in \mathbb{N}^+$ if

H_1 $K' = \langle\|M_p'\|,\|M'\|\rangle$ and $K = \langle\|M_p\|,\|M\|$ are respectively k'-l'-n' and k-l-n model element cores

H_2 $i_1,\ldots,i_r \in \{1,\ldots,k',k'+l'+1,\ldots,k'+l+n'\}$, $i_n \leq i_{n+1}$ and $J_1,\ldots,j_s \in 1,\ldots,k,k+l+1,\ldots,k+l+n$, $J_g \leq J_{g+1}$

H_3 $a_1,\ldots,a_t \in \{1,\ldots,k',k'+l'+1,\ldots,k'+l'+n'\}$, $a_n \leq a_{n+1}$ and $b_1,\ldots,b_u \in \{1,\ldots,k,k+l+1,\ldots,k+l+n\}$, $b_g \leq b_{g+1}$ so that

1) $\{a_1,\ldots,a_t\} \cap \{i_1,\ldots,i_r\} = \Lambda$
$\{b_1,\ldots,b_u\} \cap \{J_1,\ldots,J_s\} = \Lambda$

2) $\{a_1,\ldots,a_t\} \cup \{i_1,\ldots,i_r\} =$
$\{1,\ldots,k',k'+l'+1,\ldots,k'+l'+n'\}$
$\{b_1,\ldots,b_u\} \cup \{J_1,\ldots,J_s\} =$
$\{1,\ldots,k,k+l+1,\ldots,k+l+n\}$

then

A) $\|\lambda\|$ is a K',K link iff

1) $\|\lambda\|$ is an $\|M_p'\|,\|M_p\|$ relator

2) for all $\langle x',x\rangle,\langle y',y\rangle \in |M_p'| \times |M_p|, z' \in |M_p'|$

a) if $I[M_p'\times M_p](\langle x',x\rangle,\langle y',y\rangle) \neq \Lambda$ then $I[M_p'\times M_p](\langle x',x\rangle,\langle y',y\rangle) \cap \lambda(\langle x',x\rangle,\langle y',y\rangle) \neq \Lambda$.

b) if $\langle x',x\rangle \in |\lambda|$ and $IM_p'(x',z') \neq \Lambda$ then there is some $z \in |M_p|$ so that $I\,M_p(z',z) \neq \Lambda$ and $\langle z',z\rangle \in |\lambda|$.

B) For all $X \subseteq |M_p'|$, $\|\lambda\|$ is a $\langle i_1,\ldots,i_r\rangle$-K', $\langle J_1,\ldots,J_s\rangle$-K link in X iff

1) $\|\lambda\|$ is a K',K link

2) for all $h \in \{1,\ldots,r\}$, $g \in \{1,\ldots,s\}$,

   a) $|D_1(\lambda) \cap X{:}h| \neq |M_p'{:}h|$

   b) $|D_2(\lambda){:}g| \neq |M_p{:}g|$

3) $|D_1(\lambda){:}a_1,\ldots,a_t| \times |D_2(\lambda){:}b_1,\ldots,b_u| =$
$$|M_p'{:}a_1,\ldots,a_t| \times |M_p{:}b_1,\ldots,b_u|$$

4) for all $\langle x',x\rangle,\langle y',y\rangle \in |M_p'| \times |M_p|$ and $\langle\mu',\mu\rangle \in M_p' \times M_p(\langle x',x\rangle,\langle y',y\rangle)$, if $\langle\mu',\mu\rangle \in \lambda(\langle x',x\rangle,\langle y',y\rangle)$ then

$$|\{x'\}{:}a_1,\ldots,a_t| = |\{x\}{:}a_1,\ldots,a_t|$$
$$|\{y'\}{:}b_1,\ldots,b_u| = |\{y\}{:}b_1,\ldots,b_u|$$

3. INTERPRETING AND REDUCING LINKS

3.0

Fundamental to the view of empirical science expounded here is that some K', K external links may be understood intuitively as providing an empirical interpretation for at least some of the components of K. External links are links between model elements with different categories of potential model elements. That is,

$$\|M_p'\| \neq \|M_p\| .$$

Intuitively, a K', K link is an interpreting link for K, or interprets K, when models of K' serve as acceptable means of measuring or determining the values of components in potential models of K. More precisely, an interpreting $\langle i_1,\ldots,i_r\rangle$-K', $\langle J_1,\ldots,J_s\rangle$-K link is, intuitively, a link that allows us to infer something interesting about values of the components $\langle J_1,\ldots,J_s\rangle$ in at least some potential models of K from knowledge of the values of the components $\langle i_1,\ldots,i_r\rangle$ in models of K'. The laws of K are then used to say more about the values of the components $\langle J_1,\ldots,J_s\rangle$ - to claim that they satisfy additional conditions.

3.1

The concept of an interpreting link is largely a pragmatic concept. Which links are used as interpreting links is a fact about the practice of empirical science, not a fact about the formal properties of links. Thus, we can not give a purely formal characterization of interpreting links. Nevertheless, we are able to give some formal necessary conditions for links to be used as interpreting links. First, it is clear that for interpreting $\langle i_1,\dots,i_r\rangle$-K', $\langle J_1,\dots,J_s\rangle$-K links, at least some models of K' must be linked. That is,

A) $D_1(|\lambda|) \cap |M'| \neq \Lambda$.

Intuitively, members of this set are "acceptable" measuring devices or measuring situations for the components $\langle J_1,\dots,J_s\rangle$. Members of $D_1(|\lambda|)$ outside this set correlate values of $\langle i_1,\dots,i_r\rangle$ with classes of values for $\langle J_1,\dots,J_s\rangle$, but these correlations are just "meaningless numbers" - for example, readings from faulty instruments. One might think that an interpreting $\langle i_1,\dots,i_r\rangle$-K', $\langle J_1,\dots,J_s\rangle$-K link, $\|\lambda\|$, should "determine" the values of $\langle J_1,\dots,J_s\rangle$ in the sense that

B) for all $\langle x',x_1\rangle$, $\langle x',x_2\rangle \in |\lambda|$,

$|\{x_1\}: J_1,\dots,J_s| = |\{x_2\}: J_1,\dots,J_s|$.

Though many interpreting links will be determining in this sense, there appears to be no clear reason why we should demand that they be. We choose to operate with a rather broad concept of interpretation in which interpreting links need only provide information about the values of the components they interpret. That is, for some $x' \in |M'|$ that is $\|\lambda\|$-linked, when the value of $\langle i_1,\dots,i_r\rangle$ is fixed in x' not all values of $\langle J_1,\dots,J_s\rangle$ may appear in x's in $|M_p|$ that are $\|\lambda\|$-linked with x'. This is entailed by the fact that

C) $\|\lambda\|$ is a $\langle i_1,\dots,i_r\rangle$-K', $\langle J_1,\dots,J_s\rangle$-K link in $|M'|$.

3.2

There is at least one additional restriction to be placed on interpreting links. Consider the case of a K', K link in which the laws of K', together with the link, "entail" the laws of K. More precisely, $\|\lambda\|$ is such that

$$\text{for all } \langle x',x\rangle \in |\lambda|, \text{ if } x' \in |M'| \text{ then } x \in |M|. \qquad D)$$

In this case, it is not plausible to regard K' as providing an interpretation of components in K. We may not think of models for K' providing acceptable methods of "measuring" values of components in K about which the laws of K say "something more". There is nothing more to say using only the laws of K. There can be no data that fails to satisfy the laws of K. This suggests that links with this property should not count as interpreting links. That is, K', K links that interpret K should have the property that

$$\text{there exist } \langle x',x\rangle \in |\lambda| \text{ so that } x' \in |M'| \text{ and } x \notin |M|. \qquad E)$$

We shall see below, links that reduce K' to K fail to have this property. Note that $\|\lambda\|$ could have property E) and yet every $|\lambda|$-linked model of K' was linked to some model of K'.

3.3

Consider a situation symmetric to the one just considered - a K', K link $\|\lambda\|$ in which the laws of K, together with the link, entail the laws of K'. That is

$$\text{for all } \langle x',x\rangle \in |\lambda|, \text{ if } x \in |M| \text{ then } x' \in |M'|. \qquad F)$$

Were $\|\lambda\|$ to be regarded as a K interpreting link, this would mean intuitively that all "data" that satisfied the laws of K had been obtained from acceptable measurements. There could be no "bad data" that just happened to satisfy the laws of K. We might attempt to formulate all empirical theories in a way so that F) is true. In fact we do not appear to do this. We seem, rather, to formulate them so that F) is always false. Intuitively, the laws of isolated empirical theories are always formulated so that they entail nothing substantive about what counts as acceptable data for them.

$$\text{there exist } \langle x',x\rangle \in |\lambda| \text{ so that } x \in |M| \text{ and } x' \notin |M'|. \qquad G)$$

We shall see below, links that reduce K' to K fail to have this property.

3.4

We may summarize these ideas as follows.

D2 For all categories $\|\lambda\|$, K, K' $\in$ |SET|, k, k', l, l', n, n' $\in \mathbb{N}^+$ so that $\|\lambda\|$ is a K', K link:

A) $\|\lambda\|$ is external iff $\|M_p'\| \neq \|M_p\|$;

B) $\|\lambda\|$ is internal iff $\|\lambda\|$ is not external;

C) $\|\lambda\|$ is K-interpreting only if

1) $\|\lambda\|$ is external

2) $D_1(|\lambda|) \cap |M'| \neq \Lambda$

3) there exist
$\langle i_1,\ldots,i_r\rangle \in \{1,\ldots,k',k'+l'+1,\ldots,k'+l'+n'\}$
$i_n \leq i_{n+1}$
$\langle J_1,\ldots,J_s\rangle \in \{1,\ldots,k+l+1,\ldots,k+l+n\}$
$J_g \leq J_{g+1}$
so that $\|\lambda\|$ is a $\langle i_1,\ldots,i_r\rangle$-K', $\langle J_1,\ldots,J_s\rangle$-K link in |M'|

4) there exist $\langle x',x\rangle, \langle y',y\rangle \in |\lambda|$, so that:

a) $x' \in |M'|$ and $x \notin |M|$

b) $y \in |M|$ and $y' \notin |M'|$.

Note that we give only necessary conditions for K-interpreting K', K links.

3.5

Next let us consider the reduction relation on model elements. We say that model element K' reduces to model element K or, equivalently K reduces K' 'just when there is a K', K link that reduces K' to K. We then say what it is for a K', K link $\|\lambda\|$ to reduce K' to K.

First, $\|\hat{\lambda}\|$ must be a surjective functor from some full sub-category $\|M_r\|$ to $\|M_p'\|$. Intuitively, this means that every potential model in the reduced theory K' corresponds to at least one, and generally many, potential models of the reducing theory K and the potential models of the reducing theory to which they correspond - $|M_r|$ - comprise some systematically describable sub-class of the potential models of the reducing theory K. Moreover, some members of $|M_r|$ must be models for the reducing theory K and the $|\lambda|$-images of these models must be models for the reduced theory. This is the "entailment condition". It says roughly that the laws of the reducing theory, together with the reducing link, entail the laws of the reduced theory. A bit more precisely, models of the reducing theory $|\lambda|$-correspond only to models of the reduced theory. We make this precise in the following definition.

D3 For all categories $\|\lambda\|$, K',K $\in$ |SET|, if K' and K are model elements and $\|\lambda\|$ is a K',K link then $\|\lambda\|$ is a reducing link iff there is a category $\|M_r\|$ so that

A) $\|M_r\| \subseteq_f \|M_p\|$

B) $\|M_r\| \cap \|M\| \neq \|\Lambda\|$

C) $\|\hat{\lambda}\| : \|M_r\| \rightarrow \|M_p{}'\|$ and surjective

D) $\|\bar{\lambda}\|(\|M_r\| \cap \|M\|) \subseteq_f \|M'\|$

A K', K reducing link $\|\lambda\|$ reduces K' to K in the sense that: i) every potential model for K' is $|\lambda|$-related to at least one (and generally several) potential models for K (-C)); ii) some K models are $|\lambda|$-related (-B)) and; iii) if x' $\in$ $|M_p{}'|$ is $|\lambda|$-related to a model for K, then x' must be a model for K' (-D)).

3.6

Note that there may be some models for the reduced theory K' that are not $|\lambda|$-related to any models for the reducing theory K. Intuitively, we might regard the reducing link as "restricting" the class of models for the reduced model element K' in this way. The reducing theory K is "more fundamental". Any model of the reduced theory K' that does not correspond to a model for the more fundamental reducing theory K should thus be excluded because it can not be "explained" in terms of the more fundamental picture of the world provided by the reducing theory. It has been suggested (Mayr (1981)) that a more intuitively satisfactory concept of reduction could be obtained by replacing '$\subseteq_f$' with '=' in -D). This would mean intuitively that the reduction did not narrow douwn the content of the reduced theory or, still more intuitively, all claims of the reduced theory could be "deduced" from claims of the reducing theory. A precise rendering of 'deduced' here (Pearce (1982)) sharpens these intuitions. It may be that all examples of "real life" reducing links do not narrow down the content of the reduced theory. However, we think that such a "strengthening" of the reduced theory may sometimes motivate the search for reducing links. For example, one might have expected that certain thermodynamic processes that were permitted by simple equilibrium thermodynamics might be ruled out by the reduction of this theory to classical statistical mechanics. This was not the case. However, the reduction of simple equilibrium thermodynamics to quantum statistical mechanics would apparently rule out "zero entropy" states that were permitted by the thermodynamic theory.[3]

More generally, we might hope that a more fundamental understanding of underlying phenomena, like statistical mechanics and molecular genetics, might show us how to strengthen our "macroscopic" or "phenomenological" theories. Thus we choose a concept of reduction relation that allows for this possibility.

3.7

Note too that a K', K reducing link $\|\lambda\|$ that reduces K' to K can not be a K-interpreting link because D2-C-4-b) is violated. Further, a K', K link whose converse - a K,K' link - reduces K' to K can not be a K-interpreting link because D2-C-4-a) is violated. More intuitively, if a K', K link is a "reducing link" in either direction, then it can not be a K-interpreting link. Still more intuitively, reducing links can not be interpreting links. This is to be expected. Reducing links connect model elements in the same "theory family". Different members of the same theory family provide different ways of formulating logically related empirical claims. Indeed, equivalent model elements K' and K - those linked by a $\|\lambda\|$ that reduces K' to K and K to K' - are just different way of formulating essentially the same claim. Thus, members of the same theory family can not, except in a trivial way, be regarded as providing empirical interpretations for each other.

4. MODEL ELEMENT NETS

4.0

On our view, the logical structure of the whole of empirical science at any given time - the synchronic structure of empirical science - may be exhibited as a set of model elements together with the intertheoretical links among them. Formally, we may think of a model element net N as an ordered pair

$$N = \langle |N|, L \rangle$$

where |N| is a set of model elements and L is a set of binary links linking members of |N|. It is convenient to characterize model element nets in two steps. We begin by defining 'linked model element set'.

D4 For all N ∈ |SET|, N is a linked model element set iff there exist |N| and L ∈ |SET| so that

A) $N = \langle |N|, L \rangle$

B) for all K ∈ |N|, there exist k,l, n so that K is a k-l-n model element

C) for all $\|\lambda\| \in L$, there exist K', K ∈ |N| so that $\|\lambda\|$ is an K', K link.

4.1

The set L imposes a binary relational structure on |N| in an obvious way and it is somewhat more convenient to discuss the properties of model element nets in terms of this relational structure. Let L(K',K) ⊆ L be the set of all K', K links in L. Note that we may intuitively think of L(K',K) as containing just ONE link

$$\lambda[K',K] = \cap \{\|\lambda\| \in L(K',K)\}.$$

Consider the binary relation $L_r \subseteq |N| \times |N|$ that contains ⟨K',K⟩ just in case L(K',K) ≠ Λ. That is, $L_r(K',K)$ just when there is some link between K' and K. Clearly,

$$\langle |N|, L_r \rangle$$

is a binary relation structure. L_r is not transitive but we can "complete" L_r by adding to it all pairs of model elements that were linked by "chains" of links. We call this "completion" of L_r '$\mathbb{L}_r$'.

D5 For all $N \in |SET|$, if $N = \langle |N|, L \rangle$ is a linked model element set then for all K', $K \in |N|$:

A) $L(K',K) := \{\|\lambda\| \in L \mid \|\lambda\| \text{ is a } K', K \text{ link}\}$

B) $\lambda[K',K] := \bigcap \{\|\lambda\| \in L(K',K)\}$

C) $L_r := \{\langle K',K \rangle \in |N|^2 \mid L(K',K) \neq \Lambda\}$.

D) $\mathbb{L}_r := \{\langle K',K \rangle \in |N|^2 \mid$ there exist $K_1, \ldots, K_n \in |N|$ so that $K' = K_1$, $K = K_n$ and, for all $i \in \{1, \ldots, n-1\}$, $\langle K_i, K_{i+1} \rangle \in L_r\}$

4.2

Up to this point we have considered the most general possible linked model element set in which the links may be of any sort, including interpreting and reducing links but not restricted to these. Linked model element sets that may be used to describe the global structure of empirical science have some additional properties having to do with the special kinds of links we considered in Sec. 3. We will call linked model element sets with these additional properties 'model element nets'. Here we consider only additional properties of model element nets that have to do with interpreting and reducing links.

4.3

First consider interpreting links. If we think of interpreting links roughly as channels or paths for the transmission of informatition, then the relation between $L(K',K)$ and $L(K,K')$ is an important contingent fact about the logical structure of empirical science. Roughly, our conception of a K-interpreting K', K link is that it serves to transmit information F R O M K' T O K. This suggests that there should be a kind of asymmetry for interpreting links.

Information should be concieved as flowing in only ONE direction between two adjacent linked model elements. Further, it seems intuitively clear that, when K' interprets K in N, there should be no links in N that reduce K' to K. This suggests we should require that:

A) for all K', K ∈ |N|, if λ[K',K] is an interpreting link then L(K,K') = Λ.

Note that A) has the force of ruling out any kind of K, K' links when λ[K',K] is an interpreting link, not just K'-interpreting links. In the case of reducing links, we do not want anything like A) since we commonly recognize theories to be "equivalent" just when there are reducing links "going both ways" (Balzer and Sneed (1977)). However, we do want to rule out interpreting links "going the other way". Thus we might require

B) for all K', K ∈ |N|, if λ[K',K] is a reducing link then either L(K,K') = Λ or λ[K,K'] is not an interpreting link.

Generally nothing can be inferred about λ[K,K'] from properties of λ[K',K] since the definition of a linked model element set imposes no conditions relating the membership of L(K,K') and L(K',K).

4.4

We summarize these ideas in the following definition.

D6 For all N ∈ |SET|, N is an model element net iff there exist |N| and L ∈ |SET| so that

A) N = <|N|,L> is a linked model element set

B) for all K', K ∈ |N|, if λ[K',K] is a K-interpreting link then L(K,K') = Λ.

C) for all K', K ∈ |N|, if λ[K',K] is a reducing link then either L(K,K') = Λ or λ[K,K'] is not a K'-interpreting link.

4.5

A model element net may be viewed as placing restrictions on arrays of set-theoretic structures. The "content" of a model element net consists of just those arrays of structures that satisfy the restrictions imposed by the net. That is, the content of a model element net consists of structures that meet all the requirements this net imposes on "the way the world is". The fundamental intuitive ideas are these. The model classes tell us what potential models are empirically possible in the absence of links. Links tell us what combinations of potential models are empirically possible. Together they tell us what combinations of models are possible.

4.6

One natural way to move in making these intuitive ideas precise is to think of the content of a model element net as a collection of some kind f binary relation structures consisting of ordered pairs of models. These structures would be analogous to single models in the content of single model elements. Each structure in the content of the net N would be one configuration of potential models that the net admitted as empirically possible. The structure exhibited by the ordered pairs would in some way "mirror" the structure of the net N or its corresponding binary relation structure N_r. That is, the structure N_r says something about how empirically possible models for members of K must be related. However, it does not say all there is to say. There are many possible ways that binary relation structures of models could "mirror" the structure of N_r. Only some of these are plausible candidates for the content of the net N. To specify these candidates would require a significant detour from the main task of this paper. A detailed treatment may be found in Balzer et al. (1983) . We shall simply introduce some notation for the content of a net which we require for present purposes.

4.7

For model element nets N, we may let

$C_p[N]$:= the content of N.

The sub-script 'p' is used to indicate that this concept of "content" is a structured sub-class of potential models. It distinguishes this concept from the "non-theoretical content" considered below in Sec. 5. It is a class of binary relation structures consisting of models for the members of |N| that is homomorphic to N_r. On this view of the content of N, we may see how the net operates to "narrow down" the content of each of its members. For K ∈ |N|, we may take $C_p[N](K)$ to be the set of all models of K that appear in some member of $C_p[N]$. That is

$$C_p[N](K) := \{x \mid x \in |M| \text{ and there is an } X \in C_p[N] \text{ and } x \in |X|\}.$$

Alternatively, $C_p[N](K)$ is the class of all members of |M| that are linked to some model of at least one of the model elements K' in |N| that are linked with K. Clearly, $C_p[N](K)$ will be a proper sub-class of |M| just in nets in which there are members of |M| that are not linked to models of any of the model elements that are linked with K.

5. NON-THEORETICAL STRUCTURES

5.0

It is useful to distinguish those components in the potential models of K that are "interpretable" or "non-theoretical" in the net N from those that are "theoretical". Roughly, the theoretical components in the potential models of K are those components in the potential model structures that are not affected by any of K's interpreting links. The non-theoretical components are those whose values are correlated in some way, by interpreting links for K, with values of components of potential models in other model element cores.

5.1

We may make the distinction between theoretical and non-theoretical components in K precise in the following way.

D7 For all N ∈ |SET|, if N is a model element net and, for all $K = \langle \| M_p \|, \| M \| \rangle \in |N|$, if there exist k,l,n so that K is a k-l-n model element core, then, for all $i \in \{1,\ldots,k,k+l+1,\ldots,k+l+n\}$.

A) $|M_p{:}\ i|$ is K non-theoretical in N iff there is some $K' \in |\tilde{N}|$ so that

1) there exist k',l',n' so that K' is a k'-l'-n' model element

2) K' ≠ K

3) there exist
$J_1,\ldots,J_s \in \{1,\ldots,k',k'+l'+1,\ldots,k'+l'+n'\}$ so that
$\lambda[K',K]$ is a $\langle J_1,\ldots,J_s\rangle$-K', $\langle i\rangle$-K link in M'

4) $\lambda[K',K]$ is a K-interpreting link in N

B) $|M_p{:}i|$ is K theoretical in N iff $|M_p{:}i|$ is not K non-theoretical in N.

Note that this distinction between theoretical and non-theoretical components is relative to both K and the model element net N in which K appears. However, it is not a "totally global" concept since we do not have to look at all of N. We just need to look at those parts that are connected to K by interpreting links.

5.2

This definition of non-theoretical components is not quite adequate. To see why, note that components in the potential models of K do not count as non-theoretical unless they are linked "singly" to K' (though not necessarily by a determining link). A $\langle J_1,\ldots,J_s\rangle$-K', $\langle i_1,i_2\rangle$-K link does not necessarily make i_1 K-non-theoretical. Of course, the same link MAY also be a $\langle J_1,\ldots,J_s\rangle$-K', $\langle i_1\rangle$-K link, but it need not be. If it just rules out pairs of values for $\langle i_1,i_2\rangle$ while admitting all values for i_1, it is not. For example, the link between the pressure function (P) in classical hydrodynamics (CHD) and the energy (U) and volume (V) functions in simple equilibrium thermodynamics (SETH) provided by P = -DU/DV makes neither U nor V SETH-non-theoretical because it only rules out $\langle U,V\rangle$ pairs, but not U-values or V-values. Thus, by our definition this link would produce no non-theoretical components. But, this link does play an essential role in interpreting SETH and somehow the interpreting information it provides should appear as restrictions on the values of some non-theoretical components. Intuitively, this link makes the defined SETH component "thermodynamik pressure", Π:= -DU/DV, SETH-non-theoretical. Counting the defined component Π among the SETH-non-theoretical components would capture the intuition that this link is essential to the interpretation of SETH. In this case, the $\langle P\rangle$-CHD, $\langle \Pi\rangle$-SETH link is determining. But this does not seem to be essential. Generalizing, one might think that our definition of non-theoretical components should be broadened to include the possibility that defined components are non-theoretical. But, doing this would mean that we could no longer uniquely define the non-theoretical structures associated with a model element in a net as we do below in D8. Countenancing non-uniqueness here would considerably complicate the subsequent discussion. For this reason, we rest with the present, admittedly deficient, definition of non-theoretical components.

5.3

We may now define the category of non-theoretical structures or "partial potential models" $\| M_{pp}[N](K) \|$ for model element K in net N. We choose this somewhat cumbersome notation to make explicit that the concept of non-theoretical structure is net-relative.

D8 For all $N \in |SET|$, if N is a model element net and, for all $K = \langle \| M_p \|, \| M \| \rangle \in |N|$, if there exist k_p, l_p, n_p so that K is a k_p-l_p-n_p model element, then $\| M_{pp}[N](K) \|$ is the category of partial potential models for K in N iff there exist $k_{pp}, l_{pp}, n_{pp}, k_{pp} \leq k_p, l_{pp} \leq l_p, n_{pp} \leq n_p$ so that

A) $\| M_{pp}[N](K) \|$ is a category

B) $|M_{pp}[N](K)|$ is a k_{pp}-l_{pp}-n_{pp} species of structures

C) for all $i_k \in k_{pp}, i_l \in l_{pp}, i_n \in n_{pp}$ there exist $J_k \leq k_p$, $J_l \leq l_p$, $J_n \leq n_p$ so that

$$|M_{pp}[N](K) : i_x| = |M_p : J_x| \quad x \in \{k,l,n\}$$

D) for all $i \in \{1, \ldots, k_{pp}\} \cup \{k_{pp}+l_{pp}, \ldots, k_{pp}+l_{pp}+n_{pp}\}$ $|M_{pp}[N](K):i|$ is K non-theoretical in N

E) there is no $k_{pp}', l_{pp}', n_{pp}'$; $k_{pp} < k_{pp}' \leq k_p$, $l_{pp} < l_{pp}' \leq l_p$, $n_{pp} < n_{pp}' \leq n_p$ so that $M_{pp}[N](K)'$ is a k_{pp}'-l_{pp}'-n_{pp}' species of stuctures satisfying B) through D) above and $|M_{pp}[N](K)| = |M_{pp}[N](K)' : 1, \ldots, k_{pp}+l_{pp}+n_{pp}|$

5.4

The Ramsey functor -- Ram -- for K in N is just the "forgetful functor" from $\| M_p \|$ to $\| M_{pp}[N](K) \|$.

D9 For all $N \in |SET|$, if N is a model element net and, for all $K = \langle \| M_p \|, \| M \| \rangle \in |N|$, if $\| M_{pp}[N](K) \|$ is the category of partial potential models for K in N then $Ram = \langle Ram_o, Ram_m \rangle$ is the Ramsey functor for $\| M_p \|$ in N iff $Ram: \| M_p \| \to \| M_{pp}[N](K) \|$ so that for all $x \in |^p M_p|$,

$$Ram_o(x) = |\{x\}: 1,\ldots,k_{pp}+l_{pp}+n_{pp}|.$$

Note that we do not need to specify Ram_m -- the member of Ram that maps M_p-morphisms into $M_{pp}[N](K)$-morphisms -- because morphisms for species of structures are uniquely determined by the structure specification.

5.5

We may think of the laws of K as determining a sub-category of $\| M_{pp}[N](K) \|$ whose objects are just those members of $|M_{pp}[N](K)|$ that can be "filled out" with theoretical components in some way that satisfies the laws of K. We call this sub-category 'the non-theoretical content of K' and denote it by '$C_{pp}[N](K)$'. Clearly, the non-theoretical content of K in N -- $C_{pp}[N](K)$ -- is just the Ramsey functor image of the models of K.

$$C_{pp}[N](K) := \overline{Ram}(\| M \|).$$ [4]

Intuitively, $C_{pp}[N](K)$ is what K alone says about its "data". We can be a bit more precise about this when we have a clearer idea of how to specify the intended applications of K.

6. INVARIANCE PRINCIPLES

6.0

We are now in a position to say more about the properties of morphisms in the categories that appear in theory elements. Recall that we denote the M_p and M_{pp}-isomorphism relations by '$M_p^=$' and '$M_{pp}^=$' respectively and let us denote the corresponding isomorphism equivalence classes by $[y]_p$ and $[x]_{pp}$. First note that, since Ram is a functor, $M_p^=$ and $M_{pp}^=$ have the following property:

A) if $M_p^=(y,y')$ then $M_{pp}^=(Ram(y),Ram(y'))$

and since Ram is also surjective:

B) if $M_{pp}^{=}(x,x')$ then there exist y,y' so that $M_{p}^{=}(y,y')$ and $x = Ram(y)$; $x' = Ram(y')$.

Note however, that for x,x' so that $M_{pp}(x,x')$ there may also be z,z' so that $x = Ram(z)$, $x' = Ram(z')$, but not $M_{p}^{=}(z,z')$. It follows from A) and B) that, for all $y \in |M_p|$,

C) $\overline{Ram}([y]_p) = [Ram(y)]_{pp}$

From C) and the fact that $|M|$ is M_p-isomorphism invariant, it follows that $C_{pp}[N](K)$ is M_{pp}-isomorphism invariant in the sense that, for all $x \in |M_{pp}|$,

D) $x \in C_{pp}[N](K)$ iff $[x]_{pp} \subseteq C_{pp}[N](K)$.

6.1

Consider a non-theoretical structure $x \in |M_{pp}|$. Then

$$Ram<(x) \subseteq |M_p|$$

is the set of all "theoretical emendations" of x -- all the ways of filling x out with theoretical components.

$$Ram<(x) \cap |M|$$

is the set of all "interesting", K-law satisfying theoretical emendations of x. Consider $y \in Ram<(x) \cap |M|$ and the class of all members of $Ram<(x)$ that are M_p-isomorphic to y

$$[y]_p \cap Ram<(x).$$

Since $|M|$ is M_p-isomorphism invariant

$$[y]_p \cap Ram<(x) \subseteq |M|.$$

That is M_p-morphisms work in such a way that whenever y is a law satisfying theoretical emendation of x - $y \in |M|$ and $Ram(y) = x$ - every $y' \in Ram<(x)$ that is M_p-isomorphic to y is also a law satisfying theoretical emendation of x. We may then consider the quotient class

$$(Ram<(x) \cap |M|)/M_p^{=}.$$

Intuitively, each $M_p^=$-equivalence class here corresponds to one "value" for the theoretical components. Members of $M_p^=$-equivalence classes correspond, in many cases, to expressions of this value for the theoretical components in different "units" or different "coordinate systems".

6.2

This suggests that the M_p-morphisms should be related to the maximal degree of uniqueness with which theoretical components may be determined by the laws of K. In the absence of internal links or constraints, it is natural to expect that $\| M_p \|$ should be such that there is exactly one $M_p^=$-equivalence class in $\mathrm{Ram}{<}(x) \cap |M|$. That is, for all $x \in |M_{pp}|$, $y \in \mathrm{Ram}{<}(x) \cap |M|$,

$$[y]_p = \mathrm{Ram}{<}(x) \cap |M| .$$

Intuitively, all K-law satisfying theoretical emendations of x are "empirically equivalent" and the morphism concept for $\| M_p \|$ should reflect this fact. This is an additional requirement on the morphisms of categories appearing in model elements in nets. It is relative to the net in which the model element appears in that the formulation of the requirement depends on what the non-theoretical structures of the model element are and this, in turn, depends on how the model element is linked with other model elements in a net.

6.3

The situation here is somewhat more complicated in the case where K has internal links. Intuitively, the requirement on M_p-morphisms is that they describe the maximal degree of uniqueness up to which K-theoretical components may be determined by both the laws and internal links of K. In most cases, the internal links will work to provide additional restrictions on theoretical components stronger than those provided by the laws of K applied to single models. Thus, generally, M_p-morphisms satisfying this requirement will be such that

$$(\mathrm{Ram}{<}(x) \cap |M|)/M_p^=$$

is not a singleton. Space does not permit an elaboration of these further requirements on M_p-morphisms here. For a more complete discussion, from a somewhat different point of view see (Sneed (1979)).

6.4

Recall that intertheoretical links are "relators". This entails that they preserve morphism-types in the same way that functors do. Thus M_{pp}-isomorphisms will "reflect" the isomorphisms of the model elements connected to K by interpreting links in this sense. If K' is linked to K by an $\langle i_1,\ldots,i_r\rangle$-K', $\langle J_1,\ldots,J_s\rangle$-K interpreting link, then by D1-B-4), M_{pp}-isomorphic structures will be linked to M_p'-isomorphic structures in which the unlinked components in the M_p-structures have the same values. In general, theoretical components of M_p' will be among those linked to components in M_p. Roughly, K'-theoretical components will typically be linked with K-non-theoretical components in interpreting K. Thus, M_{pp}-morphisms will usually reflect the degree of uniqueness up to which theoretical components in the theories used to interpret K can be determined. This suggests, first, that it is natural to expect that the content of K be M_{pp}-isomorphism invariant in the sense of C) above. Further, it suggests that the effect of all model elements connected to K by restricting links should be M_{pp}-isomorphism invariant. That is, for all $x \in |M_{pp}|$,

$$x \in \overline{\mathrm{Ram}}(C_p[N](K)) \text{ iff } [x]_{pp} \subseteq \overline{\mathrm{Ram}}(C_p[N](K)). \qquad \text{D')}$$

Just as C) above follows from the properties of the Ramsey functor Ram and the M_p-isomorphism invariance of M, D') follows from the morphism type preserving properties of links (D1-A-2-a and -b)) and the isomorphism invariance of the laws in the linked theories. It should be noted that there appears to be no purely formal reason why the content of model elements in nets must have these invariance properties. Neither the invariance properties of the laws in individual model elements, nor the invariance properties of links between them, appear required for the concept of net content and empirical claims of model elements in the net. That these invariance properties do play a role in empirical science seems to be a contingent fact about the way this activity is actually practiced.

7. INTENDED APPLICATIONS

7.0

Let us consider how we are to regard the empirical claim of a single model element in a model element net. We have already suggested that the theory element

$$T = \langle K, I(K) \rangle$$

where

$$K = \langle \| M_p \|, \| M \| \rangle$$

claims roughly that

$$I(K) \subseteq |M|.$$

We have been purposely vague about what I(K) is and how it is specified. To begin to speak more precisely about I(K), we may note that a somewhat more plausible rendition of the empirical claim of K is that

$$I(K) \subseteq C_{pp}[N](K).$$

Here I(K) is conceived as some sub-class of the non-theoretical structures of K -- $|M_{pp}[N](K)|$. Roughly, the claim of K is that all members of I(K) can be filled out in some way with theoretical components of members of $|M_p|$ to yield a model of K -- a member of $|M|$. This seems intuitively plausible just in that it renders the empirical claim of K as a claim about its data and nothing else.

7.1

To say more about I(K), let us think about what a single model element contributes to the content of a model element net. Let us consider, for the moment, only interpreting links. A single model element K in net N has effects on the content of N that go in both "directions" from it. Among other possibilities, it may narrow down the content of all the model elements that interpret it and it may contribute to the interpretation of other model elements. Very roughly, K is used to make a claim about models of the model elements that interpret it and, in turn, model elements that it interprets make claims about its models, As a first cut, it seems natural to think of the intended applications of K has being provided by models of the model elements that are linked to K by interpreting links. But not all of these models will provide acceptable data for K. Some will be ruled out because they are not "interpreted" by model elements that are still "further back". Others may be ruled out by restrictions imposed by other model elements, besides K, that they interpret. Clearly, we do not want to include the restrictions imposed by K itself. Doing this would make K's claim trivially true. Further, it appears that we would not want to include restrictions imposed by model elements that K interprets. The reason is that the laws of K have an "indirect" effect on what these model elements rule out in the other model elements that are "behind" them. This suggests that we should think of the intended applications for K as being provided by the "net content" of interpreting model elements immediatly "behind" K in the net N. But, the "net" whose content is relevant here is not N. Rather, it is N, less everything in N that is "before" K that K interprets. Roughly, we should think of the intended applications for K as being provided by the content of a net in which K and all other model elements whose interpretation "presupposes" K have been removed.

7.2

We may make these ideas more precise in the following way. First consider the set of model elements that interpret K in N

$$<K := \{K' \in |N| \mid \langle K',K\rangle \in L_r \text{ and } \lambda[K',K] \text{ is } K\text{-interpreting}\}$$

and the set of model elements that K interprets in N

$$>K := \{K' \in |N| \mid \langle K,K'\rangle \in L_r \text{ and } \lambda[K,K'] \text{ is } K'\text{-interpreting}\}$$

A model element K in net N has "backward" and "forward" interpreting filters

$$N{<}K = \langle |N{<}K|, L< \rangle$$

$$N{>}K = \langle |N{>}K|, L > \rangle$$

associated with it, where

$$|N<K| = \{K' \in |N| \mid \langle K',K\rangle \in |L_r{}^i\}$$

$$|N<K| = \{K' \in |N| \mid \langle K,K'\rangle \in |L_r{}^i\}$$

are respectively all members of $|N|$ that are "backward" and "forward" linked to K by interpreting links in N. '$|L_r{}^i$' may be defined in the manner of D5-D) by requiring all the links in the "chain" to be interpreting links. $L<$ and $L>$ are respectively all the links in L that are interpreting links between members of $|N<K|$ and $|N>K|$. Clearly, $N<K$ and $N>K$ are model element nets. Intuitively, the interpretation of everything in the net $N>K$ "presupposes" K. We may now delete from N every model element whose interpretation may "presuppose" K to obtain

$$N \sim N>K.$$

As we have rendered it in the preceeding section, the claim of K is true of just things in the non-theoretical content of K -- $C_{pp}[N](K)$. Formally, $\lambda[K',K]$ does not pair any members of $|M_p{}'|$ with members of $|M_{pp}[N](K)|$. But we may consider

$$\lambda_{pp}[K',K] :=$$

$$\{\langle x_p{}',x_{pp}\rangle \in |M_p{}'| \times |M_{pp}[N](K)| \mid \text{there is an}$$

$$\langle x_p{}',x_p\rangle \in \lambda[K',K] \text{ so that } x_{pp} = \mathrm{Ram}(x_p)\}.$$

Suppose $\lambda[K',K]$ is a $\langle J_1,\ldots,J_s\rangle$-K', $\langle i_1,\ldots,i_r\rangle$-K link. Then, for $K' \in\ <K$, consider every member of $|M_{pp}[N](K)|$ that is $\lambda[K',K]$-linked with some member of the content of K' in $|N \sim N>K|$. That is

$$\overline{\lambda_{pp}[K',K]>}(C_p[N \sim N>K](K')).$$

Intuitively, these are members of $|M_{pp}[N](K)|$ in which information about the values of the components $\langle i_1,\ldots,i_r\rangle$ may be inferred from information about the values of the components $\langle J_1,\ldots,J_s\rangle$ in members of $C_p[N \sim N>K](K')$. That this information comes from members of $C_p[N \sim N>K](K')$ means intuitively that the information is obtained from acceptable measurement procedures -- procedures that are compatible with the laws of all the theories "relevant" to them that do not "presuppose" the theory for which they provide data. Each of the interpretors of K in N may contribute in this way to specifying K's intended applications. So we obtain

$$\bigcap \{\overline{\lambda_{pp}[K',K]>}(C_p[N \sim N>K](K')) \mid K' \in <K\}.$$

Very roughly, what K "claims" is that this information is consistent with its laws.

7.3

We have, in effect, specified necessary conditions for the intended applications of K in net N. That is, we require at least that

$$I(K) \subseteq \bigcap \{\overline{\lambda_{pp}[K',K]>}(C_p[N \sim N>K](K')) \mid K' \in <K\}.$$

It may even be plausible to regard these conditions as sufficient as well. If we regard N as including all of empirical science, we might hope that it included enough to rule out all "empirically meaningless" structures. Were this so we could replace '$\subseteq$' above by '='. We need not commit ourselves on this question here. For the sake of notational convenience, we may let

$$I[K](K) := \bigcap \{\overline{\lambda_{pp}[K',K]>}(C_p[N \sim N>K](K')) \mid K' \in <K\}.$$

Intuitively, I[N](K) is the intended applications of K as narrowly as they can be specified by N. Whether or not

$$I(K) = I[N](K)$$

we leave open.

8. REDUCING - INTERPRETATING NETS

8.0

Let us now consider model element nets in which reducing and interpreting links appear together. Let us begin by considering the net

N =

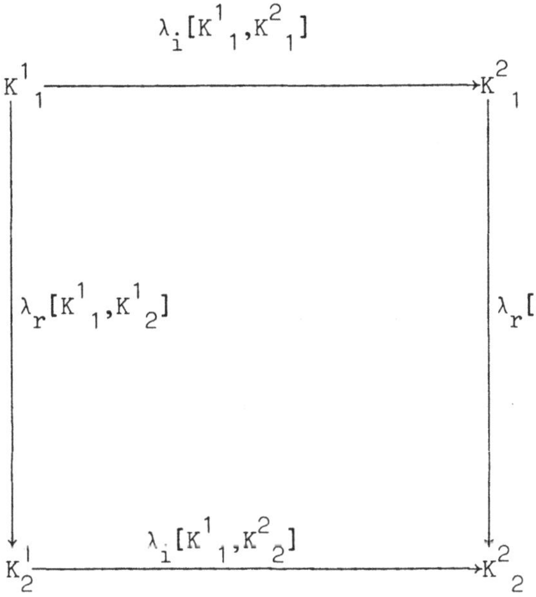

where the sub-script 'i''s indicate interpreting links and the subscript 'r''s indicate reducing links. In the non-trivial case where

$$M_p{}^1{}_1 \neq M_p{}^1{}_1 \quad \text{and} \quad M_p{}^2{}_1 \neq M_p{}^2{}_2$$

it is clear that

$$\lambda_i[K^1{}_1, K^2{}_1] \neq \lambda_i[K^1{}_2, K^2{}_2].$$

This net depicts the simplest and most obvious way we might expect interpreting and reducing links to appear together in the same net. An example of this situation is provided by classical particle kinematics (CPK = $K^1{}_2$), classical particle mechanics (CPM = $K^2{}_2$), classical rigid body kinematics (RBK = $K^1{}_1$) and classical rigid body mechanics (RBM = $K^2{}_1$).

Requiring that reducing and interpreting links be related in this way, together with the requirement that "forward linked" structures in the content of a net always be interpreted (see Balzer et al. (1983); D9-A-4-c-i) entails that $\lambda_r[K^2{}_1, K^2{}_2]$ correspond to the "strong reduction relation" of earlier treatments (Balzer & Sneed (1977/78)). Intuitively, the reduction provided by $\lambda_r[K^2_1, K^2_2]$ may be "factored" into theoretical and non-theoretical parts.

8.1

We may generalize this example in rather obvious ways. First, we might allow a multiplicity of interpreting links for each of K^2_1 and K^2_2. That is

$$<K^2_1 = \{K^j_1\} \qquad n \geq j \geq 3$$

$$<K^2_2 = \{K^j_2\} \qquad n \geq j \geq 3$$

where 'j' super-scripts indicate "corresponding" interpreting model elements. We would require that corresponding interpreting model elements either be identical or related by a reducing link. That is

$K^j_1 = K^j_2$ or $L(K^j_1, K^j_2) \neq \Lambda$ and $\lambda_r[K^j_1, K^j_2]$ is a reducing link.

We could further generalize the example by allowing the interpreting model elements for K^2_1 and K^2_2 to be "imbedded" in an interpretation net. The intuitive idea is that the interpretation net narrows down the contents of the K^j_1 and K^j_2 so that only members of these contents provide interpretations for K^2_1 and K^2_2. As in the specialization case, the appropriate interpretation nets appear to be

$$N \sim N>K^2_1 \qquad \text{and} \qquad N \sim N>K^2_2$$

and the appropriate contents are

$$C_p[N \sim N>K^2_1](K^j_1) \qquad \text{and} \qquad C_p[N \sim N>K^2_2](K^j_2).$$

8.2

Let us now consider the intended applications for K^2_2 and K^2_1 -- $I[N](K^2_2)$ and $I[N](K^2_1)$. Intuitively, we expect that each intended application of K^2_1 "corresponds via reduction" to some intended application of K^2_2. That is, we expect that every $x^2_1 \in I[N](K^2_1)$ "corresponds via reduction" to at least one $x^2_2 \in I[N](K^2_2)$. This was an independent condition imposed

in earlier treatments of reduction (Balzer & Sneed (1978/ 1979)). It is interesting to see that, in the present treatment, this condition follows from our requirements on the intended applications. To see this, we must first be precise about what we mean by 'corresponds via reduction'.
For $x^2{}_1 \in I[N](K^2{}_1)$ and $x^2{}_2 \in I[N](K^2{}_1)$, we may say that '$x^2{}_1$ corresponds to $x^2{}_2$ via reduction in net N' iff there exist

$$y^1{}_2 \in |M_p{}^1{}_2| \quad \text{and} \quad y^1{}_1 \in |M_p{}^1{}_1|$$

so that

$$1)\ \langle y^1{}_1, x^2{}_1\rangle \in \lambda_{pp}[K^1{}_1, K^2{}_1]$$

$$2)\ \langle y^1{}_1, y^1{}_2\rangle \in \lambda_r[K^1{}_1, K^1{}_2]$$

$$3)\ \langle y^1{}_2, x^2{}_2\rangle \in \lambda_{pp}[K^1{}_2, K^2{}_2]$$

The "theorem" then is that, for all $x^2{}_1 \in I[N](K^2{}_1)$, there is some $x^2{}_2 \in I[N](K^2{}_1)$ so that $x^2{}_1$ corresponds to $x^2{}_2$ via reduction in N. To see that this is true, first note that since

$$<K^2{}_1 = \{K^1{}_1\},$$

$$I[N](K^2{}_1) = \overline{\lambda_{pp}[K^1{}_1, K^2{}_1]}>(C_p[N \sim N>K^2{}_1](K^1{}_1)).$$

Further, since

$$N \sim N>K^2{}_1 = \begin{array}{c} K^1{}_1 \\ \Big\downarrow \ \lambda_r[K^1{}_1, K^1{}_2] \\ K^1{}_2 \end{array}$$

$$C_p[N \sim N>K^2{}_1](K^1{}_1) = |M^1{}_1| \cap \overline{\lambda_r[K^1{}_1, K^1{}_2]}<(M^1{}_2)$$

and

$$I[N]({K^2}_1) = \overline{\lambda_{pp}[{K^1}_1,{K^2}_1]}>(|{M^1}_1| \cap \overline{\lambda_r[{K^1}_1,{K^1}_2]}<({M^1}_2)).$$

Thus, it is clear that, for every ${x^2}_1 \in I[N]({K^2}_1)$ there exist

$${y^1}_1 \in |{M^1}_1| \quad \text{and} \quad {y^1}_2 \in |{M^1}_2|$$

so that

1) $\langle {y^1}_1,{x^2}_1\rangle \in \lambda_{pp}[{K^1}_1,{K^2}_1]$

2) $\langle {y^1}_1,{y^1}_2\rangle \in \lambda_r[{K^1}_1,{K^1}_2]$

Likewise,

$$I[N]({K^2}_2) = \overline{\lambda_{pp}[{K^1}_1,{K^2}_1]}>({M^1}_2).$$

so that

3) $\langle {y^1}_2,{x^2}_2\rangle \in \lambda_{pp}[{K^1}_2,{K^2}_2]$.

On might ask whether this would continue to be the case when ${K^1}_1$ and ${K^1}_2$ were imbedded in larger nets with interpreting links. It is rather easy to convince oneself that, so long as all interpreting and reducing links have this rectangular configuration the intended applications will be related in the way described here.

NOTES

1. Consistent with the category theoretic notation, we shall denote the class of all sets by '|SET|' and the set of all functions from X to Y by 'SET(X,Y)'.
2. '$D_i(R)$' denotes the i-th domain of the relation R.
3. I am indebted to Prof. G. Süssmann for this example.
4. When $R \subseteq X \times Y$ we use the notation 'R' to denote the corresponding sub-set of Pot(X) × Pot(Y). We extend this notation in the obvious way to relators and functors.

Joseph D. Sneed
Department of Humanities and Social Sciences
Colorado School of Mines
Golden, Colorado 80401/USA

REFERENCES

Balzer, W. and Sneed, J.D.: 1977, 1978, 'Generalized Net Structures of Empirical Theories I and II', Studia Logica 36 (3), 195-212; and Studia Logica 37 (2), 168-194.

Balzer, W. and Moulines, C.-U.: 1980, 'On Theoreticity', Synthese (44) 467-494.

Balzer, W., Moulines, C.-U. and Sneed, J.D.: 1983, 'The Structure of Empirical Science: Local and Global' to appear in Proceedings: 7th International Congress of Logic, Methodology and Philosophy of Science, Salzburg.

Bourbaki, N. (pseud.): 1968, 'Elements of Mathematics: Theory of Sets', Addison-Wesley, Reading, Mass., Chapter IV.

MacLane, S.: 1977, 'Categories for the Working Mathematician', Springer, New York.

Mayr, D.: 1976, 1981, 'Investigations of the Concept of Reduction, I and II', Erkenntnis 10, 275-294, and Erkenntnis 16, 109-129.

Pearce, D.: 1982, 'The Structuralist Concept of Reduction' Erkenntnis 18, 307-334.

Sneed, J.D.: 1979, 'The Logical Structure of Mathematical Physics, 2nd ed., Reidel, Dordrecht.

Sneed., J.D.: 1979, 'Theoretization and Invariance Principles', in: I.Niiniluoto and R. Toumela, The Logic and Epistemology of Scientific Change, (Acta Philosophical Fennica 30),North-Holland, Amsterdam, 130-178.

Gerhard Vollmer

REDUCTION AND EVOLUTION - ARGUMENTS AND EXAMPLES

1. KINDS OF REDUCTION

No other problem seems to be more central to philosophy of science than the problem of reduction. Whether a proposition, a theory, or a whole branch of science, may be reduced to another proposition, theory or discipline, is a typically metascientific question.

The concept of "reduction", however, - originally hoped to be simple and clear-cut - turned out to be complicated and ambiguous. Many authors have tried to restore clarity in this field by making conceptual distinctions. Thus, we are invited to distinguish between the following kinds of reduction:

- definability of terms (terminological or "weak" reduction) vs. derivability of propositions (deductive or "strong" reduction),[1]
- homogeneous reduction (where the concepts of the theory to be reduced are already contained in the reducing theory) vs. heterogeneous or inhomogeneous reduction (where this is not the case),[2]
- strict reduction (or reduction by deduction) vs. approximate reduction (where previous theories appear as "limiting cases" of newer ones, possibly under very special or even counterfactual conditions),[3]
- domain-combining (derivational, explanatory, or unificational) reduction vs. domain-preserving (successional or justificational) reduction,[4]
- explanation of a theory vs. explanation of its success,[5]
- reduction by deduction (or by approximation) vs. replacement (elimination, dislodgement) of theories,[6]
- reduction between theories vs. reduction between fields or branches of science.[7]

Although this list could easily be extended[8], completeness is not our task here. This contribution rather focusses on one particular distinction, namely the third one between strict and approximate reductions.

This distinction is widely used. The reason is obvious. Cases of strict reduction are quite common in structural (or formal) sciences such as logic or mathematics. In many philos-

W. Balzer et al. (eds.), Reduction in Science, 131–152.

ophers this raised the hope that similar cases of reduction would be common in empirical science too. In fact, had not scientists themselves claimed that in many historical cases a new theory implied an older one in some sense? Typical cases are "derivations" of Galileo's and Kepler's laws from Newton's.[9]

Closer scrutiny, however, reveals that historical cases of strict reduction in factual science are extremely rare. This can be shown most easily by pointing to the fact that in most cases of theory succession old and new theories are in mutual contradiction. Thus, strictly speaking, Newton's mechanics and gravitation theory lead to equations which contradict Galileo's and Kepler's laws.

These discoveries or insights prompted many philosophers of science to side with the other extreme, denying the possibility of reduction in real science altogether. Theories, then, would not be reduced to newer ones, they would rather be displaced. Such replacement would occur for reasons of comparison and rational evaluation or, even worse, for no rational arguments at all. The shortest and most frequently cited statement in this direction is Planck's dictum: "A new scientific truth does not triumph by convincing its opponents..., but rather because its opponents eventually die." This is, essentially, the view of Hanson, Kuhn, or Feyerabend.

Practicing scientists, as a rule, do not subscribe to this defeatist view. Being confronted with the impossibility of strict reduction, they prefer to talk about limiting cases, initial circumstances, counterfactual conditions, tangential embedding, "simulation" of theories, correspondence rules, etc., but are still ready to use the concept of reduction. Philosophers of science should take this - the scientist's - intuition seriously and look for adequate explications of such a weaker concept of reduction, consistent both with the scientist's use and with the historian's insistence on the near nonexistence of strict reduction in factual science.

In fact, the scene has changed within the last years. Intertheory relations have been studied more thoroughly and new concepts of reduction have been developed and applied. This trend is perfectly documented by the present volume. Even so, it still seems to be a question of trial and error, whether a special pair of theories qualifies for the application of a particular concept of reduction. It is the aim of this contribution to improve this situation. By giving arguments and examples, we shall try to show where and why strict reduction may be achieved and where approximate reduction is all we can hope for.

2. THE RELEVANCE OF REDUCTION: UNITY OF SCIENCE

Problems of reduction are relevant to different questions. One context is unity of science, another history of science. Unity of science may be interpreted as[10]

- unity of method (e.g. trial and error elimination, conjectures and refutations, hypothetico-deductive reasoning),
- unity of language (definability of terms, "weak" reduction),
- consistency (because contradictions would certainly prevent unity of science),
- unity of structure (existence of isomorphisms),
- unity of laws (derivability of propositions, "strong" reduction).

From the very beginning, we shall concentrate on the most ambitious concept of unity, on the unity of laws. Is it possible to reduce all, or many, scientific laws to some basic laws of some basic discipline, for instance, to reduce all mathematical theories to set theory, all physical theories to elementary particle physics, all natural or even all empirical sciences to physics? Is it possible to reduce sociology to psychology, psychology to biology, biology to chemistry, and chemistry to physics?

Our questions relate reduction to the aim of science: Make use of the redundancy of the world to describe it simply. This contention deserves some clarification.

In principle, the world could be chaotic, without regularities, without constants, without laws. Speaking in terms of information science or of complexity theory, the world then would be completely random, highly non-redundant, utterly complex. Planned actions, predictions, retrodictions, explanations, even short descriptions, would be impossible.

But our world is quite different. It is, though complicated enough, at least partially stable, ordered, regular, repetitive, lawful, in short redundant. And factual science may be interpreted as man's attempt to give not just some description of his world, but rather the shortest description possible. Thus, elimination of redundancy may be seen as the aim of science or, more precisely, as its cognitive aim.

Since deductive inferences don't give or add any new information - otherwise they couldn't be truth-preserving - it is obvious that unity of science in its strongest sense (namely as reduction by deduction) is not only consistent, but virtually identical with the aim of science: reduction eliminates redundancy. An ideal final stage of factual science would be a minimal, redundancy-free description of the real world. The regulative idea of such an enterprise will not only be truth, but also some kind of simplicity, of economy, of parsimony, of elegance.

Such a minimal description could be coherent, or it could consist of different independent pieces mirroring different levels of reality. The ideas of reduction and of unity in science, however, clearly side with the first alternative: One coherent world should be described by one coherent unified science. Of course, such a unified science, if it could be reached or demonstrated to be possible, would strongly affect our world-view. It would help us to determine our place in the universe. It would destroy the walls which traditionally separate the different sciences. It would stimulate an integrated (literally "psychosomatic") medicine, where body and mind are not divided artificially and treated as independent entities. Moreover, it would be quite helpful didactically. Thus, reduction in science is by no means a purely academic problem. It is related to quite general and far-reaching questions.

Are there arguments in favor of unified science? Are there reasons why unity of science should not only be welcome, but also feasible? Is the idea of unity in science nothing but wishful thinking, or are there some chances to achieve such a unity?

The best argument in support of a (deductive) unity of science, and, by the way, the only compelling one, would consist in actually performing the necessary reductions, in presenting science as a unified whole, in giving the coherent minimal description dreamt of.

However, although it can be shown that such a description would be neither unique nor identifiable as minimal[11], it is obvious that it has not been reached yet. On the contrary, the main trend in science seems to be not unification, but diversification. More and more facts are found, more levels of complexity distinguished, more theories devised, more disciplines developed. The typical scientist is a specialist. Where, then, is unity of science?

As long as the unity of science cannot be demonstrated, we will have to rely on weaker arguments, on analogies and examples, on plausibility considerations, on successful reductions and failed attempts, on non-compelling pros and cons. This is what we set out to do here. There is a good argument in favor of reduction. We shall call it the "argument from evolution" (cf. 7). We shall show that the concept of universal evolution indeed supports the idea of a unity of science, of reduction by deduction.

3. THE RELEVANCE OF REDUCTION: HISTORY OF SCIENCE

Science has its own history. The same is true of the particular branches of science. We may simply try to describe such a history. Yet we may also wish to have a more coherent account, to give explanations, to make predictions about the future of science, to set methodological standards, to discriminate good from bad science, to advise scientists how to proceed in order to be successful. For these reasons, philosophers of science try to give rational reconstructions of the history of science.

Most interesting in this respect are cases of theory succession. Suppose that two competing theories both refer to similar kinds of objects and to similar phenomena. Whereas for some time one theory is held to be true or adequate, it later falls into disgrace, is judged to be inadequate and finally replaced by its competitor. In the history of science this has happened very often. Why do scientists turn to a new theory? And even if they should have no good reasons for their preference, or no reason at all, is it possible, at least from hindsight, to justify their decision, to give arguments why in fact they were right in doing as they did (or, else, how they should have decided had all available information been properly taken into account)? Is it possible, in short, to give a rational reconstruction of the history of science?

In striving for such a rational reconstruction, we would try to compare scientific theories by some evaluative criterion and to identify one as "better" than others. Such an assignment could be made, for instance, by establishing that the new theory, T_2, is richer in content than the old one, T_1, that T_2 both explains more facts than T_1 and logically implies T_1. In such a case we would, of course, readily admit that T_2 is "better" than T_1 and that T_1 can be reduced to T_2. This is, however, not the standard case in the history of science. As a rule, a later ("better") theory will correct its predecessor, that is, both are in mutual contradiction. But if T_2 contradicts T_1, it cannot logically entail T_1, unless T_2 is inconsistent.

Thus, whatever the relation between subsequent theories, it will not be one of bare logical entailment and not one of reduction by deduction. Their relation must be more intricate, hence more difficult to find out and to explicate. What this relation precisely consists in, is still open to discussion.

It is well-known, or at least plausible, that problems of reduction with respect to the history of science are intimately connected with the question of rationality in science. Scientists are human beings, and as such they do not always behave or decide rationally. But this is not the point. Irrational decisions of individual scientists might still be compensated for by other scientists such that the scientific community would indeed behave rationally. The crucial point is whether science as a whole is a rational enterprise.

It is precisely this question which is, in view of the history of science, answered in the negative by some thinkers. For, if not even theories with the same field of intended applications can be ranked as being "better" (or "worse") than their competitors, if they cannot even be compared by some rational criterion, then the transition from one theory to another cannot rationally be justified. The decisions not only of individual scientists, but also of scientific communities, would look, by the internal standards of science, arbitrary, accidental, random.

It would not, then, make much sense to talk about the "evolution" of science or about "the evolution of physics", as Einstein and Infeld do in their common book with just this title. In using the term "evolution", we normally have in view some controlled, directed (though not goal-directed), quasi-continuous process. A sequence of irrational gestalt switches, conversions, or revolutions, would not deserve to be called an evolution. Such a view of science would also definitely contradict most scientists' intuition as to the history of science, to the comparability of scientific theories and to reduction. This paradox shows at least that reduction is not a side issue but rather a problem right at the heart of both science and its philosophy.

4. TWO TYPES OF EVOLUTION

We have distinguished two areas to which reduction is relevant, if not central: unity of science (sec.2) and history of science (sec.3). In both cases we met with the concept of evolution: universal evolution on the one hand, evolution of science on the other.

Both types of evolution are, of course, related. Factual science refers to real systems and processes and even to evolutionary processes, and the evolution of science itself is part and parcel of cultural evolution, hence of universal evolution. Moreover, both kinds of evolution are ongoing processes.

Even so, these processes are essentially different and must be carefully distinguished. The concept of universal or cosmic evolution applies to real systems, to the universe as a whole, to galaxies and quasars, to stars and planets, to continents and atmospheres, to atoms and molecules, to biomolecules and organisms, to populations and societies, to cultural and economic systems. Evolution of science, on the other hand, refers to ideas, to conceptual constructs, to hypotheses and theories, to methods, to knowledge and truth. Whereas universal evolution takes millions and billions of years, evolution of science just covers some centuries or, at most, millenia. The former is characterized by an overall growth of complexity and by causal connections between different stages, the latter by a general growth of knowledge and by specific logical relations between different parts of such knowledge.

In Fig. 1, those two types of evolution are depicted. Their difference is made conspicuous by giving them two different spatial dimensions: The evolution of real systems goes from left to right, the evolution of science from top to bottom.

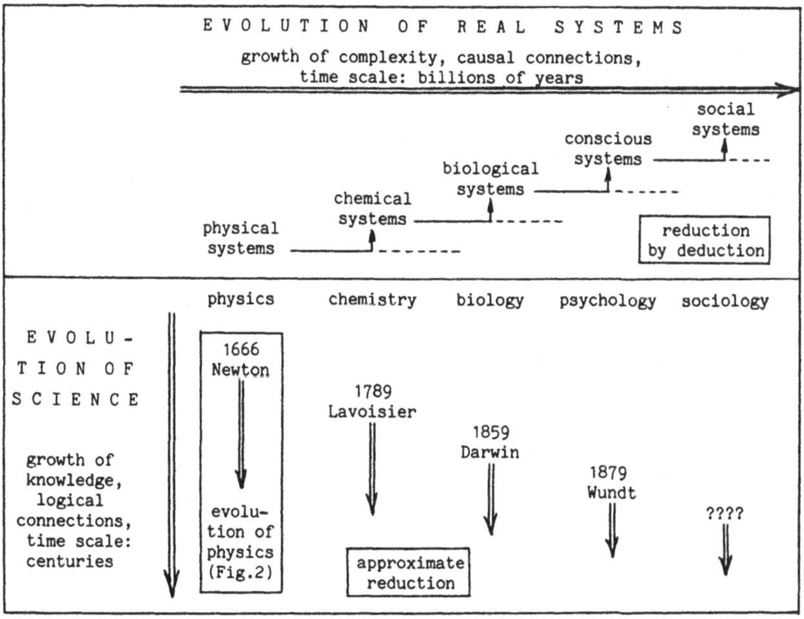

Fig. 1: Two different types of evolution: evolution of real systems and evolution of science. To these different types of evolution, different types of reduction are relevant: reduction by deduction and approximate reduction.

For all we know, cosmic evolution started with a Big Bang, possibly with quarks and leptons (or with prequarks), and gradually led to new and more complex systems, to elementary particles, nuclei, atoms, molecules, etc. That means, it started with physical systems, exclusively characterized by physical properties, and gave rise, in due time and course and wherever conditions were favorable, to chemical substances showing, in addition to their physical traits, chemical properties, to self-replicative structures, to organisms, to conscious beings, and to social systems.

Although in this process a great variety of new structures came in - new properties, new relations, new regularities, new laws - no law from a lower level was violated by systems of an upper level. Typical questions in this context would be: Do birds defy Newton's law of universal gravitation? Could the age of the earth as determined by physics (Lord Kelvin) differ from its age as estimated from biology (Darwin)? May organisms contradict the entropy law? Can free will, or mind, or consciousness, if understood as brain functions, escape the law of energy conservation? Of course not.

Organisms don't violate the laws of physics, and biology must not contradict physics. If such cases seem to occur, then the physical law supposed to be true was simply false beforehand. Cosmic evolution is an overarching, quasi-continuous, consistent, complexity-increasing process which affects all real systems.

Now, real systems, their properties, processes and interactions, are investigated, described and - if we are lucky - explained by science, in particular by the factual sciences as are, e.g., physics, chemistry, biology, psychology, sociology, history, or economy.

All these sciences have their own history. Their roots are lost in the myths and religions of prehistoric man. If, however, we characterize science by its use not only of observation, of speculation and of logic, but also of experiment and of mathematics, then science is an outcome of not more than the last three centuries. We are even able to specify, by the works of some great heroes of science, the historical junctures when those disciplines turned sciences. (This is done in Fig. 2, naming Newton, Lavoisier, Darwin and Wundt.)

Understandably enough, such specifications are always somewhat conventional or even arbitrary. We could replace Newton by Galileo, Lavoisier by Priestley, or Darwin by Lamarck. But this manoeuvre would not really change the qualitative picture, namely the succession of those disciplines. Our modern sciences came into being in the same order as the systems they investigate. This is not accidental. The more complicated systems are, the longer took it cosmic evolution to produce them and the longer took it man to handle them scientifically.

Although this sequence of origins of modern sciences has its particular interest, it is not our actual concern here. Apart from science as a whole, every science has its own history, its evolution. Thus, we talk about the evolution of physics (not of physical systems!), the evolution of biology (not about biological evolution!), etc. It is those histories where the concept of approximate reduction is most fruitful. As an exemplary case, we shall study the evolution of physics.

5. THE EVOLUTION OF PHYSICS

There are several factual sciences the history of which may be characterized by the following general trends: accumulation of knowledge, improvement of measuring precision, correction of mistakes, deepening of theories, and diversification of disciplines. Even if this description should not fit all disciplines, it certainly applies to physics. Starting with mesocosmic objects and phenomena, scientists have collected more and more information. They have replaced everyday experience by observation, intentional action, directed investigation, planned experiment, quantitative measurement, laboratory work and instrumental research. They have devised better theories providing comprehensive descriptions, adequate explanations, reliable predictions, stimulating questions. At the same time, they have established more and more branches of physics.

Some of these trends are documented in Fig. 2. The first lines exhibit objects of scientific research ("systems") and pertinent physical disciplines. Those disciplines are followed through time from top to bottom. The most important physical theories are represented by the names of their creators. Of course, only a small fraction of those remarkable men could be named.

Successive theories, referring roughly to the same area of intended applications, are connected by double lines (or arrows). Any two of them, if so connected, qualify for case studies in theory dynamics or approximate reduction. Again, there are many more cases. It is not necessary, therefore, to restrict such studies to the standard cases like Ptolemy-Copernicus, Kepler-Newton, Galileo-Newton, and Newton-Einstein.

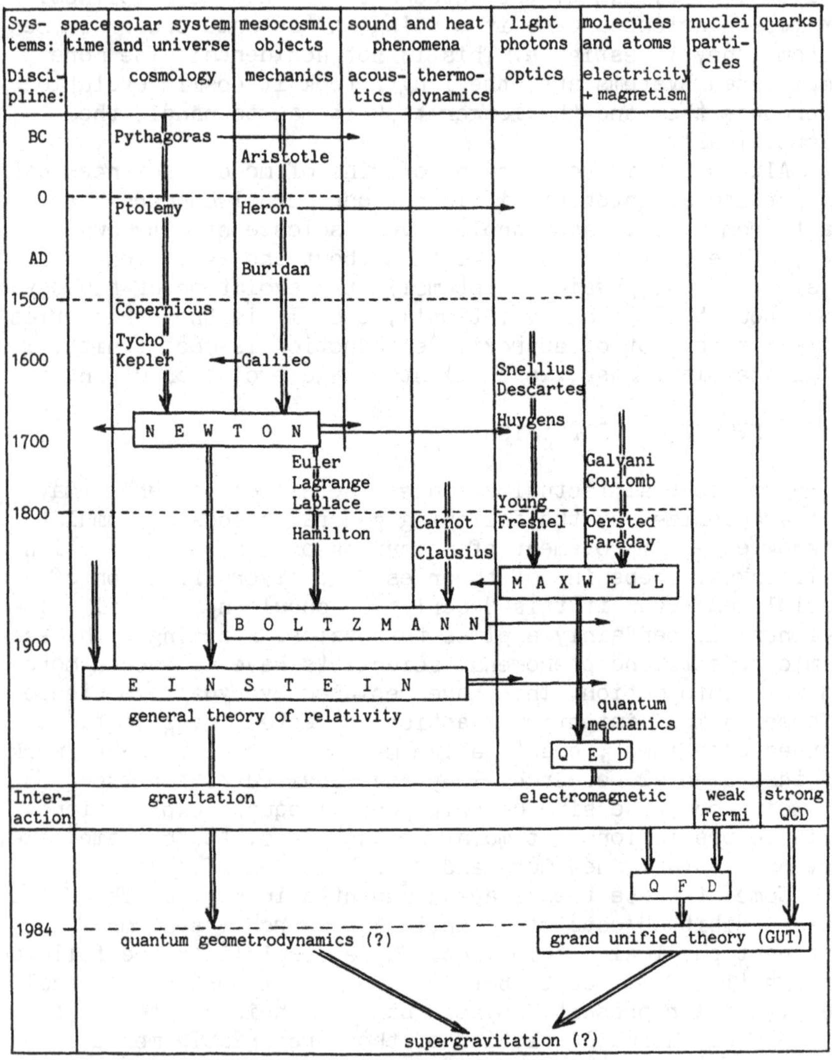

Fig. 2: The evolution of physics. Despite much diversification, there is also unification by comprehensive theories. The great unifiers of physics, Newton, Maxwell, Boltzmann and Einstein, are put into boxes where the separation of (at least) two disciplines ends. More recent unifications can hardly ascribed to single persons. Quantum electrodynamics (QED by Feynman, Tomonaga, Schwinger) unites quantum mechanics and electrodynamics. Quantum flavour dynamics (QFD by Glashow, Salam, Weinberg) integrates electromagnetic and weak interactions. Quantum chromodynamics (QCD by Gell-mann) describes strong interactions and is hoped to be united with QFD by the Grand Unified Theory (GUT) which is about to be developed and tested. Quantum theories of gravitation (QGD) and unified theories of all interactions (supergravitation) are still highly speculative.

Horizontal arrows indicate contributions of particular scientists to other fields of physics.

It is well-known that physics started with celestial and terrestrial mechanics in antiquity. The Middle Ages did not contribute too much - that's one reason for calling them the "Dark Ages". Fig. 2 just mentions Buridan reminding us of his interesting impetus theory which lies well in between Aristotle's and Galileo's theories of motion.

Modern physics, however, arose only in modern times, which seems to be a truism, but isn't. Optical phenomena were studied from 1600 on. Acoustics became a science, by being integrated into mechanics, from 1700 on. Electricity and magnetism turned sciences late in the 18th century, thermodynamics during the 19th century. Molecular and atomic physics, nuclear and elementary particle physics are late outcomes of our own century.

Thus, Fig. 2 demonstrates one of the general trends named above, the branching out of physical theories and physical knowledge. Indeed, whoever looks at physics and its history, will be impressed, if not depressed, by this process of disintegration in the body of physical knowledge. Thus, the characterization given at the beginning of this section applies perfectly well to physics.

But this is not all there is. All diversifications notwithstanding, there is still, at least in physics, another tendency, directed towards unification. Now and then, scientists succeed in formulating a comprehensive theory which integrates two hitherto separate areas or disciplines. Such achievements are always judged as great successes. They are steps towards the unity of science, at least of physics. They counteract the disintegrating effect of the growth of knowledge stressed above. They even nourish an age-old philosophical dream concerning the unity of nature.

The great unifiers of physics are Isaac Newton (uniting terrestrial and planetary physics by his mechanics and theory of gravitation), James Clerk Maxwell (integrating electricity, magnetism and optics into his electrodynamics), Ludwig Boltzmann (reducing thermodynamics to statistical mechanics), and Albert Einstein (his general theory of relativity blending space, time and gravitation into geometrodynamics).

Of course, there are more unifying theories in physics, less encompassing and less known. But completeness is not our task here. We rather want to demonstrate that the evolution of physics - and, for that matter, of all science - shows antagonistic trends, diversification and unification. Whether a fictitious final physics (or science) would exhibit a complete unification or remain diversified, is not known.

There are, however, some current trends in physics indicating that a unification of the fundamental forces of nature might indeed be possible. Electromagnetic and weak forces may be said to be united in quantum flavour dynamics, that is in the theory of electroweak interaction. Right now (1984), Grand Unified Theories (GUTs), aimed at integrating strong interactions into this puzzle, are about to be formulated and tested. And there exist already fascinating ideas about a comprehensive theory which would include gravitation, thus uniting all known interactions of nature (see Fig. 2). Thus, unification of physics is not only a most ambitious, but also a most successful research program.

Learning about these advances in fundamental physics, we might be tempted to ask with Stephen Hawking: "Is the end in sight for theoretical physics?"[13] Will cosmology and elementary particle physics finally be united, reduced to one overarching theory, to a world formula? We shall, however, not try to answer these questions.

Successful unifications pose particular problems of reduction[14]. In most cases, a unifying theory will not just combine two hitherto unconnected theories. It will rather correct them, explain their temporary success and, at the same time, allow new predictions. All this can easily be verified for the unifying theories of physics exhibited in Fig. 2, for instance for Newton's, or Maxwell's, or Einstein's theories.

The conclusion is this. The history of physics shows convergent and divergent trends. Whereas diversification may be seen as a cumulative process, unification and correction cannot be so interpreted. In either case, reduction by deduction, or strict reduction, is inapplicable because neither new nor incompatible information can be deduced from consistent knowledge. Thus, in reconstructing the evolution of science, especially of physics, only approximate reduction, if any, will be appropriate.

For the reconstruction of evolution in nature, however, the case is different.

6. THE HIERARCHY OF THE SCIENCES

There are in nature systems of different complexity: physical, chemical, biological, conscious systems (cf. Fig. 1). They are described and sometimes explained by sciences of decreasing generality, but of increasing complexity: physics, chemistry, biology, psychology.

It would not be adequate, however, to characterize those different sciences as having disjoint object classes, e.g. to view physics as investigating solely inanimate systems, whilst biology were studying living systems, that is organisms. If this characterization were correct, physics and biology could not contradict each other, not even in principle. And psychology, inquiring into systems with mental attributes, consciousness, self-consciousness and thought, couldn't possibly contradict biology or neurophysiology.

And yet, in factual science they do just that. Sometimes at least, different disciplines are thought to be inconsistent with each other. Darwin, a biologist, and Kelvin, a physicist, came to quite different results with regard to the age of the sun. Some thinkers suspected the law of entropy increase to be inapplicable to organisms (Planck) or even to be false in biology (du Noüy). Eccles admits that, for interactionism to be true, the law of energy conservation must be violated during the interactions between mind and brain.

It is therefore much more adequate to view the factual sciences as narrowing down their fields of interest to ever smaller domains. Physics investigates all real systems, be they dead or alive, inanimate or minded. It does not care, however, for all properties of those systems. Chemistry then specializes to systems which show chemical reactions, to atoms and molecules, to polymers and even larger systems. Some of those systems show additional properties, e.g. self-replication, heredity, death. They are called living systems and as such are the objects of biology. Again, psychology concentrates on some peculiar organisms, namely conscious systems, and studies their peculiar behavior. Thus, the object classes

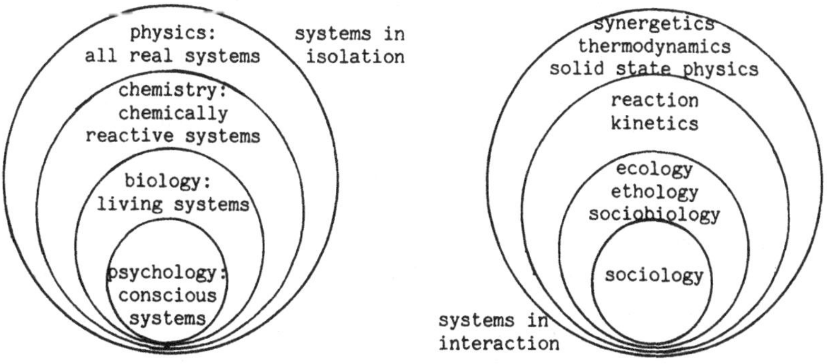

Fig. 3: The hierarchy of factual sciences

of those sciences are best characterized by set inclusion. This is illustrated in Fig. 3.

The case of sociology is somewhat deviating and therefore slightly more complicated. In principle, it may be asked for all real systems how they interact with each other and with other real systems and what cooperative phenomena there are. This is done for physical systems by physics itself, e.g. by thermodynamics, by solid state physics, or by synergetics. For chemical systems, it is also done by the science of chemistry itself, for living systems by genuine biological disciplines such as ecology, etholgy and sociobiology. For conscious systems, however, especially for man, this is not done by the pertaining discipline, e.g. anthropology, but by an autonomous science, sociology. (This peculiar role might be one of the reasons for the persisting uncertainty about the scientific status of sociology.) Similar considerations would apply to the science of history. The disciplines investigating systems in interaction are also presented in Fig. 3.

Having established such a hierarchical view of the factual sciences and their objects, we are now ready to recognize their evolutionary coherence.

7. THE ARGUMENT FROM EVOLUTION

Not only have all real systems their individual history, they are, in a sense, further united by a common history. Not only do all complex systems consist of smaller and simpler parts, they also originated from other (mostly simpler) systems in the course of evolution. The sequence of quarks, elementary particles, nuclei, atoms, molecules, macromolecules, organelles, unicells, multicellar organisms, plants, animals, men and other systems not only mirrors a scala naturae, a ladder of increasing complexity, but at the same time one of evolutionary succession, the order of their coming into being. Science has begun to recognize evolution as a universal process.

This cosmic process is mainly governed by causal laws, at least above the quantum level. Earlier states of real systems are causally connected with later states. This applies not only to processes of decay or of mere maintenance, but also to the emergence of new systems. It applies to all kinds of evolution, to cosmic processes, organismic evolution, human history, individual development, artificial synthesis.

Now, philosophical analysis has shown that scientific theories map causal relations (between events) onto logical relations (between statements). If this is true, it should be possible to remodel the processes of evolution in our scientific theories. And if complex systems originated from simpler ones in an at least partially deterministic way, it should be possible to derive at least some of the properties and laws of those complex systems from our knowledge about simpler systems. Thus, the facts of cosmic evolution provide an argument for the possibility of reduction by deduction. They support the hope that such an evolutionary reduction should be feasible. In fact, they even give us a hint at how to proceed: Try to mirror evolutionary processes in scientific theories, try to mirror the evolutionary unity of nature by a deductive unity of science.

For this evolutionary reduction no approximate reduction would be needed, provided our pertaining scientific theories are correct. (For a discussion of this questionable presupposition see sec. 8.) Nature doesn't use approximations. The later, the complex system should be deductively explained from its earlier, simpler subsystems. Complete knowledge about parts should suffice for the explanation of wholes.

The origin of life may serve as an example. If our evolutionary outlook is correct, there have been aeons of cosmic evolution, when living systems couldn't and didn't exist. Since there are now, at least on this tiny spot of the universe, living organisms, a transition from non-living to living matter must have taken place. What is more, the seeming gap between non-living systems and living organisms must have been bridged from the inorganic side, since before the existence of organisms there simply were no systems to which biological laws could apply.

But if the physical and chemical properties of matter were sufficient to produce particular self-replicating systems, that is life, then physical and chemical laws should also be sufficient to explain their emergence. That means that the gap between biochemistry and biology should be bridged by deductions from biochemistry. The fascinating discipline of biogenetics is just exploring this area. (It goes without saying that the heuristic path to such reduction laws might be found by going the other direction.)

Obviously, the "argument from evolution" supports a reductionist point of view. Even so, evolution does neither prove the unity of the world nor guarantee the unity of science. But this is neither new nor devastating. As long as there is no perfect knowledge with humans anyway, we shall (have to) content ourselves with conjectural knowledge and with good arguments. And the argument from evolution seems to be a good one.

There are further arguments, pointing into the same direction (reduction by deduction), but showing weaker thrust. We shall call them "the argument from synthesis" and "the argument from development".

The argument from synthesis points to the fact that we are able to synthesize complex things from simple ones, repeatedly, reliably, predictably. We are able to plan such synthetic processes, to control them, to change them, to use them. In this manner, physicists have synthesized heavy elements, Wöhler has synthesized organic compounds from inorganic material, Miller has synthesized biomolecules from abiotic substances, and biochemists synthesize virus genomes from purely enzymatic substrates. And again, we are convinced that such processes are governed by causal laws. Therefore, it should be possible to map them into concepts and theories, to describe and to explain them theoretically, to reduce later stages to previous ones.

The argument from development draws on the fact that organisms develop from undifferentiated stages to highly complex ones, from small cells to large and mature individuals. This development occurs again according to causal laws. Should it not then be possible to mirror this causal process in a theory of ontogenetic development? To explain final stages from previous ones? Again, causal connections in nature would be mirrored by logical connections in theory, relations of production by relations of deduction.

Nevertheless, although both arguments are similar in structure to the argument from evolution, they are not as compelling as the latter. Synthesis and development are not the best models for the processes of universal evolution, for several reasons.

For one, synthetic processes are planned and carried through by intelligent beings (scientists). The activity of such an outside ruler might be a decisive element in the course of synthetic processes. And there is no ruler in evolution. It is precisely the task and aim of science to dispense with such an omnipresent, omniscient and omnipotent outside authority.

For another, although developmental processes don't exhibit such a conscious ruler, they still are directed and controlled by a genetic program. And there is no program in cosmic or in biological evolution. It is precisely the achievement of science, for instance of Darwin's theory, to demonstrate how evolution could get along without such a program. (This difference is, by the way, partly responsible for the inadequacy of Haeckel's biogenetic "law", according to which ontogenetic development should be a condensed recapitulation of phylogeny.)

Thus, we might be tempted to read the argument from synthesis as saying "Given the actions of an intelligent ruler, the products of synthesis can be reduced to the initial conditions." and the argument from development as saying "Given a clever enough program, the outcomes of development can be reduced to its inputs." But this is not enough. The argument from evolution is not a candidate for such a cryptotheistic or cryptovitalistic interpretation. It simply and unconditionally claims: "The results of evolution can be reduced to previous conditions." This is by far the strongest claim. And that is why we should direct our attention to the argument from evolution.

Even so, we should not ignore the difficulties lurking here.

8. DIFFICULTIES FOR "EVOLUTIONARY" REDUCTION

By the argument from reduction, universal evolution is interpreted as suggesting the possibility of reduction by deduction and as supporting an ambitious unity of science program. There are, however, several problems we should be aware of.

One problem arises from the fact that even our best theories are not ultimate truths. They might be false and often are. It is, however, only to correct theories that the argument from evolution fully applies. If system S_2 evolved from system S_1, then theory T_2, referring to S_2, would logically entail theory T_1, referring to S_1. If, however, either T_1 or T_2 fail to be correct, then it is unlikely that there will be logical entailment between T_1 and T_2.

Let's again take biology as an example. Living systems have evolved from non-living ones. According to the argument from evolution, a fictitious ideal physics would be able to explain the origin of life. Moreover, it would entail all laws of an equally fictitious ideal biology. As long as biology is not in this perfect state, our perfect physics could not imply such an imperfect biology. That means that even in evolutionary contexts we shall often have to put up with approximate reductions.

This insight is bound to damp our optimism concerning strong reductions, but does not render the concept of evolutionary reduction worthless. First and foremost, it remains a <u>regulative idea</u> we might use and strive for even without finally reaching it. Second, evolution has run through innumerably many steps. Science might succeed in pursuing them, some accurately, some approximately. Anyway, the progress of science will be piecemeal. We should not ask to find out all at once. For some evolutionary steps, reduction by deduction might already be adequate. It is certainly worthwhile to check this. Third, it will be quite helpful to be clear about where and why the concept of reduction by deduction is not adequate. We have seen that in general it does not apply to the evolution of science (unless some scientific progress just consists in reconstructing an evolutionary step). This insight enjoins any philosopher of science to make his choice: Either he insists on studying strict reduction. Then he should direct his attention to evolutionary processes and try to achieve evolutionary reduction. Or he sticks to problems of theory succession and will have to face the difficulties of approximate reduction. This is, at least a clear and instructive alternative.

There are more difficulties to evolutionary reduction. In going "upstairs" from physics to chemistry, to biology, to psychology, not only new properties arise, but also new problems for evolutionary reduction. One of these difficulties is complexity itself. Although complexity is a creative element in evolution, it is also a handicap in science. Thus, we may have at our disposal all basic information necessary to describe a system and still not be able to derive, predict or explain its properties from the available equations and data. Our failure might be due to our unability to perform the necessary computations.

The relation of chemistry to physics may serve as an example. Most scientists are convinced that atoms and molecules are ruled exclusively by the known principles of electrodynamics and quantum mechanics as are Maxwell's equations, Schrödinger's (or Dirac's) equation, and Pauli's exclusion principle. As to these basic principles, our knowledge seems to be complete. Even so, physicists have so far not been able to deduce the behavior of an uranium or even an oxygen atom from these laws. Most atoms and all the more all molecules are too complicated for such "ab initio" computations.

This problem is by no means peculiar to the relation (and hoped-for reduction) of chemistry to physics. It arises for every stage of evolution, on every level of complexity, in every branch of science. It occurs not only with relations between different sciences, but already inside physics. In many cases, we will have to rely on approximations, on statistical methods, on mean values, on additional assumptions.

This does not mean, however, that we have to fall back to approximate reduction which is constitutive for theory succession. In coping with the problems of complexity, we are not necessarily thrown on contradictions, counterfactual conditions, unrealistic assumptions. Although our initial information might not suffice to yield the hoped-for reduction, but call for supplementary assumptions, this auxiliary information can be perfectly consistent with all our basic laws and facts.

Thus, we have traced out a further kind of reduction, neither strict (as reduction by deduction) nor contradictory (as in theory succession). This type of reduction is relevant when evolutionary reduction is possible in principle, but, due to complexity, impossible in practice. Maybe this kind of reduction truly deserves to be called "approximate" reduction.

Complexity is still a rather general concept. It is possible to specify the impediments for evolutionary reduction more in detail. Particular kinds of complex systems are, for instance, chaotic systems, feedback systems, systems with random elements, etc. We shall not discuss them in detail, but just list them. Fig. 4 presents different sciences, their objects of inquiry, some new properties relevant on that particular level, and the additional problems combined with them.

Science	Objects of inquiry	New properties	Additional problems
Physics	all systems, e.g.: particles, fields atoms many-body systems	cooperative effects	complexity strong causality violated ("chaotic" systems)
Chemistry	molecular systems	auto-catalysis	cyclic (cybernetical) causality
Biology	living systems	self-replication and heredity, mutations	constitutive role of chance uniqueness, individuality
	higher organisms	pain and suffering	ethical problems of experimenting
Psychology	conscious systems man	reflection inner aspect	observational effects mind-body problem
Sociology	social systems		small number of relevant systems

Fig. 4: Problems for evolutionary reduction

9. SUMMARY

The concept of reduction is ambiguous. Whereas in formal science strict reduction is the rule, in empirical science it is the exception. In the latter, approximate reductions are much more frequent.(1) Even so, there are in factual science cases where strict reduction (or reduction by deduction) applies or may, at least, function as a regulative idea. This can be shown by reference to different areas where problems of reduction arise, namely unity of science (2) and history of science (3). In both contexts, we encounter evolutionary processes, evolution of real systems (universal evolution) on the one hand, evolution of science on the other.(4) For the latter, physics may serve as an example. Despite all diversification, the evolution of physics shows theory succession, correction and unification of theories. Here approximate reduction will be adequate.(5) The case of universal evolution is different. The hierarchy of the sciences is based on the increasing complexity of real systems.(6) This sequence, in turn, reflects the creative process of universal evolution. And if complex systems originated from simple systems according to causal laws, then it should be possible to mirror these evolutionary processes in our scientific theories and to explain the complex from the simple. This "argument from evolution" suggests that strict reduction between different disciplines and sciences should be possible, at least in principle.(7) Even so, there is no room for naive optimism. Although our argument may serve as a heuristic cue pointing to the right strategy, the problems posed by the complexity of most real systems are still hard enough.(8)

Zentrum für Philosophie und Grundlagen der Wissenschaft
Justus-Liebig-Universität Gießen

NOTES AND REFERENCES

1. R. Carnap: 1938, 'Logical Foundations of the Unity of Science', Int. Enc. Unified Science I, 1. Reprinted in H. Feigl and W. Sellars (eds.): 1948, Readings in Philosophical Analysis, New York, pp. 408-423.
2. E. Nagel: 1961, The Structure of Science, Routledge & Kegan Paul, London, pp. 339-341. Nagel is relying on previous work from 1949. The distinction between homogeneous and inhomogeneous reduction is also made in L. Sklar: 1968, 'Types of Inter-Theoretic Reduction', Brit. J. Phil. Science

18, 109-124, p.110, and in K. Friedman: 1982, 'Is Intertheoretic Reduction Feasible?' Brit. J. Phil. Science 33, 17-40, p. 20.

3. The concept of approximate explanation was hinted at, but not worked out, in C.G. Hempel: 1965, Aspects of Scientific Explanation. The Free Press, New York, p. 344. See also E. Scheibe: 1973, 'The Approximative Explanation and the Development of Physics', in P. Suppes et al. (eds.), Logic, Methodology, and Philosophy of Science IV, North Holland, Amsterdam, pp. 931-942. R.M. Yoshida: 1977, Reduction in the Physical Sciences, Dalhousie University Press, Halifax.
4. T. Nickles: 1973, 'Two Concepts of Intertheoretic Reduction' J. Phil. 70, 181-201. Nickles also uses awkward indices: $reduction_1$, $reduction_2$. Making the same distinction, Wimsatt calls them "explanatory" reduction and "successional" (or "intra-level") reduction. See W.C. Wimsatt: 1976, 'Reductive Explanation: A Functional Account', in R.S. Cohen et al. (eds.): 1976, PSA 1974, Reidel, Dordrecht, pp. 671-710.
5. Sklar[2], p. 112.
6. T.S. Kuhn: 1962, The Structure of Scientific Revolutions, The University of Chicago Press, Chicago. - P.K. Feyerabend: 1962, 'Explanation, Reduction, and Empiricism', in H. Feigl and G. Maxwell (eds.): 1962, Scientific Explanation, Space and Time, University of Minnesota Press, Minneapolis, pp. 28-97. - K.R. Popper: 1974, 'Scientific Reduction and the Essential Incompleteness of All Science', in F.J. Ayala and T. Dobzhansky (eds.), Studies in the Philosophy of Biology, Reduction and Related Problems, Macmillan, London, pp. 259-284. - According to Kuhn, Feyerabend and even Popper, the relevant cases of reduction in factual science are cases of displacement. A similar point is made by M. Spector: 1978, Concepts of Reduction in Physical Science, Temple University Press, Philadelphia, and refuted in R. Yoshida: 1981, 'Reduction as Replacement', Brit. J. Phil. Science 32, 400-410.
7. L. Darden and N. Maull: 1977, 'Interfield theories', Phil. Science 44, 43-64.
8. More attempts at classifying different concepts of reduction are made in K.F. Schaffner: 1967, 'Approaches to Reduction', Phil.Science 34, 137-147.
9. M. Born: 1949, Natural Philosophy of Cause and Chance, Clarendon Press, Oxford, and Dover, New York 1964, pp. 129-134, even claims that Newton's law of gravitation may be "derived" or "deduced" from Kepler's laws, not only vice versa.

10. For a comparative account of those concepts of unity of science see G. Vollmer: 1984, 'The Unity of Science in an Evolutionary Perspective', Proc. 12th Conf. on the Unity of the Sciences (Chicago, Nov. 1983), Int. Cultural Foundation Press, New York (in press).
11. That the minimality of a description cannot be proven in every case, is stressed in G.J. Chaitin: 1975, 'Randomness and Mathematical Proof', Sci. American 232, May 1975, 47-52.
12. See also G. Vollmer: 1977, 'Theoriendynamik und Ablösung einer Theorie durch eine neue (bessere): Simulation statt Erklärung', in G. Patzig, E. Scheibe and W. Wieland (eds.), Logik, Ethik, Theorie der Geisteswissenschaften (XI. Dt. Kongreß für Philosophie, 1975), Meiner, Hamburg, pp. 493-499.
13. S. Hawking: 1980, Is the End in Sight for Theoretical Physics? Cambridge University Press, Cambridge.
14. See, e.g., R.L. Causey: 1976, 'Unified Theories and Unified Science', in R.S. Cohen et al. (eds.), PSA 1974, Reidel, Dordrecht. - P. Kitcher: 1981, 'Explanatory Unification', Phil. Science 48, 507-531.

David Pearce and Veikko Rantala

LIMITING CASE CORRESPONDENCE BETWEEN PHYSICAL THEORIES

However sceptical one's attitude to the unrestricted validity of the so-called Correspondence Principle (henceforth: CP), there is no denying that CP has played a significant role, not only in the actual development of physical theories, but also in the construction of general methodologies of science. The concept of intertheory correspondence is thus important for both descriptive and normative aspects of methodology, where it figures prominently in the context of discovery as well as in the context of justification.

The central idea of CP that one physical theory may 'contain' another as a limiting case, or that a later theory approximates its predecessor when certain limits are taken, is deceptively simple to state. It is equally easy to produce ready-made examples from the history of science that persuasively fit the pattern of such a relationship. Indeed, there is a rather widespread agreement that a relation of correspondence exists, for example, between Van der Waal's equation and the ideal gas law, or between Newton's gravitation theory and Kepler's laws, or between substantial portions of special relativistic mechanics and classical mechanics, and so on. This much is usually readily conceded. But there remains precious little agreement concerning in exactly what this correspondence relation consists; and this is where the real problems begin in earnest.

So long as limiting relations between theories are discussed only in intuitive and vague terms, there seems to be little hope for reaching a consensus about precisely what constitutes a correspondence relation, and which pairs of actual theories exemplify it. Consequently, it is impossible under these terms to reach any informed judgement about the validity of CP as a general principle governing scientific development; and

W. Balzer et al. (eds.), Reduction in Science, 153–185.

even in the 'paradigm' cases of correspondence, such as those just mentioned, there will inevitably remain a large and shady area of uncertainty and controversy about whether the theory changes in question are continuous or discrete, or whether they count as instances of reduction, intertheoretic explanation, or whatever.

Fortunately, physicists and philosophers of science have lately come to acknowledge the need for a sharper and more penetrating analysis of the correspondence relation. And a number of recent works have offered us more detailed and precise explications of correspondence, have studied in a formal setting the actual relations obtaining between selected pairs of physical theories, and have looked in greater depth at the role of correspondence and CP in various methodologies of science.[1]

In an earlier paper, (1983a), we proposed a new method for treating limiting relations between physical theories, based on the tools of nonstandard analysis that were first applied in this context in Rantala (1979). Besides offering an exact reconstruction and logical analysis of the correspondence relation - which we believe to be formally adequate - this method seems to us to shed new light on a number of the more problematic logical and philosophical questions that have long been associated with theory change, correspondence, and reduction in general. With the aid of this method we were able to reconstruct the logical relation holding between certain key fragments of classical and (special) relativistic particle mechanics; and success here led us to conjecture that other instances of correspondence in science would admit a similar characterisation.

The present paper, which is intended as a sequel to our (1983a), is in two parts. The first gives an informal presentation and defence of our approach to correspondence; we concentrate on some of its chief points of contrast (and similarity) as compared with other methods currently available. In Part II we revert to the framework of our (1983a) in order to illustrate our approach with a further case study: we sketch a proof of the fact that certain axiomatised versions of Newton's celestial mechanics and Kepler's laws of planetary motion stand in the relation of (limiting case) correspondence. In itself this fact is of course well-known, but the present reconstruction incorporates several important aspects of the relation that are usually left out of consideration.

I

Put in a nutshell, our main thesis about (limiting case) correspondence is that it can be conceived as an exact, logical relation between theories, on a par with, and indeed in a suitable sense a special case of, intertheoretic reduction in general. What we mean by this is that if T and T' are theories, and T corresponds to T' in some limit, then there is a direct and straightforward logical link from T' to T. What we do not mean is that there is an informal and possibly vague mathematical connection between T' and T that is mediated by counterfactual hypotheses, mathematical limits and approximations, and idealising conditions, as is usually assumed. Such 'mediating' features are indeed characteristic of correspondence at the naive or informal level of analysis, and making their presence explicit is an important first step in studying the relation. Our claim, however, is that at a deeper level of analysis, where the connection of T to T' is suitably formalised, the link has a simple and direct, logical form which, moreover, throws into sharper relief certain aspects of the continuity between T and T' that are obscured on the traditional view and have been rightly questioned as a result.

Before examining how all this is possible, we may do well to reflect for a moment on the classical account of reduction. Its central feature is that reduction is construed as a logical, deductive relation between theories: if theory T is reducible to theory T' there should be a set A of auxiliary assumptions such that T' together with A deductively entails T. In short:

(1.1) $T' \;\&\; A \vdash T.$

This is only of course a rough and ready reduction schema that omits many of the finer details. But it already captures the core of the relation as being deductive, and it shows how one may naturally subsume reduction (at least formally) under the general heading of (D-N) explanation. (A finer characterisation would distinguish in A between special conditions required for the reduction, and correspondence or bridge principles that may be needed to connect the languages of T and T').

Even a cursory glance at the above schema shows that it does not accurately represent the majority of those intertheoretic relations in science that we would normally like to term 'reductive'. The main difficulties, which emerge already in the examples mentioned, is that usually T follows only approximately from T' (or T' entails only a corrected version of T), and the limiting assumptions and special conditions required for the deduction may be strictly false or incompatible with T' itself. The problems facing (1.1) as an adequate general picture describing reduction and correspondence are too well-known to need documenting here in any detail. Suffice it to remark that the difficulties in obtaining a simple deductive link between the reducing and reduced theories persuaded some philosophers to abandon the classical account of reduction altogether, and it led many others to give up the idea that limiting relations between supplanting and supplanted theories in science could in general be explanatory. Those philosophers who have retained the basic intuition of the standard account of reduction fall roughly into two categories: those who try to incorporate limits and approximations into an essentially deductive schema, and those who focus on the mathematical side of the approximation process in order to render it in precise, quantitative terms. Let us briefly survey each of these approaches in turn.

Deductive approaches to correspondence

These may be characterised by their retention of schema (1.1) subject to a slight modification so as to allow for the possibility that the corresponded or reduced theory T follows only approximately from the corresponding or reducing theory T', i.e. when certain parameters of T' approach their limiting values. In other words, (1.1) may be replaced by

(1.2) $T' \,\&\, A \vdash^{*} T$

or by

(1.3) $T' \,\&\, A \vdash T^{*}$

where, in the first expression $\vdash^{*}$ stands for approximate entailment, and in the second T* denotes an approximate counterpart of T.

(1.3) has turned out to be the most popular kind of ammendment to (1.1). It was accepted by Hempel (1965), Putnam (1965) and Nagel (197o), in particular, and it has found its way into many subsequent, more sophisticated accounts of reduction. The other version, (1.2), is in some respects preferable, but it suffers from the drawback that ordinary logic contains no notion of approximate inference. Attempts to construct such a 'logic' of approximate inference (see, e.g., Czarnocka and Zytkow (1982)) seem to lead in effect to a form of (1.3) in which a single entailment relation is replaced by a sequence of distinct entailments in which a mathematical parameter appearing in A approaches a given limit and the inferred theory T* gets closer and closer to the original theory T.

It seems rather remarkable, even with the benefit of hindsight, that a schema such as (1.3)could ever have been seriously proposed as a response to the challenge of producing an adequate deductive account of reduction in cases where limits and approximations are inextricably involved. For, in the first place, in order to obtain T* as a consequence of T' together with auxiliary assumptions usually requires that parameters occurring in those assumptions take values that are counterfactual or else forbidden by T' itself. So, when limits are actually taken, rather then merely approached, the entailing conjunction will be either false or inconsistent. And, secondly, the schema leaves completely open the sense in which T* is supposed to approximate T. How is one to determine in general when an approximation is 'sufficiently' close?

It is in the very nature of the concept of approximation that the borderline between 'quite close to' and 'rather far from' is vague; and it requires just as much clarification as the notion of reduction that is to be explicated. Naturally, one could select some conventional standard of 'proximity' based on empirical considerations: for instance, by choosing A in (1.3) to be such that the inferred theory T* is empirically indistinguishable from T given usual allowance for errors of measurement. But the problem here is that such standards of accuracy depend heavily on the theories involved and on the particular empirical situations to which they are being applied; they will have to be revised in each different context.

It seems clear, therefore, that even as a first step (1.3) is hardly a viable alternative to (1.1). On the contrary, in its usual formulation it replaces a well-defined logical relation of entailment and appeals instead to an unclear and problematic notion of approximation.

Recently, philosophers have paid closer attention to the difficulties involved in adhering to (1.3) as a general schema of reduction. Krajewski (1977) and the Poznan School in Poland have tried to provide a coherent expression of the logical relation obtaining in correspondence.[2] Their approach takes stock of the process of idealisation in science, whereby one of the senses in which theories progress is that they gradually remove idealising assumptions and replace them with more concrete or factual assumptions. Thus, on Krajewski's account, Newtonian mechanics reveals idealising assumptions that are present in Kepler's laws, e.g. that the ratio of the mass of the sun to the mass of a planet is infinite, that planet-planet interactions have no influence on overall planetary motion, and so forth. Consequently, though the correspondence relation of T' to T is thought to be essentially deductive, the connection between the theories is mediated by the drawing of idealising and factualising assumptions. In a simplified form Krajewski's view of correspondence can be pictured in the following way:

(1.4)

$$\begin{array}{ccc} T' & \dashrightarrow & T \\ \updownarrow & & \updownarrow \\ T'^{*} & \longrightarrow & T^{*} \end{array}$$

Here, the bottom arrow is a logical implication, and the remaining arrows denote approximate relations. Suppose that the earlier theory T consists of a functional law F(x)=0 which 'corresponds' to the law F'(x)=0 of T'. T is said to be re-interpreted in the light of T'. The re-interpretation is denoted by T* and comprises the addition of a numerical parameter p with the property that F(x)=0 in T* if p is equal to a suitable limiting value, say zero. In contrast, the parameter p is strictly positive in T', since there p>0 implies F'(x)=0, and F' is assumed to be different from F. From T' we may pass to a so-called 'abstracticized' version T'* of T' in which we may add to F'(x)=0 the condition p=0 and under this assumption deduce T* as a consequence. I.e.

(1.5) $$T'^{*} \,\&\, (p=0) \vdash T^{*}.$$

In (1.4) the passage from T to T* is one of idealisation, T and T' are taken to be mutually incompatible, and the limiting condition p=0 conflicts with T'. However, in (1.5) we achieve ordinary entailment from consistent hypotheses because the constraint p>0 is removed from T' and no longer appears in the 'abstracted' theory T'*.

Krajewski's account of the correspondence relation is of course much richer than this thumbnail sketch suggests, and it seems to clarify several interesting logical and methodological features of the problem. Nevertheless, a number of objections to it may be raised. For one thing, Krajewski makes no attempt to analyse further the approximate relation that holds between T and the inferred theory T*; and, secondly, his approach is exclusively syntactical, so that, in particular, no semantical analysis of possible meaning-changes occurring in the transition from T to T' is forthcoming. Indeed, Krajewski abruptly dismisses the latter problem by simply claiming that concepts common to the two theories undergo no essential shift in meaning, whereupon a syntactical description of the correspondence relation is held to be fully adequate. In other words, a term like 'mass' is supposed to have the same sense in, say, both classical and relativistic mechanics; only the empirical laws involving mass happen to be different in the two theories. It is hardly necessary to emphasise that this assumption has been subject to severe and prolonged criticism in the recent philosophy of science. Finally, though there is a clear sense in which (1.4) improves upon (1.1) or (1.3), there is quite a heavy price to pay for it: from the logical point of view the correspondence relation becomes much more complex than originally envisaged. This fact does not really count as an objection to the correctness of (1.4) or (1.5), but it does invite us to reconsider whether a logically simpler relationship might not be materially adequate.

In at least two important directions the approach of Krajewski and the Poznan school has recently been improved. Niiniluoto (1983) has successfully combined certain features of the idealisation and concretisation process in the development of theories with a thorough quantitative analysis of approximative relations between scientific laws. One of his chief aims is to render intelligibly a version of methodological realism, viz. of the thesis that science progresses through a series of theories possessing ever-increasing degrees of verisimilitude. For this purpose Niiniluoto employs a state space conception of physical theories, similar to that of van Fraassen (1970). In this setting the numerical equations of a physical theory may be represented by surfaces in a suitable function space, and the distance between physical laws can be 'measured' by assigning a metric to the space. Armed with a well-defined concept of approximation and 'convergence to a limit', Niiniluoto is then able to reconstruct a sophisticated picture of the correspondence relation that comprises an account of how the laws of T and T' are in close proximity (in the limit) and, analagously, how their intended models - considered as 'factual' and 'idealised' (counterfactual) structures to which T and T' apply - are related.

Niiniluoto's proposals lead to a detailed, but quite complex, account of correspondence and theory-development. It remains within the spirit of the classical, deductive view of reduction, but in addition to the usual logical apparatus it requires of course an accompanying mathematical theory of approximations. One slight difficulty for this approach might be that the metric-space characterisation of convergence and approximation is not the most general one and might be potentially too restricted to handle all possible cases of limiting relations between scientific theories. A second potential drawback - which incidentally applies equally to some other, more general mathematical representations of approximation - is that the choice of a suitable metric plays a crucial role here. For, physical laws that may be 'close', and therefore good approximations to eachother, under one characterisation of 'distance' may turn out to be rather far apart, and therefore poor approximations, under another, equally 'reasonable' characterisation.

Non-deductive approaches to correspondence

Niiniluoto's use of the state space representation of physical theories in several respects bridges the gap between deductive and non-deductive approaches to correspondence. For, the latter all share the common feature that the concept of a physical law as a syntactic or linguistic (or propositional) entity is dropped in favour of a structural representation of laws, generally by means of classes of mathematical 'models'. In other words, the key feature of a theory is taken to be a class of mathematical structures that directly represent the actual or idealised events and processes that the theory is supposed to describe. One advantage is that, as mathematical objects, such structures may easily be embedded in a suitable topological space, so that the neighbourhood of any model is precisely delineated and limits, imprecisions and approximations can be dealt with in the usual analytical way. Naturally, the correspondence relation is then defined primarily as a model-theoretic link, (or something that may be transformed into one), and in this sense the connection is not straightforwardly, deductive, though only some authors explicitly deny that the resulting model of reduction is inferential.[3]

There are different ways of approaching correspondence from a structural perspective. Ludwig (1978) and his followers conceive the models of a physical theory in Bourbaki species-of-structures style, and speak of the 'embedding' of one theory in another. On the other hand, structuralists (e.g. Moulines (1980)) adopt Sneed's (1971) reconstruction of physical theories as sets of models defined by set-theoretic predicates, and refer to the 'approximate reduction' of one theory to another. However, for our present purpose these differences in terminology and framework are marginal, for there is a common core of assumptions and methods at work here.

In essence, what it means in this approach for a theory T to be reducible to a theory T' is that a functional relation exists between their respective classes of models, viz. a mapping from models of T' into models of T that satisfies certain further properties. This mapping is in some respects a sematical counterpart to the deductive relation between the theories' laws asserted on the classical view of reduction. Thus, models of the reduced theory 'correspond' to, or are correlated with, models of the reducing theory - their inverse images under the mapping. Now, approximations may be handled using the Weil-Bourbaki concept of 'uniformity', where the space of models of each theory is assigned a uniform structure of a particular kind. And limit processes - such as the tending of some parameter of T' to a special value - may be described by sequences of models the limits of which may lie in the so-called *completion* of the uniform space. Thus, when the reduction or correspondence relation is approximative, models of T 'correspond' to converging sequences of models of T', the limits of such sequences being not themselves models of T'. For example, a given model of Keplerian planetary orbits may correspond to a Cauchy converging sequence of Newtonian orbits, the limit of which is not itself Newtonian, i.e. not a model of T'. However, the approximate reduction of T to T' can be characterised as an exact reduction of T to a theory $\hat{T}'$ (called the completion of T') whose models are defined by taking a suitable completion of the space of models of T' with respect to a uniformity (Mayr (1981)).

This view of correspondence thus differs from the deductive account, first, in that the central feature of the relation is a structural mapping, and, secondly, that this relation is defined as a mathematical connection between T and a modified or approximated version $\hat{T}'$ of T', rather than as a syntactical connection between T' and a corrected or approximated version T* of T.[4]

Now, the functional link between models of $\hat{T}'$ and T may happen to induce a deductive relation between the laws of those theories when considered linguistically. But T cannot generally be said to be inferred from T', or a consistent extension of T', because $\hat{T}'$ has models that are not models of T', i.e. when looked at syntactically the completion of a theory is not generally an extension of that theory. In short, on this approach we may have to give up the idea that the laws of theories satisfying the correspondence relation are deductively connected, and settle instead for a purely mathematical relation between their models; a relation that is moreover relativised to a suitable choice of topology and uniformity. Notice that, though the notion of 'proximity' defined for a uniform space in more general than the corresponding idea in a metric space, the same problem of 'choice' affects both. Notice too that in this framework two quite distinct (though related) concepts of reduction are characterised: ordinary (strict) reduction is a sort of limiting case of approximate reduction (as the degree of approximation required tends to zero).

Correspondence and nonstandard models

Any acceptable formal explication of an intertheoretic relation must achieve at least two things: it should respect as far as possible the usual intuitive understanding of the relation - in this case of the idea that one theory approximates another in the limit - and it should be materially adequate in the sense of accurately describing at least some accepted instances of the relation found in actual scientific practice. Beyond this general requirement, an acceptable account of how a theory T corresponds in the limit to a theory T' should also provide grounds for answering the following types of question:

(1) What is the syntactic connection between T and T' ?

(2) What is the semantic connection between T and T' ?

(3) In what sense, if any, is the connection logically definable ?

(4) How is the limiting or approximate character of the relation formally reflected ?

(5) How does the relation fit into the broader pattern of reductive (or nonreductive) intertheory relations ?

(6) In what sense, if any, does T' explain (the successful part of) T; or, how are the empirical successes of T absorbed by the later theory T' ?

(7) In the case of physical theories connected by correspondence, what is the relation, if any, between their symmetries or invariance properties ?

There are of course other methodological aspects of correspondence that an adequate explication should bring out, however these seven seem to amount to a fundamental shortlist of desiderata. Most accounts of correspondence known to us tend to focus on only some of these questions, at the expense of the others.[5]

We can best describe our own approach by considering a simple example that will serve as a prototype. Imagine that we have two theories each consisting of a single law describing the functional dependence of a real-valued physical quantity Q on n real-valued physical quantities $P_1, \ldots, P_n$, and suppose that in T' Q additionally depends on a further numerical parameter $p > 0$. Thus we have two laws of the form:

$$Q(x) = f(P_1(x),\ldots,P_n(x))$$
$$Q(x) = f'(P_1(x),\ldots,P_n(x),p).$$

Now suppose that T' approaches T in the limit as p tends to zero, i.e. for all x,

(1.6) $$\lim_{p \to 0} f'(x,p) = f(x)$$

We assume here that the extension of Q in the two theories is different and that the limiting value p = 0 is incompatible with T'. Now, we could depict the convergence of T' to T by the usual ε, δ technique. But if we assume that the mathematical bases of the theories are constituted by nonstandard analysis we can modify the form of (1.6) (which is of course an abbreviation for a sequence of expressions). By employing a nonstandard extension $*\mathbb{R}$ of the reals $\mathbb{R}$ we can make use of the <u>infinitesimals</u> and <u>infinite reals</u> contained in $*\mathbb{R}$ and the property, well-known from A. Robinson's theory of nonstandard analysis, that for any finite, nonstandard real r in $*\mathbb{R}$ there is a standard real number s in $\mathbb{R}$ such that their difference is infinitesimal. Such an s is usually called the <u>standard part</u> of r and denoted by st (r). Given quite natural assumptions (e.g. continuity), we can then rewrite (1.6) in the form:

(1.7) st(f'(x,p)) = f(x), for all infinitesimal p.

In other words, a converging sequence of values of f' is replaced by a single condition on p under which for all x the value of Q in T' and T is 'nearly the same'. Moreover, this condition is compatible with T' under the assumption that relevant nonstandard entities are admitted alongside the standard ones. If we now consider the models of the two theories, we see that the essential feature of their relation is that certain (standard) models of T are connected, by (1.7), with certain (nonstandard) models of T', rather than with sequences of models of T' or with limit models that lie outside T'. And here we depart markedly from the other approaches to correspondence considered so far.

Though the above example is extremely simple, the method can be generalised for arbitrary theories;

the main ideas remain in principle the same. We assume that the classes of models of any corresponding theories T and T' are defined so as to admit structures whose mathematical parts may be standard or nonstandard in a suitable sense - e.g. if the reals are involved we allow that $\mathbb{R}$ or $^*\mathbb{R}$ may belong to the models. And we can generalise the standard part operation so that it applies to any space of mathematical entities endowed with a Hausdorff topology, and yields for any suitable nonstandard model $\mathfrak{M}$ of T' a model $^r\mathfrak{M}$ which can be regarded as a <u>standard approximation</u> of $\mathfrak{M}$ in the sense that the values of all physical functions inside $^r\mathfrak{M}$ are just the standard parts of the corresponding functions inside $\mathfrak{M}$. To return to our example, if we take a nonstandard model $\mathfrak{M}$ of T' that is 'nearly standard', then the extension of the quantity Q in $^r\mathfrak{M}$ will be infinitely close to the extension of Q in . What is more - and this is the key point - from condition (1.7) it follows that $^r\mathfrak{M}$ is a (standard) model of T. Thus, in the case of a correspondence of T to T', the standard part operation can be characterised as a function mapping a subclass of models of T' into models of T. Viewed quite generally, therefore, the semantical side of a (limiting case) correspondence is described by a function F:K'-> K, where K and K' are respectively definable subclasses of models of T and T', and for each $\mathfrak{M}$ in K', F ($\mathfrak{M}$) is a standard approximation of $\mathfrak{M}$.

To complete the picture at the syntactical level, suppose that the language or vocabulary of T is given by a similarity type τ , and that of T' by a type τ'. Furthermore, let the relevant collections of models K and K' be defined as elementary (axiomatisable) classes in some logic L. Then, syntactically, the connection between T and T' will be expressed by a translation, say I, that maps L (τ)-sentences into L(τ ')-sentences in such a way that truth is preserved (relative to F); i.e. for all models $\mathfrak{M}$ in K' and L(τ)-sentences φ ,

(1.8) $$\mathfrak{M} \models_L I(\varphi) \Leftrightarrow F(\mathfrak{M}) \models_L \varphi$$

where $\models_L$ is the truth relation for L. So, for example, if σ, ϑ' and ϑ are L-sentences axiomatising the class K and the classes of models of T' and T,

in that order, then it follows from (1.8) that
(1.9)

$$\vartheta', \sigma \models_L I(\vartheta).$$

In other words, T' together with auxiliary assumptions σ entails a translation of (the axiom of) T. More generally, we can say that the translations of the laws and consequences of T can be inferred from a consistent extension of T' obtained by conjoining the relevant limiting conditions and special assumptions (σ) to the axioms of T'.

Clearly (1.9) is a close cousin to the original deductive schema of reduction expressed by (1.1). The fact that translation plays a crucial role in our picture of correspondence does not as such distinguish it from other types of reductive relations. Translation is after all a natural, and often indispensable, feature of reduction contexts. Limiting relations between physical theories are distinguished rather by the manner in which the model-theoretic correlation F is characterised by means of the standard part operation. However, a special property of the limiting relation between theories is that the translation I is not as a rule an identity mapping on terms that are common to the vocabulary of each theory. More accurately, in the example considered an atomic expression of T involving the relation Q will be translated by I into a complex formula of T' containing other terms besides. Here, I includes in effect a description of the process of taking standard parts. In this sense the earlier theory T can be said to be re-interpreted in the light of the later theory T', and some terms of the former may undergo a change of meaning in the transition to the new theory. However, explanations accomplished by the supplanted theory are, in a well-defined sense, taken over by its successor, for they may be translated into statements that, in the limit, remain valid in the new theory. Thus we can make clear sense of the well-accepted idea that, in this type of theory change, the later theory corrects and improves upon its forerunner whilst showing how it was nevertheless approximately right within agreed limits.

The method just sketched not surprisingly looks

somewhat more complicated when all the finer details are laid forth, such as is required for any fully-fledged application. However, once the basic logical technicalities are dispensed with, the procedure is quite straightforward. What is more, the fundamental structure of the correspondence relation then has an extremely simple form which permits a uniform treatment of intertheory relations of various types within a single, basic 'model'; and at the same time this representation is highly informative, especially regarding logical features of the correspondence relation. One obvious advantage of this method comes directly from its use of nonstandard analysis: Robinson's theory clearly affords a much more elegant treatment of mathematical limits than the standard one, and corresponds much closer to our ordinary, 'prescientific' intuitions.

A further advantage of nonstandard analysis here is that, as compared with other accounts of correspondence, this approach has a more 'logical' flavour to it. Naturally, we have to exploit certain topological properties of the underlying mathematics of a physical theory, in order to define the relevant nonstandard models. But from the standpoint of elemantary logic all of the nonstandard extensions are structurally similar, so that all of them which contain the relevant nonstandard entities lead to equivalent descriptions of the way in which the appropriate physical quantities converge in the limit. For this central purpose, therefore, it does not matter which of such nonstandard extensions of the underlying mathematical basis we choose. So we are not faced with the problem of having to select one mathematical concept of 'proximity' rather than another (e.g. one particular metric or uniformity). Obviously, in the typical cases where the ordinary analysis of real numbers is involved, it is convenient to work with the usual topology on $\mathbb{R}$.

The above remarks suggest the general lines along which we propose to answer the basic questions (1)-(6); for fuller details we refer to our (1983a) and to Part II below. But the method also supplies an account of point (7) that we can briefly consider here.

The symmetries (or invariance properties) of a

physical theory are usually represented by a group of transformations, such as the Galilean group in classical mechanics, the Poincaré group in special relativity, and so on. Model-theoretically, a transformation can be regarded as a certain mapping (or relation) between structures under which the class of models of the theory is closed (i.e. under which the laws of the theory are preserved). The method of nonstandard analysis can also be applied here: one may study the relation between certain limiting or infinitesimal transformations in T' and their natural counterparts in T.

Now, several philosophers of physics have claimed that in a correspondence relation the symmetries of T and T' are somehow preserved or correlated.[6] This need not be taken to mean that the transformation group of each theory is the same, but it can more loosely be interpreted as saying that the algebraic *structure* of the symmetries is preserved. In our framework, the latter idea can be approximated with the aid of a commuting diagram. Suppose that the correspondence of T to T' is captured by the pair of maps <F,I> as above, that $\mathfrak{M}'$ and $\mathfrak{N}'$ are models in the domain K' of F, and that $\mathfrak{N}'$ is connected to $\mathfrak{M}'$ by a transformation h ' belonging to the symmetry of T', i.e. $h': \mathfrak{M}' \rightarrow \mathfrak{N}'$; then the preservation thesis implies the claim that there is a symmetry h of T such that $h: F(\mathfrak{M}') \longrightarrow F(\mathfrak{N}')$, or in other words that the following diagram commutes:

(1.1o)

$$\begin{array}{lccc} T' & \mathfrak{M}' & \xrightarrow{\quad h' \quad} & \mathfrak{N}' \\ & F \downarrow & & \downarrow F \\ T & \mathfrak{M} & \xrightarrow{\quad h \quad} & \mathfrak{N} \end{array}$$

More completely, we can represent a (sub)symmetry of a theory by a *category* comprising an invariant collection of models (the 'objects') together with a collection of maps between them (the 'morphisms'). Then, given a correspondence <F,I> of T to T', we can associate with each theory a suitable subsymmetry (category) based on the correspondence, and formulate the general thesis that F can be extended to a *functor* between the categories just defined.

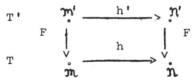

Though we consider it an open problem to what extent this thesis is true for all cases of correspondence in physics, we have pointed out by two case studies that there is some truth in it. In any case, the thesis provides an exact reconstruction of the idea that the algebraic structure of symmetries is preserved in a correspondence relation.

All in all, the syntactic and semantic properties of the maps F and I in a correspondence relation, together with the possible functorial connection they establish between the appropriate symmetries of the theories, establish an important sense in which the passage from T to T' is continuous and even cumulative. If this pattern of correspondence is indeed a characteristic feature of scientific change generally, then the method sketched here yields a useful measure of the continuity of scientific development.

II

At this juncture we end our informal discussion of the properties of the correspondence relation, and turn instead to an actual case study. First we should make clear what our notion of a 'theory' is, and how logics can be applied to it; then we repeat the general definition of correspondence given in our (1983a).

We work with a structural, or model-theoretic, concept of a physical theory that we have introduced and defended in our (1983). Though by a 'logic' we understand any of the logics admitted in general model theory (or abstract logic), for present purpose we require only first order logic, $L_{\omega\omega}$, and the infinitary logic, $L_{\omega_1\omega}$, admitting countable disjunctions and conjunctions. Our logical notation is fairly standard, and we assume that the reader is familiar with the rudiments of ordinary (many sorted) model theory.

By a theory we mean a structure

$$T = \langle \tau, N, M, R \rangle$$

such that τ is a (many sorted) similarity type

(symbols plus sorts) fixing the vocabulary of the theory; N is a collection of models of type τ (i.e. $N \subseteq Str(\tau)$ called the admitted structures for T; $M \subseteq N$ is called the class of models of T; R is a collection of relations of the form $\mathfrak{R} = \{< \mathfrak{M}_1, \ldots, \mathfrak{M}_n; h> \ldots\}$ ... where h is a sequence of relations between various domains of $\mathfrak{M}_1, \ldots, \mathfrak{M}_n \in N$; the collections N,M,R are set-theoretically definable (in an appropriate metatheory), and N,M can be defined in some logic L, i.e. they can be axiomatised by L-sentences in the type τ. Such a logic L is called adequate for T.

Thus a theory is essentially determined by a vocabulary, a collection of admitted structures which fix the 'conceptual structure' of T, a collection of models representing (semantically) the 'laws' of T, and a collection of relations between admitted structures that may include, e.g., morphisms, symmetries, constraints, or other types of relations that are in some sense characteristic of the theory. We assume that a theory possesses adequate logics in the sense that the relevant model-classes can be defined as L-elementary classes for some logic L; but we allow in general that any logic L may be applied to T in a metascientific context providing that the vocabulary of T is admitted by the logic in question. We assume as usual that to any logic L can be associated a collection Str_L of admitted structures, a collection $Sent_L$ of sentences, and a truth relation $\models_L \subseteq Str_L \times Sent_L$.

To characterise intertheory relations in a uniform manner, it is convenient to begin with a very general reductive relation that we shall call correspondence. Let $T = <\tau, N, M, R>$ and $T' = <\tau', N', M', R'>$ be theories, and let L,L' be logics that are adequate for T,T' respectively.

(2.1)

A correspondence of T to T', relative to <L,L'> is a pair <F,I> of maps:

(i) $F: K' \rightarrow K$ onto, where K,K' are non-empty subclasses of M,M' resp.;

(ii) $I: Sent_L(\tau) \longrightarrow Sent_{L'}(\tau')$; such that

(iii) K is definable in L, and K' in L';

(iv) $\forall \mathfrak{M} \in K', \forall \varphi \in Sent_L(\tau)\ (F(\mathfrak{M}) \models_L \varphi \iff \mathfrak{M} \models_{L'} I(\varphi))$

Clearly, condition (iv) ensures that truth is preserved Here F is called the structural correlation and I the translation. If L=L', we call <F,I> simply an L-correspondence of T to T'; for notational simplicity we shall assume this identification in what follows.

We next distinguish three special cases of correspondence. The first two are straightforward generalisations of concepts of the same name familiar in logic and methodology, the third is the central notion of this paper which we have been discussing in Part I above.

(2.2)

<F,I> is an L-interpretation of T in T' if (2.1) holds together with the property K'=M';

(2.3)

<F,I> is an L-reduction of T to T' (or embedding of T in T') if (2.1) holds with K=M;

(2.4)

<F,I> is an limiting L-correspondence of T to T' if (2.1) holds together with (i) K' is a class of nonstandard models of T' satisfying certain finiteness properties; (ii) $\forall \mathfrak{M} \in K'$: $F(\mathfrak{M})$ is a standard approximation of $\mathfrak{M}$.

Here we cannot give a precise definition of (i) and (ii) of (2.4), for which we refer to our (1983a). However, the basic idea as presented in Part I above is that any model $\mathfrak{M}$ of a theory T can be described as a structure

$$\mathfrak{M} = <\mathfrak{B}; \ldots\ldots>$$

where $\mathfrak{B}$ is a structure representing the underlying mathematics of the theory, e.g. $\mathfrak{B}$ may be a model of real or complex analysis. In that case, $\mathfrak{M}$ is called standard if $\mathfrak{B}$ is a standard model of analysis, containing the ordinary real or complex numbers, operators and so forth. And $\mathfrak{M}$ is called nonstandard if $\mathfrak{B}$ contains infinitesimals, infinite 'reals', 'reals' that are infinitesimally close to standard reals, etc.; i.e. if $\mathfrak{B}$ is a suitable nonstandard extension of the standard model. Let $\mathfrak{M}' = <\mathfrak{B}'; \ldots>$ be a nonstandard model of T. For suitable reals, vectors, real- and vector-valued functions in $\mathfrak{B}'$ one may then define their standard parts in much the usual way, and further, call a standard model $\mathfrak{M} = \langle \mathfrak{B}; \ldots . \rangle$ a standard approximation of $\mathfrak{M}'$ when $\mathfrak{B}'$ is a nonstandard extension of $\mathfrak{B}$, and each function, relation and individual in $\mathfrak{M}$ is the standard part of the corresponding component of $\mathfrak{M}'$. The relevant definitions can be generalised for arbitrary theories by means of nonstandard topology (see our (1983a)). Thus, in (2.4) K' is to be a class of nonstandard models of T' that possess standard approximations, and F is a mapping that forms the standard approximation of each member of K'.

The above discussion concerning symmetries can be generalised in a natural way so as to take account of the relation of R to R'. For reasons of space we shall not repeat the appropriate modification here, but it consists essentially in the idea expressed by (1.1o) where $\{\langle \mathfrak{M}', \mathfrak{N}'; h' \rangle \ldots\}$ is a relation in R' (e.g. a symmetry of T') that corresponds via an extension of F to a relation $\{\langle \mathfrak{M}, \mathfrak{N}; h \rangle \ldots\}$ in R (e.g. a symmetry of T).

Correspondence in Celestial Mechanics

We now proceed to study the relation between Newton's theory of gravitation and Kepler's theory of planetary motion. Basically this study will be quite analogous, and in many respects even similar, to our analysis of particle mechanics in (1983a). Thus we shall omit many details to avoid unnecessary repetition. We use E. Scheibe's formulations of these theories (Scheibe, 1973); naturally they are reconstructions of the original ones. They are invariant under the Galilean transformation. We shall show below that there is a limiting case correspondence of (reconstructed) Kepler's theory to an 'inessential' extension of Newton's theory. For that purpose, we shall pick a mathematical result from Mayr (1981), where Scheibe's formulation is also used.

First, let $\mathfrak{B}$ be a model of analysis in the

sense of our (1983a). Let B be the domain of $\mathfrak{B}$ and τ_0 its type. Now let

$$\mathfrak{M} = < \mathfrak{B}; P, T, s, k, d >$$

be a three-sorted expansion of $\mathfrak{B}$ of type $\tau \supset \tau_0$, with P and T as new domains, s a function $s: P \times T \to B$, and k, d individuals from B and P repectively. The intuitive meaning is that P is a set of bodies (planets, stars), T is an interval of time, s states the position of a body p at time t, d denotes a distinguished body (the sun), and k is Kepler's constant. If $a \in B$, let $a_{\mathfrak{B}} = \{ b \in B | b \in^{\mathfrak{B}} a \}$.

We now define M_0 to be the class of all models $\mathfrak{M}$ of type τ satisfying:

(2.5) $\mathfrak{B}$ is a (standard or nonstandard containing infinitesimals) model of analysis;

(2.6) P is a finite, non-empty set;

(2.7) T is a interval on $\mathbb{R}_{\mathfrak{B}}$;

(2.8) $K \in (\mathbb{R}+)_{\mathfrak{B}}$;

(2.9) $s: P \times T \longrightarrow (\mathbb{R}^3)_{\mathfrak{B}}$;

(2.10) s is twice differentiable on T;

(2.11) For all $p \in P$ such that $p \neq d$ and all $t \in T$,

(i) $s(p,t) \neq s(d,t)$;

(ii) $D^2 s(d,t) = \bar{0}$;

(iii) $$D^2 s(p,t) = -k \frac{s(p,t) - s(d,t)}{|s(p,t) - s(d,t)|^3}$$

(iv) $$\frac{1}{2} |Ds(p,t) - Ds(d,t)|^2 - k |s(p,t) - s(d,t)|^{-1} \in (\mathbb{R}^-)_{\mathfrak{B}}.$$

Let M be the least class of models of type τ including M_0 and closed under isomorphism. Elements of M are called <u>models of Kepler's theory of planetary motion</u> (KP). The theory itself may be defined as

$$KP = <\tau, Str(\tau), M, \mathcal{G}>$$

where $\mathcal{G}$ is the class of all Galilean transformations in M. A Galilean transformation may be characterised by a parameter u in the following way. Let $\mathfrak{M}$ as above be a model in M_0 and $u \in (\mathbb{R}^3)_{\mathfrak{B}}$. Consider a transformation $t \to \bar{t}$, $s \to \bar{s}$, such that for all $t \in T$, $p \in P$, $\bar{t} = t$ and $\bar{s}(p,\bar{t}) = s(p,t) - ut$. Then the parameter u defines a Galilean transformation $g: \mathfrak{M} \to \bar{\mathfrak{M}}$, where $\bar{\mathfrak{M}}$ is a model $< \mathfrak{B}; P, T, \bar{s}, k, d> \in M_o$. We can therefore specify the class of all Galilean transformations by

$$\mathcal{G} = \{< \mathfrak{M}, \bar{\mathfrak{M}}; g> \mid \mathfrak{M} \in M \ \& \ g: \mathfrak{M} \to \bar{\mathfrak{M}} \text{ is a Galilean trans.}\}$$

Next consider models of the form

$$\mathfrak{M} = < \mathfrak{B}; P, T, s, m >$$

where $m: P \to B$, and the other components are as before; here m specifies the 'mass' of a body p. Let ρ be the type of $\mathfrak{M}$, and let W_0 be the class of all models of type ρ satisfying (2.5)-(2.7), (2.9), (2.1o) and the following:

(2.12) $m: P \to (\mathbb{R}^+)$;

(2.13) $\forall p, p' \in P \ (p \neq p')$, $\forall t \in T$, (i) $s(p,t) \neq s(p',t)$;

$$\text{(ii) } D^2 s(p,t) = - \sum_{\substack{q \in P \\ q \neq p}} m(q) \, \frac{s(p,t) - s(q,t)}{|s(p,t) - s(q,t)|^3} .$$

Let W be the least class of models of type ρ including W_0 and closed under isomorphism. Its elements are <u>models of Newton's theory of gravitation</u> (NG). If $\mathcal{H}$ corresponds to the Galilean transformations in W, this theory is given by

$$NG = <\rho, Str(\rho), W, \mathcal{H}>.$$

We now define a trivial extension of NG by just expanding the language, without any new axioms. Let $\tau' = \rho \cup \{\underset{\sim}{k}, \underset{\sim}{d}\}$ be the expansion of ρ obtained by adding individual constants of the sort of B and P respectively. Let

$$M_0' = \{\mathfrak{M} \in Str(\tau') \mid \mathfrak{M} \upharpoonright \rho \in W_0\} ,$$

$$M' = \{\mathfrak{M} \in Str(\tau') \mid \mathfrak{M} \upharpoonright \rho \in W \} ,$$

and let $\mathcal{G}'$ be the Galilean transformations in M'. We next consider the theory

$$NG' = <\tau', Str(\tau'), M', \mathcal{G}'>.$$

Now, it is clear from the discussion in our (1983a) that the classes M and W can be defined in the infinitary logic $L_{\omega_1\omega}$, i.e. the above sets of axioms can be formalised by sentences of $L_{\omega_1\omega}(\tau)$ and $L_{\omega_1\omega}(\rho)$ respectively (here it is not necessary to write out these sentences in full). Similarly, the class M' is defined by the same $L_{\omega\omega}(\rho)$-sentence as W. Hence $L_{\omega\omega}$ is adequate for both KP and NG' in the sense indicated earlier.

To describe the correspondence of Kepler's theory to this inessential extension of Newton's theory we must first specify the relevant subclasses K and K' of M and M' for which the structural correlation is to be defined. Essentially, we shall require that K is the class of all standard models of KP in which the distance between any two different 'particles' (here 'heavenly bodies') has a lower limit $a \neq 0$, i.e. the bodies cannot be arbitrarily close to eachother. And we shall take K' to be a class of nonstandard models of NG' satisfying the same condition (with $a \in \mathbb{R}$, i.e. standard), certain finiteness and other properties to guarantee the existence of standard approximations, together with the conditions that k is equal to the mass of the Sun and that the masses of all planets are infinitesimal. K and K' can be shown to be $L_{\omega_1\omega}$-definable. Next we describe a mapping F from K' onto K which forms standard approximations of the Newtonian models, essentially by letting the masses of all planets go to zero. It can then be seen that F induces a translation, I, from sentences in the language of KP into sentences in the language of NG' such that (2.1)(iv) holds; and thus that $\langle F,I \rangle$ is an $L_{\omega_1\omega}$-limiting correspondence of KP to NG'. Furthermore, F preserves the structure of the appropriate Galilean transformations.

Structural correlation.[7] let K_o be the class of all standard models $\mathfrak{M} = \langle \mathfrak{U}; P, T; s, k, d \rangle$ in M_o satisfying:

(2.14) There is $a \in \mathbb{R}^+$ such that for all $p, p' \in P$ such that $p \neq p'$ and all $t \in T$, $|s(p,t) - s(p',t)| > a$.

Let K'_o be the class of all nonstandard models $\mathfrak{M} = \langle {}^*\mathfrak{U}; P, T; s, m, k, d \rangle$ in M'_o satisfying (8.4) of our (1983a),[8] and

(2.15) There is $a \in \mathbb{R}^+$ such that for all $p, p' \in P$ such that $p \neq p$ and all $t \in {}^eT$, $|s(p,t) - s(p',t)| > a$;[9]

(2.16) s, Ds, and D^2s are finite on ${}^eT$;[10]

(2.17) $m(d)$ is finite and non-infinitesimal;

(2.18) $m(p)$ is infinitesimal for all $p \neq d$;

(2.19) ${}^rs$ is twice differentiable on ${}^rT$ and $D({}^rs) = {}^r(Ds)$, and $D^2({}^rs) = {}^r(D^2s)$;[11]

(2.20) $k = m(d)$;

(2.21) For all $p \in P$ such that $p \neq d$ and all $t \in {}^eT$, $\frac{1}{2}|Ds(p,t) - Ds(d,t)|^2 - k|s(p,t) - s(d,t)|^{-1} \in \mathbb{R}^-$.[12]

Let $\mathfrak{M} = \langle {}^*\mathfrak{U}; P, T; s, m, k, d \rangle$ be in K_o'. Then the following model of type τ is uniquely determined:

$${}^r\mathfrak{M} = \langle \mathfrak{U}; P, {}^rT; {}^rs, st(k), d \rangle.$$

Again, this model is a <u>standard approximation</u> of $\mathfrak{M}$ in the sense defined in section 3 of our (1983a).

<u>Lemma 1.</u> For all $\mathfrak{M} \in K_o'$, ${}^r\mathfrak{M} \in K_o$.

<u>Proof.</u> We need only consider (2.11):(ii)-(iv).

<u>ad</u> (ii): It follows from (2.15) and (2.18) that for all $t \in {}^rT$, each addendum on the right hand side of the equation

$$D^2s(d,t) = \sum_{\substack{q \in P \\ q \neq d}} m(q) \frac{s(d,t) - s(q,t)}{|s(d,t) - s(q,t)|^3}$$

is infinitesimal. So for all $t \in {}^rT$,

$$D^2s(d,t) \approx \bar{0}.$$

Thus

$${}^r(D^2s)(d,t) = st(D^2s(d,t)) = \bar{0},$$

whence by (2.19)

$$D^2({}^rs(d,t)) = \bar{0}.$$

<u>ad</u> (iii): As above, we see by (2.15)-(2.20) that

for all $p \in P$, $t \in {}^{r}T$,

$$D^2({}^{r}s(p,t)) = st(-m(d) \frac{s(p,t)-s(d,t)}{|s(p,t)-s(d,t)|^3})$$

$$= -st(m(d)) \frac{st(s(p,t))-st(s(d,t))}{|st(s(p,t))-st(s(d,t))|^3}$$

$$= -st(k) \frac{{}^{r}s(p,t)-{}^{r}s(d,t)}{|{}^{r}s(p,t)-{}^{r}s(d,t)|^3} .$$

ad (iv): This follows from (2.21) together with the second condition of (2.19). □

Lemma 2. For all $\mathfrak{M} \in K_o$, there is $\mathfrak{M}' \in K'_o$ such that ${}^{r}(\mathfrak{M}') = \mathfrak{M}$.

Proof. Let $\mathfrak{M} = \langle \mathfrak{A}; P,T; s,k,d \rangle$ be in K_o and let ${}^{*}\mathfrak{A}$ be as before. Since there does not seem to exist a straightforward and direct proof, we use an approximation result from Mayr (1981). It follows from this result that the following holds:

(2.22) For every $\varepsilon \in R+$, there is a model of Newton's theory of gravitation, $\langle \mathfrak{A}; P,T; r,m,k,d \rangle$, such that $m(d) = k$ and for all $p \in P$, $t \in T$, $p \neq d \Rightarrow m(p) < \varepsilon$ and $|r(p,t)-s(p,t)| < \varepsilon$.

This result can be handled in the following way. Consider the set $(\mathbb{R}^3)^T$; i.e., the set of all functions from T into $\mathbb{R}^3$.

(2.22) can be rewritten as follows:

(2.23) For every $\varepsilon \in \mathbb{R}+$, there are $r_p, r_{p'}, \ldots \in (\mathbb{R}^3)^T$ and $m_p, m_{p'}, \ldots \in \mathbb{R}+$ (for all $p, p', \ldots \in P$) such that for all $t \in T$,

(i) r_p is twice differentiable on T ($p \in P$);

(ii) $r_p(t) \neq r_{p'}(t)$ ($p \neq p'; p, p' \in P$);

(iii) $$D^2 r_p(t) = - \sum_{\substack{q \in P \\ q \neq p}} m(q) \frac{r_p(t)-r_q(t)}{|r_p(t)-r_q(t)|^3} \quad (p \in P);$$

(iv) $m_d = k$;

(v) $m_p < \varepsilon$ ($p \neq d; p \in P$);

(vi) $|r_p(t) - s_p(t)| < \varepsilon$ ($p \in P$). [13]

Since $T \in A$ and $\mathbb{R}^3 \in A$, also $(\mathbb{R}^3)^T \in A$, whence it has a name in τ_o. Furthermore, the conditions (i)-(vi) are expressible by an $L_{\omega\omega}(\tau_o)$-formula, whence (2.23) can be translated into an $L_{\omega\omega}(\tau_o)$-sentence.[14] Then by the transfer principle (see our 1983a) and by (2.23) the following result holds in $*\mathfrak{U}$:

(2.24) For every $\varepsilon \in \hat{\mathbb{R}}+$, there are functions $*r_p, *r_p, \ldots : \hat{T} \to \hat{\mathbb{R}}^3$ and $m'_p, m'_p, \ldots \in \hat{\mathbb{R}}+$ (for all $p, p', \ldots \in P$) such that for all $t \in \hat{T}$,

(*i) $*r_p$ is twice differentiable on $\hat{T}$ ($p \in P$);

(*ii) $*r_p(t) \neq *r_{p'}(t)$ ($p \neq p'; p, p' \in P$);

(*iii) $$D^2 *r_p(t) = -\sum_{\substack{q \in P \\ q \neq p}} m'(q) \frac{*r_p(t) - *r_q(t)}{|*r_p(t) - *r_q(t)|^3} \quad (p \in P);$$

(*iv) $m'_d = k$;

(*v) $m'_p < \varepsilon$ ($p \neq d; p \in P$);

(*vi) $|*r_p(t) - *s_p(t)| < \varepsilon$ ($p \in P$).

Now, let ε be a given infinitesimal number in $\hat{\mathbb{R}}+$ and let $*r_p, m'_p$ (for each $p \in P$) be functions and 'reals' corresponding to ε, as indicated in (2.24). Let $s' : P \times \hat{T} \to \hat{\mathbb{R}}^3$ and $m' : P \to \hat{\mathbb{R}}+$ be the functions defined by

$$s'(p,t) = *r_p(t),$$
$$m'(p) = m'_p.$$

Then the model

$$\mathfrak{M}' = \langle *\mathfrak{U}; P, \hat{T}; s', m', k, d \rangle$$

is a nonstandard model of Newton's theory of gravitation such that the following holds:

(2.25) For all $p \in P, t \in \hat{T}, s'(p,t) \approx s(p,t) = {}^*s_p(t)$.

That $\mathfrak{M}' \in K'_o$ and $\mathfrak{M} = {}^r(\mathfrak{M}')$ is obvious on the basis of the conditions (iv),(v),(2.25) and the fact s and its relevant derivatives are continuous, ${}^r(s') = s$, and $st(k) = k$.[15] □

It is not hard to see that we can establish results concerning appropriate categories which are quite analogous with the theorems (8.6 and 8.8) concerning classical and relativistic particle mechanics proved in our (1983a). First, it is clear that K_o is closed under the Galilean transformations and K'_o under the <u>finite</u> transformations (in which the parameter u is finite). Secondly, if $g: \mathfrak{M} \to \bar{\mathfrak{M}}$ ($\mathfrak{M} \in K'_o$) is a Galilean transformation induced by conditions like those mentioned above, then

$$\tilde{t} = t;$$
$$\tilde{{}^r s}(p,\tilde{t}) = {}^r s(p,t) - st(u)t \quad (t \in {}^r T)$$

induces a Galilean transformation ${}^r g: {}^r\mathfrak{M} \to {}^r\bar{\mathfrak{M}}$ (${}^r\mathfrak{M} \in K_o$).

Thirdly, if K and K' are the least classes of models of type τ and τ' respectively, including K_o and K'_o respectively and closed under isomorphism, they are $L_{\omega_1\omega}$-definable. Finally, in analogy with Theorem 8.8 of our (1983a), the following result holds:

<u>Theorem 3</u>. Let G be the class of all Galilean transformations of models in K, and let G' be the class of all finite Galilean transformations of models in K'. Then $\tilde{K} = \langle K,G \rangle$ and $\tilde{K}' = \langle K',G' \rangle$ are categories, and the mapping $F: K' \cup G' \to K \cup G$ defined

by: $F(\mathfrak{M}) = {}^r\mathfrak{M}$ ($\mathfrak{M} \in K'$), $F(g) = {}^r g$ ($g \in G'$), is a functor $F: \underset{\sim}{K}' \rightarrow \underset{\sim}{K}$.

Translation.

Since we again use the logic $L_{\omega_1\omega}$, and since $F \upharpoonright K'$, where F is as in Theorem 3, is a 'limiting mapping' as in our (1983a), we can establish a truth-preserving translation $I: Sent_{L_{\omega_1\omega}}(\tau) \rightarrow Sent_{L_{\omega_1\omega}}(\tau')$ quite analogously with the case of classical and relativistic mechanics. The only difference is that the formula $\vartheta(t)$ of (1983a)§8 now has to say that t is in ${}^rT$ (instead of ${}^oT$; c.f. note 11). As before the translation essentially provides a syntactic description of the process of forming standard approximations.

Thus it follows that <$F \upharpoonright K'$,I> is an $L_{\omega_1\omega}$-limiting correspondence of KP to NG' which preserves the categorial structure of the symmetries involved.

David Pearce
Institut für Philosophie
Freie Universität
Habelschwerdter Allee 30
1000 Berlin 33
BRD

Veikko Rantala
Department of Philosophy
University of Helsinki
Unioninkatu 40 B
00170 Helsinki 17
Finland

NOTES

1. For definitions and case studies of the correspondence relation, see the works cited below, and various contributions to the present volume. For recent views concerning the role of CP in scientific methodologies, see especially Post (1971), Krajewski (1977), Nowak (1980) and Zahar (1983).
2. See, e.g. Nowakova (1975), Nowak (1980).
3. For example Stegmüller (1979). See also the introductory remarks of Moulines in his contribution to this volume.
4. For further details on this approach to correspondence the reader is referred to the papers by Ludwig, Mayr, Scheibe and Schmidt in this volume.
5. Niiniluoto (1983) covers all but point (7), at least in passing. Other works tend to fare less well on this score.
6. Post (1971) and Redhead (1975) have looked at this question in some detail. The remarks made in the remainder of this paragraph are amplified somewhat in our (1983a).
7. At this point we need to introduce some further notation and nomenclature, mostly drawn from our (1983a). Henceforth, we work with a fixed, standard model of analysis $\mathfrak{A}$ of type τ_o, with domain A where A is the <u>superstructure</u> over the set of natural numbers $\mathbb{N}$; and with a fixed, but arbitrary, nonstandard extension $*\mathfrak{A}$ of $\mathfrak{A}$, with domain *A. $*\mathfrak{A}$ is of the form

$$*\mathfrak{A} = \langle *A, *\in, a\rangle_{a\in A}.$$

As usual, for any element of $a \in *A$, we set $\hat{a} = \{b\in *A \mid b *\in a\}$, and for any relation Q of elements of A that is $L_{\omega\omega}$-definable in $\mathfrak{A}$, we let *Q be the corresponding relation of $*\mathfrak{A}$ defined by the same formula. In particular, $\mathbb{R}\in A$ and $*\mathbb{R}\in *A$, and the other sets and relations required for nonstandard analysis are also elements of *A. For example, the relations $\approx$ and st are $L_{\omega_1\omega}$-definable in $*\mathfrak{A}$, where for $a,b\in *\mathbb{R}$, '$a\approx b$' reads 'a is <u>infinitesimally close</u> to b' and 'st(a) = b' reads 'b is the <u>standard part</u> of a' (cf. our (1983a §§3-5). If T is any interval on $\hat{\mathbb{R}}$, we denote its 'finite part' by ^{e}T. And if T is non-infinitesimal, and ${}^{e}T$ non-empty, we define the standard part of T by

$$ {}^{r}T = T \cap \mathbb{R}. $$

Assume that a function F: $T \to \hat{\mathbb{R}}$ is finite on ${}^{r}T$, then its standard approximation ${}^{r}f$: ${}^{r}T \to \mathbb{R}$ is defined by putting

$$ {}^{r}f(t) = st(f(t)). $$

These and other definitions given in our (1983a) can be straightforwardly adapted to handle nonstandard vector analysis. By a vector we mean a 'generalised n-tuple' (for any $n \in \omega$): $*\langle a_1, \ldots, a_n \rangle \in \hat{\mathbb{R}}^n$. It is standard (i.e. belongs to $\mathbb{R}^n$ and hence to A) just in case its components are standard, that is, if $a_1, \ldots, a_n \in \mathbb{R}$. Clearly, a nonstandard vector belongs to *A. We use the familar notation for the usual operations on standard vectors. From Robinson's transfer principle (cf. our 1983a (3.2)) it follows that these operations and the ordinary rules for them are straightforwardly generalisable for the nonstandard vectors, since they are $L_{\omega\omega}(\tau_0)$-definable in $\mathfrak{A}$. Similary, all the required definitions concerning the reals can be generalised for vectors. In particular, we say that a vector is infinite, finite, or infinitesimal if all its components have the corresponding property, and that $a \in \hat{\mathbb{R}}^n$ is infinitely close to $b \in \hat{\mathbb{R}}^n$ if a-b is infinitesimal. It should also be clear what it means that a function F: $T \to \hat{\mathbb{R}}^n$ has a limit in $a \in T$, or that f is differentiable, continuous, etc., on T. Likewise, its standard approximation ${}^{r}f$ is defined as above, bearing in mind that the standard part of a (finite) vector is always a standard vector. In what follows, we let $\vec{0} = \langle 0,0,0 \rangle$.

8. I.e. that T is a non-infinitesimal interval such that ${}^{e}T$ is non-empty.

9. For Lemma 1 below it is not sufficient to require merely that $a \in \hat{\mathbb{R}}^{+}$.

10. Actually the finiteness of D^2s on ${}^{e}T$ follows from the finiteness of s and the conditions (2.13):(ii), (2.15), (2.17), (2.18). Naturally,(2.19) is only possible on the assumption (2.16).

11. ${}^{r}s$ is a simpler form of standard approximation than the function ${}^{o}s$ required in our (1983a), since

here we do not need to deal with infinitesimal, non-zero time transformations.

12. Again, notice that it would not be sufficient to take $\hat{\mathbb{R}}^-$ instead of $\mathbb{R}^-$.

13. For each $p \in P$, s_p is the function on T defined by $s_p(t) = s(p,t)$.

14. If we choose for each $p \in P$ two variables $\underset{\sim}{x}_p$ and $\underset{\sim}{\chi}_p$ of the sort of A, and a similar variable $\underset{\sim}{x}$ for ε, such that all the variables are distinct, then the $L_{\omega\omega}(\tau_0)$-sentence is of the form

$$\forall \underset{\sim}{x}(\underset{\sim}{x} \in \underset{\sim}{\mathbb{R}}^{+} \Rightarrow \exists \; \underset{\sim}{x}_p | p \in P \quad \exists \; \underset{\sim}{\chi}_p | p \in P \; (\bigwedge_{p \in P} (\underset{\sim}{x}_p \in (\underset{\sim}{\mathbb{R}}^3)^{\underset{\sim}{T}} \wedge \underset{\sim}{\chi}_p \in \underset{\sim}{\mathbb{R}}^{+}) \wedge \forall t \varphi)$$

where φ is a formalisation of (2.23):(i)-(vi).

15. Their continuity follows from (2.10) and (2.11): (i),(iii).

REFERENCES

Czarnocka, M. and Zytkow, J. (1982): 'Difficulties with the reduction of classical to relativistic mechanics', in Krajewski, W. (ed.), Polish Essays in the Philosophy of the Natural Sciences, D.Reidel, Dordrecht.

van Fraassen, B. (197o): 'On the extension of Beth's semantics of Physical theories', Philosophy of Science 37, pp.325-339.

Hempel, C. (1965): Aspects of Scientific Explanation, Free Press, New York.

Krajewski, W. (1977): Correspondence Principle and the Growth of Science, D. Reidel, Dordrecht.

Ludwig, G. (1978): Die Grundstrukturen einer physikalischen Theorie, Springer-Verlag, Berlin-Heidelberg-New York.

Mayr, D. (1981): 'Approximative reduction by completion of empirical uniformities', in Hartkämper, A. and Schmidt, H.-J. (eds.), Structure and Approximation in Physical Theories, Plenum Press, New York.

Moulines, C.-U. (198o): 'Intertheoretic approximation the Kepler-Newton case', Synthese 45, pp.387-412.

Nagel, E. (197o): 'Issues in the logic of reductive explanations', in Kiefer, H. and Munits, M. (eds.), Mind, Science, and History, SUNY Press, Albany.

Niiniluoto, I. (1983): 'Theories, approximations,

and idealisations', Invited Address, 7th. International Congress for Logic, Methodology and Philosophy of Science, Salzburg.
Nowak, L. (198o): _The Structure of Idealization: Towards a Systematic Interpretation of the Marxian Idea of Science_, D. Reidel, Dordrecht.
Nowakova, I. (1975): 'Idealization and the Problem of Correspondence', _Poznan Studies in the Philosophy of Science and Humanities_ 1, pp.65-7o.
Pearce, D. and Rantala, V. (1983): 'New foundations for metascience', _Synthese_ 56, pp.1-26.
Pearce, D. and Rantala, V. (1983a): 'A logical study of the correspondence relation', _Journal of Philosophical Logic_
Pearce, D. and Rantala, V. (1983b): 'The logical study of symmetries in scientific change', in Weingartner, P. and Czermak, H. (eds.), _Epistemology and Philosophy of Science_, Vienna.
Post, H.R. (1971): 'Correspondence, invariance and heuristics: in praise of conservative induction', _Studies in History and Philosophy of Science_ 2, pp.213-255.
Putnam, H. (1965): 'How not to talk about meaning', in Cohen, R. and Wartofsky, M. (eds.), _Boston Studies in the Philosophy of Science_, Vol. 2, Humanities Press, New York.
Rantala, V. (1979): 'Correspondence and nonstandard models: a case study', in Niiniluoto, I. and Tuomela, R. (eds.), _The Logic and Epistemology of Scientific Change_ (Acta Philosophica Fennica 3o), North-Holland, Amsterdam.
Redhead, M. (1975): 'Symmetry in intertheory relations', _Synthese_ 32, pp.77-112.
Scheibe, E. (1973):'The approximative explanation and the development of physics', in Suppes, P., Henkin, L., Joja, A. and Moisil, G. (eds.), _Logic, Methodology and Philosophy of Science IV_, North-Holland, Amsterdam.
Sneed, J.D. (1971): _The Logical Structure of Mathematical Physics_, D.Reidel, Dordrecht.
Stegmüller, W. (1979): _The Structuralist View of Theories_, Springer-Verlag, Berlin-Heidelberg,New york.
Zahar, E. (1983): 'Logic of discovery of psychology of invention ?', _The British Journal for the Philosophy of Science_ 34, pp.243-262.

Dieter Mayr

CONTACT STRUCTURES, PREDIFFERENTIABILITY AND APPROXIMATION

1. INTRODUCTION

For a constructive-axiomatic approach to general relativistic spacetime, a mathematical theory of contact structures has been developed (cf. Mayr (1983)). Roughly, contact structures characterize at a very basic level a concept of tangentiability related to mathematical objects like sets, relations or maps. And - as a consequence of the general level - contact structures transport the concept of differentiability from normed vector or Banach spaces to suitable uniform spaces (pregeodesic contact spaces).

Any filter base over the diagonal of a product set combined with a certain regularity condition generates a so-called cone structure. Graphically, two curves are in the n-cone at a fixed point, if the sine of the angle between their tangents is smaller than 1/n. The cone structure canonically induces a so-called contact uniformity on the set of subsets, relations or maps which are defined in a neighborhood of the fixed point. In the case of maps the elements of the Hausdorff completed contact uniformity are called pre-jets expressing the analogy to the well-known jets (cf. Bourbaki (1971) 12.1). Restricted to a special subclass of (pregeodesic) subsets the Hausdorff completion may be imagined as a generalized version of a tangent space.

After the introduction of a suitable set of "translations", that is a transitive set of maps whose actions may be interpreted as generalized parallel transport, so-called preaffine maps are determined by an invariance property of the transported pre-jets. And consequently, the fundamental idea of differentiability - the local approximation of maps by affine maps (cf. Bourbaki (1971) 1.2) - canonically leads to a concept of predifferentiability and a generalized calculus works intrinsically without the matching properties of a Banach space. In this view, contact structures lay down the mathematical core for generalizing the grounds of differential topology and geometry.

The approximative correspondence of physical theories

W. Balzer et al. (eds.), Reduction in Science, 187–198.

represents a nice field of application for contact structures. The usual approximations of points or maps are sometimes not fully adequate to reproduce all aspects of an approximative correspondence (cf. for instance the contribution of H.-J. Schmidt in this volume). But additional approximations, e.g. of differentials, already presuppose Banach structures which are not intrinsically related to the empirical domains of the theories. To avoid the Banach machinery we will use contact structures to perform more complex, so-called contact approximations. The adequacy and necessity of this approximation type may be clarified by the following examples:

(1) In a rectangular triangle the hypotenuse can be approximated by a sequence of steplines whose steps become smaller and smaller. If the other sides of the triangle have unit length, then all steplines have length 2 in contrary to the length $\sqrt{2}$ of the hypotenuse. Moreover, one can easily find approximating curves whose lengths tend to infinity.

(2) To approximate the orbit $p: \mathbb{R} \to \mathbb{R}^3$ of a particle one often uses the so-called ε-tube of the particle

$$T_p(\varepsilon) := \{(t,x) \in \mathbb{R} \times \mathbb{R}^3 \mid \vee s \in \mathbb{R}: \ |t-s| < \varepsilon \wedge |x-p(s)| < \varepsilon\}.$$

Obviously, if $\varepsilon > 0$ tends to zero, all curves, passing through the ε-tube, approximate the orbit. But therein is a tremendous set of lines with physically bizarre properties, for instance curves with arbitrarily high velocities and accelerations. And if we deal with charged particles, a sequence of permanently stronger radiating particles may even approximate a non-radiating particle.

Clearly, these absurd consequences simply follow from the pure approximations of points - an approximation of order zero, as it were. Approximations of additional structures, for example differentials, may exclude those paradoxical situations. In most cases of approximative reduction, the intuitive approximation idea automatically includes the convergences of the differentials (up to a certain order). Emphasizing this plausible statement we have to face a more or less ordered hierarchy of approximations which determines the approximative comparison of theories. And finally, in order to cover more general types of theories we use the basic frame of contact structures.

2. CONTACT STRUCTURES

The starting point in the theory of contact is the concept

of tangentiability which generalizes the known germ equivalence of sets. We recall that two subsets A,B of a set X are said to be germ equivalent wrt. a filter $\mathcal{F}$ iff there is a filter element $F \in \mathcal{F}$ with $F \cap A = F \cap B$ (cf. Bourbaki (1966) I, § 6.9). To illustrate our idea of tangentiability we give a simple example. Curves are called tangential, if the angle between their tangents is smaller than any positive real. And angles may be determined by distance relations: if one goes n steps along one side of the angle and meets after k steps the second side, then k/n is an approximative estimation of the angle. Consequently, we have to specify the concept of steps with constant "length" whereby the last term should only be viewed as a generalization of the metric distance. The most fundamental way to perform steps in a set is described by the well-known relation product defined on the Cartesian product of the set (cf. Bourbaki (1968) R, § 3.10). To copy the idea of constant length we will use a global criterion of a generalized length comparison expressed by the structure of a filter base on the product set (in contrast to the germ filter defined on the set only). Let $\Delta_X := \{(x,x) | x \in X\}$ be the diagonal of X×X.

<u>2.1 Definition:</u> $\mathcal{N} \subset \mathrm{Pot}(X \times X)$ is said to be a filter base over Δ_X iff

(i) $\mathcal{N} \neq \emptyset$.
(ii) $\bigwedge N \in \mathcal{N}: \Delta_X \subset N$.
(iii) $\bigwedge M,N \in \mathcal{N} \bigvee L \in \mathcal{N}: L \subset M \cap N$.

If N is any entourage, i.e. an element of $\mathcal{N}$, we may understand $(x,y) \in N$ by the idea that "x and y are N-close". Obviously, the defined filter base is a first part of a fundamental system which generates a uniformity (cf. Bourbaki (1966) II, § 1.1). With suitable uniformities metrics can be constructed whose distances are compatible with the "constant length" of the entourages (cf. Bourbaki (1966) IX, § 1.4). In this sense we talked about a "generalization of the metric distance".

The product of the "relations" $N,M \in \mathcal{N}$ is defined by

$$N \circ M := \{(x,z) \in X \times X | \bigvee y \in X: (x,y) \in M \wedge (y,z) \in N\}.$$

The relation product characterizes the steps which are N- or M-close. The n-fold product of an entourage represents the result of any n steps or in other terms, of any chain of length n. As usual, the N-neighborhood of an $x \in X$ is the set

$$N(x) := \{y \in X \mid (x,y) \in N\}$$

and all neighborhoods form a subbase of a topology on X, the so-called $\mathcal{N}$-topology (cf. Bourbaki (1966) I, § 2.3). Thus, the set

$$N^n(x) := \{y \in X \mid (x,y) \in N^n\}$$

is a neighborhood of x in which we can reach any point y by n steps from x to y. Now, we are ready to define our central concept of tangentiability; let $A, B \in \mathrm{Pot}(X)$ and $\mathcal{N}$ be a filter base over Δ_X.

2.2 Definition:
(1) A weakly touches B at $x \in X$ iff

$$\bigvee k \in \mathbb{N} \bigwedge n \in \mathbb{N} \bigvee N_n \in \mathcal{N} \bigwedge N \in \mathcal{N},\ N \subset N_n \bigwedge a \in A:$$
$$a \in N^n(x) \rightarrow \bigvee b \in B: (a,b) \in N^k.$$

(2) A and B are said to be weakly tangential at $x \in X$ iff A weakly touches B at $x \in X$ and vice versa.
(3) If $k = 1$ we obtain the non-weak versions of tangentiability.

The idea of the concept of touch is easily illustrated. A curve (weakly) touches a surface, for example, if the angle between the tangent and the tangential plane is arbitrary small, or in our language, if the ratio k/n of steps numbers - running n-times along the curve and crossing the surface after k steps - tends to zero. In Euclidian spaces there are special filter bases (so-called Euclidian systems), such that the defined tangentiability coincides with the usual (Euclidian) one.

Clearly, germ equivalent sets are tangential, if the corresponding filters are compatible, i.e. if the germ filter is a neighborhood filter wrt. the $\mathcal{N}$-topology. Moreover, the weakly tangential relation is an equivalence relation in contrast to the tangential relation. To compensate for the structural lack we introduce a special regularity condition. Let $\mathcal{N}$ be again a filter base over Δ_X.

2.3 Definition: $\mathcal{N}$ is said to be t-regular iff

$$\bigvee k \in \mathbb{N} \bigwedge M \in \mathcal{N} \bigvee N \in \mathcal{N}: N^2 \subset M \subset N^k.$$

By the first inclusion $\mathcal{N}$ becomes - apart from a symmetry condition - a fundamental system of a uniformity on X. But more characteristic is the second inclusion; it enforces a kind of metric similarity between the neighborhoods together with a smooth convergence structure. In the Euclidian $\mathbb{R}^n$, for example, the fundamental systems defined by the entourages

$$N_\varepsilon := \{(x,y) \in \mathbb{R}^n \times \mathbb{R}^n \mid |x-y| < \varepsilon\},$$

whereby ε runs through $\mathbb{R}_+$, $(0,1)$ or $\{1/n^i \mid n \in \mathbb{N},\ i \in \mathbb{N} \text{ fixed}\}$, respectively, are all t-regular with $k \geq 2$ (Euclidian systems). However, if ε is an element of $\{1/n! \mid n \in \mathbb{N}\}$, the convergence is too fast and the t-regularity fails.

By a t-regular filter base the tangential relation becomes an equivalence relation and coincides with the weakly tangential relation. Moreover, t-regularity is a uniform invariant and is preserved by transitions to product sets and restrictions to suitable ($\mathcal{N}$-convex) subspaces (cf. Mayr (1983) A.4).

Now, we define our central concept of contact. Let $\mathcal{F}_x(X)$ be the set of local maps in X defined in a neighborhood of $x \in X$, $\mathcal{N}$ t-regular and $f,g \in \mathcal{F}_x(X)$.

<u>2.4 Definition:</u>

(1) f and g are in the n-cone at $x \in X$, $C_x^{\mathcal{N}}(f,g;n)$, iff

$$\vee\, N_n \in \mathcal{N} \wedge N \in \mathcal{N},\ N \subset N_n \wedge y \in N^n(x):\ (fy,gy),(gy,fy) \in N.$$

(2) f and g are in contact at $x \in X$, $C_x^{\mathcal{N}}(f,g)$, iff

$$\wedge\, n \in \mathbb{N}:\ C_x^{\mathcal{N}}(f,g;n).$$

Two real functions are in the n-cone at $x \in \mathbb{R}$ (wrt. an Euclidian system), for example, if the sine of the angle between their tangents at x is smaller then 1/n. Again, the contact is an equivalence relation which generalizes the germ equivalence of maps (cf. Bourbaki (1966) I, § 6.10). But what is more, the cone structure canonically induces a uniformity on the set of local maps by the following entourages.

<u>2.5 Definition:</u>

$$C(n) := \{(f,g) \in \mathcal{F}_x(X) \times \mathcal{F}_x(X) \mid C_x^{\mathcal{N}}(f,g;n)\}$$

$$C_x(\mathcal{N}) := \{C(n) \mid n \in \mathbb{N}\}.$$

$C_x(N)$ is a countable totally ordered fundamental system of the so-called contact uniformity. To realize the mathematical position of our contact structures we recall the well-known norm contact defined, for example, on the set $F_x(\mathbb{R}^m,\mathbb{R}^n)$ of local maps from $\mathbb{R}^m$ in $\mathbb{R}^n$ (cf. Bourbaki (1971) § 1.1).

2.6 Definition: f and g of $F_x(\mathbb{R}^m,\mathbb{R}^n)$ are in norm contact at $x \in \mathbb{R}^m$, $C_x^{|\cdot|}(f,g)$, iff

$$\bigwedge \varepsilon>0 \bigvee U_x \bigwedge y \in U_x: |fy-gy|_n<\varepsilon|x-y|_m.$$

whereby U_x is a neighborhood of x. The fundamental meaning of the norm contact is given by the following definition (cf. Bourbaki (1971) § 1.2.1).

2.7 Definition: $f \in F_x(\mathbb{R}^m,\mathbb{R}^n)$ is said to be differentiable at $x \in \mathbb{R}^m$ iff there is a (continuous) affine map $a: \mathbb{R}^m \to \mathbb{R}^n$ with $C_x^{|\cdot|}(f,a)$.

The linear and convex structures of the Euclidian spaces state an important connexion between the two contact concepts. If we fix m = n, for the present, and take an Euclidian system (or more general a t-regular fundamental system generated by a norm), then the two contacts coincide. Thus, differentiability may be redefined by our contact structure which only depends on the system N. And we can say even more. We recall to the translation-invariant uniformity (cf. Schaefer (1971) I, § 1.3) which is generated by any neighborhood base of the zero element. Using only topological and algebraical (i.e. first of all normfree) properties we can define a so-called contact base of the zero element which generates the same contact relation as any t-regular system of a norm. (cf. Mayr (1983) A.3). Thus, differentiability is completely redefined without norms, just in the sense of Dieudonné's statement:"...this definition [of contact] only depends on the topologies..." (cf. Dieudonné (1969) VIII, 1. p.149).

Let us turn back to the general case. The elements

$$\hat{f}_x \in J_x(X) := \widehat{F_x(X)}$$

of the Hausdorff completion of $F_x(X)$ (wrt. the contact uniformity) are called pre-jets to emphasize the analogy to jets (more precisely: 1-jets) which are contact classes of local maps in a manifold (cf. Bourbaki (1971) § 12.1). Pre-jets are represented by minimal Cauchy filters, for example by neighborhood filters (cf. Bourbaki (1966)II, § 3.2).

If N generates a Hausdorff topology, there is a punctual operation

$$\hat{f}_x: \{x\} \to \{y\} \ , \quad \text{whereby } fx = y,$$

because any map in a C(n)-small filter element of $\hat{f}_x$ transports x to y. Consequently, x is said to be the source and y the end (or target) of the pre-jet. Thus, the pre-jet space divides into separate open subspaces with different ends:

$$J_x(X)_y, \quad y \in X.$$

In general, uniformities may have different t-regular fundamental systems which generate non-equivalent contact structures. To show the differences we present a comparability criterion on the basic level of the filter bases M,N over Δ_x.

2.8 Definition:
(1) N is said to be t-finer than $M, M \underset{t}{\leq} N$, iff

$$\forall\, l \in \mathbb{N}\ \forall\, M_o \in M \wedge M \in M,\ M \subset M_o\ \forall\, N \in N:\ N \subset M \subset N^l.$$

(2) M and N are said to be t-comparable, $M \underset{t}{\sim} N$, iff

$$M \underset{t}{\leq} N \wedge N \underset{t}{\leq} M.$$

Obviously, t-comparability implies (by the first inclusion) the equivalence of the filter bases and, in the case of fundamental systems, their uniform equivalence. But in the main, t-comparability enforces the coincidence of the corresponding contact structures. Let us abbreviate the notion of a contact space by

$$(X,N)$$

whereby N implicitely represents all tangential, cone and contact relations (generated by the t-regular N). Consequently, two contact spaces are isomorphic if there is a bijection which maps the filter bases into t-comparable systems. In the case of a common base set we specify the automorphisms by the following characterization. Let N be a filter base over Δ_x.

2.9 Definition: f: X → X fulfils a Lipschitz condition iff
$\forall\, l \in \mathbb{N} \wedge N \in N:\ (f\times f)\, N \subset N^l.$

In the special case of fundamental systems defined by norms the Lipschitz conditions turns into a global version of the usual Lipschitz continuity. Moreover, the Lipschitz condition is preserved under the composition of bijective maps and consequently, the bijections which ensure together with their inverses a Lipschitz condition (Lipschitz automorphisms), form the so-called Lipschitz group, a subgroup of the group of uniform automorphisms. Finally, a Lipschitz automorphism maps a filter base into a t-comparable filter base and thus the Lipschitz group is a contact automorphism group.

3. PREDIFFERENTIABILITY

In the following we will sketch a generalized differential calculus based on contact structures. For this purpose we consider the set $\mathcal{F}_x(X,Y)$ of local maps $f: U_x \to Y$ from an x-neighborhood of the contact space $(X,\mathcal{M})$ in the contact space $(Y,\mathcal{N})$. In a first step we will specify a correspondence between the systems $\mathcal{M}$ and $\mathcal{N}$ which determines equal "entourage distances" of $\mathcal{M}$ and $\mathcal{N}$. In the Euclidian case the correspondence is canonically given by the real numbers of the norm distances (cf. for instance definition 2.6).

3.1 Definition: The map $\kappa: \mathcal{M} \to \mathcal{N}$ is said to be a contact correspondence iff $\mathring{\kappa}$ is isotone and $\kappa\mathcal{M} \underset{t}{\sim} \mathcal{N}$.

Hereby is $\mathring{\kappa}$ the extension of κ defined by $\mathring{\kappa}(M^n) := (\kappa M)^n$ on the finite products of the $\mathcal{M}$-entourages. There is a slightly weaker condition: instead the isotony of $\mathring{\kappa}$ it already suffices a pre-isotony of κ

$$\bigvee r \in \mathbb{N} \bigwedge L,M \in \mathcal{M} \wedge m,n \in \mathbb{N}: L^{n\cdot r} \subset M^m \to (\kappa L)^{n\cdot 2} \subset (\kappa M)^m$$

and all remains true; but in the present framework we content ourselves with the simpler definition 3.1. Isotony and t-comparability are logically independent. For example, isotony even does not imply the equivalence of the filterbases $\kappa\mathcal{M}$ and $\mathcal{N}$ in general. If $\lambda: \mathcal{N} \to \mathcal{L}$ is a further contact correspondence, $\lambda\circ\kappa$ is a contact correspondence too. However, a contact correspondence does not induce a reverse correspondence or a contact isomorphism (X and Y are not bijective in general). In the case of Euclidian systems $\mathcal{M}$ and $\mathcal{N}$ defined by the norms of $\mathbb{R}^m$ and $\mathbb{R}^n$ (e.g. $\varepsilon \in (0,1)$), the map $\kappa: \mathcal{M} \to \mathcal{N}$, $M_\varepsilon \mapsto N_\varepsilon$ is an example of the mentioned canonical contact correspondence.

Now, we are ready to formulate a contact concept on the

set of local maps from X in Y. Let $f,g \in F_x(X,Y)$ and $\kappa: M \to N$ a contact correspondence.

3.2 Definition:
(1) f and g are in the n-cone at x (wrt. M and κ), $K_x^{M,\kappa}(f,g;n)$, iff

$$\bigvee M_n \in M \wedge M \in M, M \subset M_n \wedge y \in M^n(x): (fy,gy),(gy,fy) \in \kappa M.$$

(2) f and g are in correspondence-contact (in short: cor-contact) at x, $K_x^{M,\kappa}(f,g)$, iff $\wedge\, n \in \mathbb{N}: K_x^{M,\kappa}(f,g;n)$.

Obviously, the general idea of the cor-contact agrees with that in definition 2.4, but now the distance of the images fy and gy is measured by the corresponding system κM. And again, the entourages
$K(n) := \{(f,g) \in F_x(X,Y) \times F_x(X,Y) \mid K_x^{M,\kappa}(f,g;n)\}$, $n \in \mathbb{N}$, constitute a fundamental system $K_x(M,\kappa)$ of the so-called cor-contact uniformity on $F_x(X,Y)$. If X = Y and $M \underset{\tau}{\sim} N$, the cor-contact coincides with the contact (2.4) for any contact correspondence. In the Euclidian case the cor-contact agrees with the norm contact (2.6), a canonical correspondence of the Euclidian systems pre supposed.

Let us abbreviate by $J_{x_o}(X,Y)$ the Hausdorff uniform space associated with the cor-contact uniformity (cf. Bourbaki (1966)II, § 3.8). The elements $\tilde{f}_{x_o} \in J_{x_o}(X,Y)$, which we call pre-jets again, are completely represented by the neighborhood filters on $F_x(X,Y)$. To get a correspondence between pre-jet spaces with different sources we have to work - like in a manifold - with a kind of connection. For this purpose we fix a set of Lipschitz automorphisms which determine - similar like translation - a generalized parallel transport.

3.3 Definition: A subset $L_{x_o}(X)$ of Lipschitz automorphisms on (X,M) is said to be minimal x_o-transitive iff
$\wedge\, x \in X \overset{1}{\vee} \varphi \in L_{x_o}(X): \varphi x = x_o$.

The condition only ensures a unique transport. Further conditions, for example a continuous dependence from the starting point of the maps or a group structure, specify the idea of parallel translation. The fixing of $x_o \in X$ is not essential; if $L_{x_o}(X)$ is minimal x_o-transitive, then

$$L_x(X) := \{\varphi_o^{-1}\varphi \mid \varphi \in L_{x_o}(X)\} \text{ with } \varphi_o \in L_{x_o}(X),\ \varphi_o x = x_o$$

is minimal x-transitive for any $x \in X$. Let us call the triples

$$(X, M, L_{x_o}(X)), \quad (Y, N, L_{y_o}(Y))$$

pregeodesic contact spaces. The term "pregeodesic" emphasizes the possibility of characterizing at any $x \in X$ a family of subsets which may be called pregeodesic, because any two set-pre-jets, i.e. equivalence classes wrt. the tangential relation on subsets (cf. definition 2.2(3)), with sources in such a subset coincide via the "translation" maps (autoparallel).

The map

$$\phi_{xy}: J_x(X,Y)_y \to J_{x_o}(X,Y)_{y_o}, \ \tilde{f}_x \to \tilde{F}_{x_o} := (\widetilde{\psi f \varphi^{-1}})_{x_o}$$

with $\varphi \in L_{x_o}(X)$, $\varphi x = x_o$ and $\psi \in L_{y_o}(Y)$, $\psi y = y_o$ describes the transport of the pre-jets for any $x \in X$ and $y \in Y$. As a consequence of the pregeodesic structure, ϕ_{xy} turns out to be an isomorphism of the uniform spaces. On the union of all pre-jets

$$J(X,Y) := \bigcup_{x \in X} J_x(X,Y)$$

we introduce - via the parallel transport - a new uniformity by the entourages

$$J(n) := \{(\tilde{f}_x, \tilde{g}_{x'}) \in J(X,Y) \times J(X,Y) \mid (\tilde{F}_{x_o}, \tilde{G}_{x_o}) \in \tilde{K}(n)\}, n \in \mathbb{N}$$

whereby $\tilde{K}(n)$ is an entourage of the associated Hausdorff uniformity wrt. the cor-contact uniformity at x_o. The global uniformity offers a first possibility for comparing pre-jets with different sources and ends.

Now, we will take a brief look into the differential calculus of pregeodesic contact spaces. The usual concept of differentiability mainly rests on normed spaces, their affine maps and a contact relation which determines the linear approximation. Linear maps are preserved by differentiation, i.e. a linear map coincides with its own differential. With this invariance property we define a generalized sort of affine map by the use of the translation maps.

<u>3.4 Definition</u>: A map $f: X \to Y$ is said to be preaffine iff

$$\forall \tilde{g}_{x_o} \in J_{x_o}(X,Y)_{y_o} \wedge x \in X: fx = y \to \phi_{xy}(\tilde{f}_x) = \tilde{g}_{x_o}.$$

Let L(X,Y) be the set of preaffine maps from X in Y. L(X,Y) is independent from x_o and y_o, because any preaffine map (wrt. x_o,y_o) is preaffine wrt. any x,y if we change the translations from $L_{x_o}(X)$ to $L_x(X)$ and $L_{y_o}(Y)$ to $L_y(Y)$. In Euclidian spaces affine maps coincide with preaffine maps if t-regular norm systems and Euclidian translations are presupposed. Thus, in analogy to linear space calculus we can introduce the concept of predifferentiability.

3.5 Definition:

(1) $f \in F_x(X,Y)$ is said to be predifferentiable at $x \in X$ iff there is a continuous preaffine map $a \in L(X,Y)$ with $K_x^{M,\kappa}(f,a)$.

$\delta f_x := a$ is called the predifferential of f at x.

(2) f is said to be predifferentiable iff f is predifferentiable at any $x \in X$. The map $\delta f\colon X \to L(X,Y),\ x \mapsto \delta f_x$ is called the total predifferential of f.

With the above mentioned presuppositions predifferentiability coincides with differentiability in Euclidian spaces. Predifferential maps are continuous and the composite mapping theorem holds. Moreover, the set of preaffine maps is uniformized by the cor-contact uniformity in such a way that the induced topology coincides - in the case of Euclidian spaces - with the usual operator norm topology (cf. Dieudonné (1969) V. 7). Thus, continuous predifferentiability and predifferentials of higher order may be defined. Furthermore, premanifolds and generalized versions of fibre bundles are definable. We feel that the theory of pregeodesic contact spaces may generalize not only the differential calculus, but also differential topology and geometry.

Finally, let us turn back to the contact approximation mentioned in the introduction. We first present a general version, other conceptions using predifferentiability may be easily obtained. The idea of a contact approximation of maps, for example, was characterized by a double approximation: beside the approximation of the maps an approximation of the differentials or pre-jets is additionally performed. We determine the second approximation by a uniform structure on the set of the so-called total pre-jets

$$\tilde{f}\colon X \to J(X,Y),\ x \mapsto \tilde{f}_x \quad \text{with} \quad f \in F_x(X,Y).$$

Obviously, total pre-jets correspond to total differentials in an analogous way as pre-jets to differentials. With the subset $\sigma \subset \mathrm{Pot}(X)$ the so-called graph uniformity is generated by the entourages

$$G(M,n;S) := \{(\tilde{f},\tilde{g}) \mid \wedge s\in S \vee t\in S: (s,\tilde{f}_s;t,\tilde{g}_t),(s,\tilde{g}_s;t,\tilde{f}_t)\in M\times J(n)\}$$

whereby $M \in \mathcal{M}$, $n \in \mathbb{N}$ and $S \in \sigma$. If $\mathcal{M}$ includes the diagonal (discrete topology), the graph uniformity turns into the well-known uniformity of σ-convergence, for example into the uniformity of pointwise, compact or uniform convergence (cf. Bourbaki (1966) X , § 1.3). If we interpret pre-jets again as generalized tangents the graph uniformity compares the tangents by parallel transport. If X is a Hausdorff space, the graph uniformity generates a Hausdorff topology in which the tangents tend together similarly to the approximation of differentials. Combined with the usual approximations of order zero, the approximative reduction of theories can be more adequately captured.

Dieter Mayr
Sektion Physik der Ludwig-Maximilians-Universität
Theresienstraße 37, D-8000 München 2

REFERENCES

Bourbaki, N.: 1966, 'Elements of Mathematics. General Topology', Herman, Paris.

Bourbaki, N.: 1968, 'Elements of Mathematics. Theory of Sets', Herman, Paris.

Bourbaki, N.: 1971, 'Éléments de Mathématique. Variétés différentielles et analytiques' Herman, Paris.

Dieudonné, J. 1969, 'Foundations of Modern Analysis', Academic Press, New York.

Mayr, D.: 1983, 'Zur konstruktiv-axiomatischen Charakterisierung der differenzierbaren Struktur in der allgemeinen Raumzeit durch das Prinzip der approximativen Reproduzierbarkeit', Habilitationsschrift, presented to the Fachbereich Physik der Universität Marburg.

Schaefer, H.H.: 1971, 'Topological Vector Spaces', Springer, New York.

Heinz-Jürgen Schmidt

TANGENT EMBEDDING - A SPECIAL KIND OF APPROXIMATE REDUCTION

1. INTRODUCTION

1.1

This paper presents a definition of an intertheoretic relation which is based on G. Ludwig's concept of "approximate embedding".

If two theories are characterized by the same vocabulary, but different laws, we will ask in which sense the two sets of solutions (or models) of the respective laws are "close" together. Looking at simple examples (viz. v.d. Waals/ideal gas law, spherical/plane geometry, Kepler's/Galileo's law of free fall) yields the intuition that the sets of solutions are in tangent contact with respect to some identification map. This intuition will be formally developed into the definition of "proximorphisms" and a "degree of tangency λ".

As a more advanced example, the relation between classical and relativistic spacetime is analyzed in terms of tangent embeddings, and proximorphisms with $\lambda = 2$ are constructed in this case.

So far, tangent embedding is purely mathematically defined and typically generates a symmetric relation between reductive pairs of theories. In order to account for the asymmetry between reduced and reducing theories one has to consider physical observations, sets of data, and "blurred" sets of data corresponding to deviations between observation and theoretical prediction. For the full concept of tangent embedding it will be required that the difference in the degree of blurring (or "error") between the two theories can be explained by the "error of embedding". This will be a quantitative refinement of the usual conditions on the domains of empirical successes and failures.

1.2

It is a successful strategy in science to split involved

W. Balzer et al. (eds.), Reduction in Science, 199–215.

objects into simpler constituents. Applied to the problem of reduction this would amount to constructing a reduction relation as a product of intertheoretic relations (IR) which can be analysed more easily, as has been proposed by G. Ludwig (cf. his contribution to this volume). One special kind of IR, which he calls 'embedding', is restricted to the case where the two theories have the same vocabulary and interpretation, but different laws. In order to apply the concept of embedding, one usually has, in a first step, to reconstruct the two theories under consideration in a unified language, or, at least, one has to define the basic concepts of the "poor" theory in terms of the "rich" theory. In many cases the problem of approximation only arises within the second step when comparing different laws, whereas the first step may be formulated as an exact IR. Thus it seems reasonable to confine ourselves to the special case of 'approximate embedding'.

1.3

Different approaches to intertheoretic approximation may be classified according to whether they are concerned with the "approximation of laws" or the "approximation of solutions". Sometimes the adopted view simply depends on the underlying theory concept, let it be a "statement-view" or a "non-statement-view", whereas J. Ehlers (1984) (from which I borrow these very intuitive terms) considers both kinds of approximation. I shall analyse almost exclusively the "approximation of solutions", for the following reasons. If s_1 is a solution of the law L_1, say $s_1 \in L_1$, and, analogously, $s_2 \in L_2$, it may well happen that s_1 and s_2 are both approximate descriptions of the same physical situation (and hence $s_1 \approx s_2$), which then represents a successful application of both theories. Approximation of solutions is therefore of genuine interest for comparing theories. But at present one is neither able to derive it in general as a consequence of an approximation between the respective laws, nor is one forced to presuppose such an approximation. An interrelation between the two kinds of approximation would (amount) to a deep metascientific result, but for the time being we should remain content with analysing them independently. In the following three examples we shall study the approximation of solutions in detail.

2. EXAMPLES

2.1 Van-der-Waals vs. Ideal Gas Law

According to the limit-law approach, van-der-Waals law

$$(P + \frac{N^2 a}{V^2})(V - Nb) = NRT \tag{1}$$

reduces to the ideal gas law

$$PV = NRT \tag{2}$$

if the parameters a,b go to 0. This statement alone is not enough to give an adequate account of the real situation. For a given substance, a and b have fixed values depending only on the choice of the units, which is a matter of convention, not of physics. But there are certain regions in (P,V,T)-space where both laws are approximately satisfied, and a measurement of a triple (P,V,T) within this region could be regarded as a successful application of both laws. In order to give a geometric visualization let us pass to dimensionless quantities

$$p = Pb^2/a, \quad n = Nb/V, \quad t = TRb/a \tag{3}$$

and rewrite (1) as

$$p = nt + n^2(t/(1-n)-1), \tag{4}$$

where the first term represents the ideal gas approximation. Both laws can be visualized as tangent surfaces in $\mathbb{R}^3$ which touch along the line n = p = 0 (cf. fig. 1). Note that the "touching region" (or "focus") itself is nonphysical, which is a typical feature. The fact that the two sets of solutions of the laws are tangent subsets is not accidental but reflects the degree of approximation: the region in (p,n,t)-space where van-der-Waals' correction is smaller than ε is "large" when measured in units of ε. Of course, this region increases if we consider another substance with smaller a,b, which may be viewed as another aspect of the formal limit statement a,b → 0. However, it seems not to be the primary aspect, since the van-der-Waals/ideal gas law approximation would also make sense if there existed only a single substance in the world.

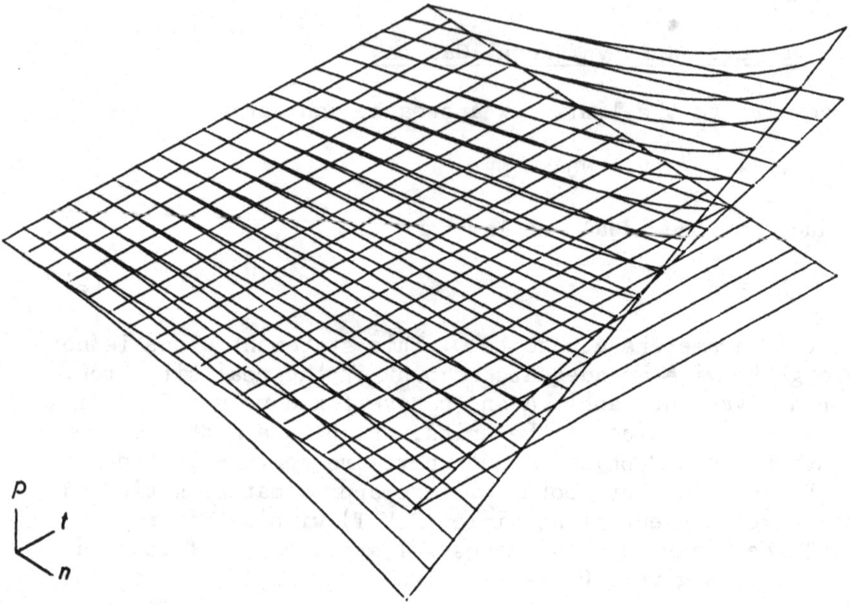

Fig. 1

This figure[1] shows 3 different surfaces in (n,t,p)-space, where (in natural units):

- n mole density
- t temperature
- p pressure.

Besides the plane $p = 0$, the graph of the ideal gas law $p = nt$ is represented by a family of straight lines corresponding to constant t. The slope increases with t. Any such line is the tangent at $n = 0$ of a van-der-Waals curve $p = nt - n^2(t/(1-n)-1)$ with constant t. This shows that the two graphs of the resp. gas laws are tangent surfaces. Note that for $0 < n < 1-t$ the van-der-Waals pressure is lower than the ideal pressure due to the attractive regime of intermolecular forces, whereas for high density or temperature the repulsive regime prevails (cf. Kuipers (1983)).

2.2 Spherical vs. plane geometry

A sphere X and a plane Y are different metric spaces, say (X,d) and (Y,D), which are not even locally isomorphic. Any local map ϕ: X $\to$ Y conserves distances only approximately:

$$D(\phi x,\phi y) \approx d(x,y). \tag{5}$$

The maximal deviation ε cannot be zero, but there are "good" maps ϕ such that $\varepsilon \to 0$ if the domain of ϕ is more and more restricted to some spot on X, the focus of ϕ. For example, if ϕ is the projection of the sphere into the plane tangent at the focus, then the azimuthal distances are conserved and the radial ones distorted according to $\theta \approx \sin\theta$, where θ is the polar angle. Thus ε is of order θ^3, if θ measures the radius of the domain of ϕ. Such a statement yields no exact bounds of error, but quick estimates like $\varepsilon \sim (\frac{500}{6000})^3 \cdot 6000\text{km} \sim 4$ km for a map of Germany.

Obviously the exponent 3 is not peculiar to the chosen map ϕ, but characteristic for the pair of metric spaces (X,Y).[2]

2.3 Free fall: Kepler vs. Galileo

Consider the Kepler ellipse of a body moving in the earth's gravitational field. Let x(t),z(t) be the Cartesian coordinates in the plane of movement of the bodies position at time t such that z(t) has a maximum at t = 0. Hence in lowest order approximation

$$\begin{aligned} x(t) &= x(0) + vt, \\ z(t) &= z(0) - gt^2/2, \end{aligned} \tag{6}$$

which is Galileos law of free fall. This approximation turns out even better than one would expect on the grounds of the above derivation. Kepler's "exact" solution is most conveniently written in terms of the <u>eccentric anomaly</u> E as

$$\begin{aligned} x &= b \sin E \\ z &= a \cos E - e \\ t &= (aE - e \sin E)b/C, \end{aligned} \tag{7}$$

where a,b,e are the usual parameters of the ellipse and C/2 is the (constant) area per time.[3] Expanding these functions into power series in E (and suitably identifying the constants of the two laws) shows that Galileo's is actually a second order approximation for x(t) and a third order one for z(t) with respect to E and, hence, to t. [This is not peculiar to the r^{-2} force law, but holds for all conservative fields of force which admit motions analytic in t.] Even in this simple example it would be difficult to calculate exact bounds for the error of approximation. However, for most purposes it suffices to know the approximate error, which is given by the next nonvanishing term in the t-power series. If we set

$$r = z(0) = a - e, \quad X = vt, \; Z = gt^2/2 \tag{8}$$

these terms come out as

$$\Delta x = -XZ/3r \quad \text{and} \quad \Delta z = Z^2/3r + X^2Z/4r^2. \tag{9}$$

Again, a Kepler orbit and a Galileo orbit can be considered as tangent subsets of the set of events $\mathcal{E}$. But what we should look for are tangent sets of solutions, i.e. tangency between the set K of Kepler orbits and the set G of Galileo orbits. These sets are specified by the condition that the value of g (resp. $\mu \equiv aC^2/(a^2-e^2)$) remains constant. Hence we need a concept of distance between orbits. Let v_0 be any velocity, d the Euclidean metric on $\mathcal{E}$ induced by the quadratic form $x^2 + y^2 + z^2 + (v_0t)^2$ and $\bar{d}$ the corresponding (quasi-) metric defined for subsets $\alpha,\beta \subset \mathcal{E}$ via

$$d_1(\alpha,\beta) = \sup_{p\in\alpha} \inf_{q\in\beta} d(p,q), \quad \bar{d}(\alpha,\beta) = \sup\{d_1(\alpha,\beta), d_1(\beta,\alpha)\}. \tag{10}$$

Consider orbits with a finite (but not fixed) time interval. A typical focus γ will be the degenerate orbit consisting only of the returning point (maximum of z(t)). Then, it is possible to show that for all $\alpha \in K$, such that $\bar{d}(\alpha,\gamma) \leq \lambda$, there exists a $\beta \in G$ such that $\bar{d}(\alpha,\beta) \lesssim \lambda^3$ and vice versa. (Use (7) and consider the case $v \gg v_0$ separately.) The symbol "$\lesssim$" indicates the replacement of the exact error by the approximate error obtained by the series expansion.

3. PROXIMORPHISMS AND DEGREE OF TANGENCY

In this section I shall propose a concept of approximation of structures which covers the examples studied in the last section and applies to a more general framework as well. This concept is a purely mathematical one and relates two structures (X,s) and (Y,t) of the same type. This means that (X,s) resp. (Y,t) are structures of species Σ resp. θ with the same "echelon construction scheme", say

$$s \subset L(X_1,\ldots X_n,\mathbb{R}) \equiv S, \qquad (11)$$
$$t \subset L(Y_1,\ldots Y_n,\mathbb{R}) \equiv T,$$

but possibly different proper axioms $\alpha(X,s)$ resp. $\beta(Y,t)$. Instead of rephrasing the relevant definitions (cf. Bourbaki (1968), IV) I shall indicate how the above examples fit into this frame. The X_i, $i = 1,\ldots,n$, are refered to as the "base sets" and s as the "structural term".

A metric is a relation between pairs of points and numbers, hence the typification is in both cases of the form $s \subset X \times X \times \mathbb{R}$ and only the proper axiom distinguishes between plane and spherical geometry.

In the gas law case, let $(X_i;A_i,M_i)(i = 1,2,3)$ be typified as $\mathbb{R}$-linear spaces, i.e.

$$A_i \subset X_i \times X_i \times X_i,\ M_i \subset \mathbb{R} \times X_i \times X_i, \qquad (12)$$

where A_i stands for "addition" and M_i for "multiplication with reals". Further, let $\gamma \subset X_1 \times X_2 \times X_3$, which is intuitively the graph of the gas law, if X_1 denotes pressure, X_2 density of moles, X_3 temperature. $s = (A_1,M_1,A_2,M_2,A_3,M_3,\gamma)$ has then an obvious typification $s \subset L(X_1,X_2,X_3,\mathbb{R})$, and $\alpha(X,s)$ resp. $\beta(X,s)$ say that the X_i are one-dimensional linear spaces possessing bases such that the coordinates of γ satisfy the ideal resp. van-der-Waals law (4). The Galileo/Kepler case may be reconstructed similarly. The base set E has to be endowed with some additional structure ν which turns it into, say a Newtonian spacetime (see, for instance, W. Balzer's article in this volume). Clearly, the typification of K is $K \subset P(E)$ (power set) and $s = (\nu,K)$. The formulation of the respective laws of free fall can be given in a (ν-admissible) coordinate system and has to distinguish between numbers which are constant only for one orbit and those constant for all orbits.

The first lesson to draw from these examples is that we should abandon "obvious" identifications of base sets. The geometric example clearly shows that identifying points in either geometry is already a matter of approximation and, further, that it is by no means unique. The identification of the spaces of events or of pressure, density, temperature in the other examples (by using the same letter) should more properly be reformulated in terms of partial isomorphisms, partial in the sense that only parts of the structural terms are involved. Isomorphisms between structures of the same type should thus turn out as special or degenerate instances of a more general concept of "proximorphism" between such structures. Let us recall the definition of an isomorphism between the two structures (X,s) and (Y,t) introduced above:

$$f \in \mathrm{iso}(x,s;Y,t) \text{ iff } f = \langle f_i : X_i \to Y_i \rangle_{i=1,\ldots,n} \text{ such that all } f_i \text{ are bijections and } F[s] = t, \text{ where } F = L(f_1,\ldots,f_n,1) \text{ is the canonical lift to echelon sets.} \tag{13}$$

A first attempt to weaken this notion of isomorphism could be to replace $F[s] = t$ by $F[s] \approx t$ and (not necessarily equivalent) $F[s'] = t'$ by $F[s'] \approx t'$ (set-theoretic complement). The symbol $\approx$ means "being close", which has to be explicated in terms of metrics on the respective echelon sets, or, more generally, in terms of uniform structures. But a glimpse at the examples shows that this definition would be too insensitive since as a rule the two sets of solutions s,t will not be "globally close" but only "locally close" when compared via F. Hence it appears natural to restrict F to a local map in such a way that the approximation $\approx$ will be at least as good as a given ε-margin (if metrics are used). I think that this is essentially the concept of approximate embedding proposed by G. Ludwig in his (1978). p.98

In most examples it is possible to choose ε arbitrarily small, giving rise to a family (germ) F_ε of more and more restricted maps. Typically the diameter δ of the domain of F_ε also tends to 0 if $\varepsilon \to 0$ it thus converges to what we have called the "focus" of an approximation, which may well consist of improper or degenerate solutions. It seems difficult and not very interesting to construct explicitely the germ F_ε, which would require exact estimates and subtle inequalities. Physicists usually have to be content with order-of-magnitude-calculations or the first few terms in a series expansion, the convergence of which is, if at all, proved some decades later.

Hence not the germ F_ε itself but some of its asymptotic properties for $\varepsilon \to 0$ should be regarded as characteristic of the comparison of the two structures. Here we shall concentrate on comparing the rates of convergence of $\varepsilon \to 0$ and $\delta \to 0$. If ε converges quicker than δ, say as δ^λ, $\lambda > 1$, (symbolically $\varepsilon = O(\delta^\lambda)$), we shall speak of <u>tangent</u> sets of solutions, λ being the <u>degree of tangency</u>. This was the case in the above examples.

Another point of minor importance[4] is that I prefer to reduce the germ F_ε to a single object. Instead of considering an exact mapping F_ε and then blurring it, one may from the first consider a blurred isomorphism, i.e. a relation, which becomes finer and finer in the neighborhood of the focus, and relates s and t. Both properties will be formulated in different definitions. We shall write $sF \equiv \{y \in T | \exists x\ S, xFy\}$etc.

D1: Let (X,s), (Y,t) be structures of the same type, $s \subset L(X_1,\ldots,X_n,\mathbb{R}) \equiv S$ and $t \subset L(Y_1,\ldots,Y_n,\mathbb{R}) \equiv T$. A relation $F \subset S \times T$ is called a <u>proximorphism</u> iff $t \subset sF$ and $t' \subset s'F$ (set-theoretic complement).[5] If, moreover, there exist $f_i \subset X_i \times Y_i$, $i = 1,\ldots,n$, such that $F \supset L(f_1,\ldots,f_n,1)$, F will be called <u>standard</u>.

The existence of proximorphisms for any (X,s),(Y,t) is trivial, take $F = S \times T$. But the existence of proximorphisms with good "focussing properties" is non-trivial; this will be made precise in the following. To this end we suppose two metrics σ,τ to be given on S resp. T, which are compatible with the "uniformities of physical imprecision" in the sense of Ludwig (1978). Let d be any metric on a set D. A subset $\beta \subset D$ of the form

$$\beta = \{y | y \in D, d(x,y) \le \alpha\}, x \in D, \alpha \ge 0 \tag{14}$$

will be called a <u>ball</u> with radius $\alpha = d(\beta)$. Let $A \subset D$ be any subset, then we define

$$\begin{aligned} d_{in}(A) &= \sup\{d(\beta) | \beta \subset A, \beta \text{ ball}\}, \\ d_{out}(A) &= \inf\{d(\beta) | \beta \supset A, \beta \text{ ball}\}, \end{aligned} \tag{15}$$

including the value ∞. A proximorphism F is (at most) ε-imprecise at $x \in S$ if $\tau_{out}(xF) < \varepsilon$. This will be the case in a subset of S which may be characterized by its inner radius

$$\delta_F(\varepsilon) \underset{\text{def}}{=} \sigma_{in}\{x|x \in S, \tau_{out}(xF) < \varepsilon\}. \tag{16}$$

This means: if the blurred map F is restricted to some ball with radius $< \delta_F(\varepsilon)$, then it will be at most ε-imprecise. According to our previous considerations we have to investigate the way $\delta_F(\varepsilon)$ tends to zero, if $\varepsilon \to 0$. Especially interesting is the case $\varepsilon \sim \delta^\lambda$, which leads us to the following

D2: Let F be any proximorphism,

$$\lambda_F \underset{\text{def}}{=} \limsup_{\varepsilon \to +0} \log \varepsilon / \log \delta_F(\varepsilon)$$

is called the degree (of tangency) of F. If F is locally a map, we set $\lambda_F = +\infty$. We note that the relational product of two proximorphisms of degree $\geq \lambda$ is not always of the same degree, but only if the foci are duly related.

In the above examples we have $\lambda = 2$ for the gas law case and $\lambda = 3$ for the cases of geometry and free fall. Several objections might be raised against our definition, some of which we shall discuss shortly.

Our definition depends to a certain extent on the choice of the metrics, and non-equivalent metrics may generate the same uniformity. If the uniformity is considered as the basic structure, the numerical value of our λ appears as a mere convention. In this context the following immediate proposition should be mentioned.

Proposition: If there are constants A,B,C,D,p > 0 such that the metrics σ_1, τ_1 on S resp. T satisfy $A\tau^p \leq \tau_1 \leq B\tau^p$ and $C\sigma^p \leq \sigma_1 \leq D\sigma^p$, then $\lambda_F(\sigma,\tau) = \lambda_F(\sigma_1,\tau_1)$.

Hence we have a large class of pairs (σ,τ) which give the same λ. Perhaps it is possible to characterize a fundamental system of entourages which generates a class of "tangentially equivalent" metrics. At least in the examples the choice of this class is rather obvious; which may suggest that some structure exists between the metric and uniformity descriptions or imprecision.[6] However, the choice of the metric becomes more delicate in cases where the focus lies at infinity.

Another objection would be that the λ value may be misleading when comparing two approximations. p = nt and

$p = 0$ are both $\lambda = 2$ approximations of the van-der-Waals law. And even worse: with respect to a finite error, a $\lambda = 2$ approximation may be better than a $\lambda = 3$ one. Consider the Newtonian motion of a planetary system. Its Kepler approximation neglects a small force, hence it will of degree $\lambda = 2$, but its "Galileo approximation" (constant force for each planet) will be of degree $\lambda = 3$, as we have seen in a similar case. Thus we have to go back to the sets $\{x | x \in S, \tau_{out}(xF) < \varepsilon\}$ in order sensibly to compare approximations with finite error.

Let us add two remarks relevant for the next section. If the structural term consists of m components $s = (s_1,\ldots,s_m)$, it will sometimes be convenient to regard a proximorphism as a tuple $F = (F_1,\ldots,F_m)$, especially if different components of F have different degrees of tangency. Although the notion of "proximorphism" merges the identification map with the blurring, in concrete examples it may be simpler to construct it in the following way. Let $(X,s),(Y,t)$ be two structures of the same type and $f_i \subset X_i \times Y_i$, $i = 1,\ldots,n$, as in the above definition D1, and τ a metric on T. For every $\xi \in t$, let $\varepsilon(\xi)$ be the smallest number such that $\tau(\xi,\xi') \leq \varepsilon(\xi)$ implies $\xi' \in L(f_1,\ldots,f_m,1)s$, analogously for $\xi \in t'$. Then the relational product of $L(f_1,\ldots,f_m 1)$ and $\{(\xi,\xi') | \tau(\xi,\xi') \leq \varepsilon(\xi)\}$ will be a proximorphism $F \subset S \times T$ by construction, and will be called the proximorphism generated by $(f_1,\ldots,f_m,1)$, τ and (s,t). This definition particularly applies if X and Y are identified for the sake of convenience.[7]

4. G-SPACETIME VS. PEM-SPACETIME

In this section we shall consider the somewhat more involved example of Galileo (G) vs. Poncaré-Einstein-Minkowski (PEM) spacetime. We choose a representation which is based on qualitative, physically fundamental concepts, namely a set P of processes, endowed with two binary relations: inclusion $<$ and congruence $\square$. The axiom $\alpha(P;<,\square)$ (resp. $\beta(P;<,\square)$) is simply:

> $(P;<,\square)$ is isomorphic to the lattice of bounded, open subsets of $\mathbb{R}^4$, where congruence is defined via the imhomogenous Galileo (resp. Poincaré) group without reflections G_{GAL} resp. G_{POI} (17)

A more intrinsic characterization of PEM spacetime in terms

of (extended) processes can be found in Mayr (1979,1983), where also the reconstruction of events and the Minkowski metric is carried out. The congruence relation there is contained in the basic structure of a group of "reproductions".

Obviously, a given G-spacetime and a PEM-spacetime are isomorphic as lattices, but not with respect to the congruence relation. So we may identify the $(P,<)$-part of both structures and consider a partially ordered set with two congruence relations $\square$ (G) and $\boxdot$ (PEM) (mnemotechnical: point carré). Choose any Euclidean metric on $\mathbb{R}^4$ and consider the resulting metric d on P and the (supremal) metric $\tilde{d}$ on $P \times P$. Let $F \subset (P \times P) \times (P \times P)$ be the proximorphism generated by $\tilde{d}$ and $(\square,\boxdot)$ and let us calculate λ_F in the neighbourhood of the focus (p_o,p_o) given by a degenerated process p_o with coordinate representation $\{(\lambda,0,0,0)|0 \le \lambda \le 1\}$ (the first coordinate is the time-coordinate). Clearly, the degree of tangency λ_F does not depend on the chosen Euclidean metric and of p_o. In order to

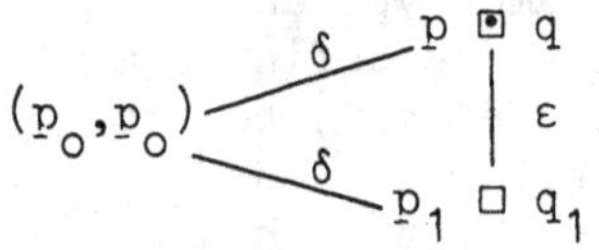

Fig. 2

calculate λ_F we have to choose $\varepsilon > 0$ and $\delta > 0$ such that within the δ-ball around $(p_o p_o)$ every PEM-congruent pair $p \boxdot q$ is ε-close to a pair $p_1 \square q_1$ and then to relate ε and δ (cf. note 5). Since translational and rotational congruences are identical in both cases, we have essentially to compare boosts (special Galileo resp. Lorentz transformations) in two dimensions, say

$$\varphi = \begin{pmatrix} 1 & 0 \\ \tanh\tau & 1 \end{pmatrix} \in G_{GAL} \quad \text{and}$$
$$\psi = \begin{pmatrix} \cosh\tau & \sinh\tau \\ \sinh\tau & \cosh\tau \end{pmatrix} \in G_{POI} \tag{18}$$

If $\tilde{d}(p_o,p_o;p,q) = \sup\{d(p_o,p),d(p_o,q)\} \le \delta$ and $q = \psi[p]$, we may infer $\tau \le \delta$. For $p_1 = p$ and $q_1 = \varphi[p]$ we have $\tilde{d}(p,q;p_1,q_1) = d(q,q_1) \le \sup_{\vec{x}\in p} \|\varphi\,\vec{x} - \vec{x}\| \lesssim \varepsilon$. This supremum will be attained at some event $\vec{x} = (t,\delta x)$ with $t,x \sim 1$. Expanding φ and ψ into a power series with respect to τ yields:

$$\varepsilon \sim \left\| \begin{pmatrix} x\tau\delta + t\tau^2/2 + \ldots \\ x\tau^2\delta/2 - t\tau^3/6 + \ldots \end{pmatrix} \right\| \sim \delta^2. \tag{19}$$

Since these estimates cannot be improved by an order of magnitude, we may conclude:

$$\lambda_F = 2 \tag{20}$$

The construction of a proximorphism F with this degree of tangency depends on an observation which is physically important and well-known (cf. Ludwig (1974), p. 369): Our physical world appeared to be non-relativistic for a long time not only because the usual velocities are so small compared with the velocity of light c, but also because the usual processes have so small a spatial extension if compared with their duration times c.

5. APPROXIMATE REDUCTION

So far, tangent imbedding is only a relation between mathematical structures and typically a symmetric relation. In order to relate it with the (asymmetric) reduction of theories, we have to refer to physical experiences. Our exposition will be provisional and largely informal.

First we assume that the components of a structure (X,s) of species Σ have a physical interpretation, the base sets X_i, i = 1,...,n, as different sorts of physical objects and the components of the structural term $s = (s_1,\ldots,s_m)$ as physical (non-theoretical) predicates and relations. The above examples may serve as an illustration.

Further, let G be the fundamental domain of applications of a physical theory PT (cf. Ludwig, this volume). For any application $g \in G$ let $A(g)$ be the corresponding set of data or observational report, consisting of a quantifier-free sentence in the language of the mathematical theory MT_Σ employing Σ (extended by individual constants). Intuitively, it typifies certain physical objects and states some relations which hold between them and, perhaps, some real numbers. If $A(g)$ is inconsistent with the axiom $\alpha(X,s)$ of Σ, some weakened version $N(g)\ A(g)$ may still be consistent. Here $N(g)$ is a tuple of "imprecision sets" which measures the quality of the application g: if $N(g)$ remains small, g will be a success of PT, if $N(g)$ has to be chosen large, it will be a failure.

Now consider two theories PT_1 and PT_2 which are almost equal up to the axiom of the species of structures, which reads $\alpha_1(X,s)$ resp. $\alpha_2(Y,t)$. Hence $N_1(g)$ and $N_2(g)$ may be different, although we assume for both theories the same G and for all $g \in G$ the "same" $A(g)$, i.e. $A_1(g)$ and $A_2(g)$ are only artificially distinguished by the use of different letters. In order to define a reduction relation suppose each $g \in G$ is assigned a standard proximorphism $F(g)$ between (X,s) and (Y,t) of degree $\geq \lambda$. Intuitively, the g-dependence accounts for the possibility of focusing $F(g)$ on the objects occurring in g, for instance adapting the focus p_0 in the spacetime example to the frame of reference in which the experiments are performed. This allows us to translate $A_2(g)$ into a statement $F(g)\ A_2(g)$ in MT_{Σ_1} which is implied by $A_2(g)$. Roughly speaking, we translate $y \in t$ into $y \in sF$, which is $\exists x \in s : xFy$, and similarly for the negation. (Recall that $t \subset sF$ and $t' \subset s'F$ holds by definition of a proximorphism.) More precisely, this is done for each component of s_ν occurring in $A_2(g)$ by using the relevant component $F_\nu(g)$ of $F(g)$. It will suffice to consider a simple example with self-explanatory notation. Let $A_2(g)$ report on some triangle in spherical geometry, viz.

$$d(a,b) = 1,\ d(b,c) = 1,\ d(c,a) = 1. \tag{21}$$

Then $F(g)\ A_2(g)$ would read: $\exists a_1,a_2,b_1,b_2,c_1,c_2,\alpha,\beta,\gamma$:

$$\begin{aligned} D(a_1,b_1) &= \alpha \text{ and } (a_1,b_1,\alpha)\ F(a,b,1),\\ D(b_2,c_1) &= \beta \text{ and } (b_2,c_2,\beta)\ F(b,c,1),\\ D(c_2,a_2) &= \gamma \text{ and } (c_2,a_2,\gamma)\ F(c,a,1). \end{aligned} \tag{22}$$

Clearly, $F(g)\ A_2(g)$ is consistent with $\alpha_1(X,s)$ if $A_2(g)$ is consistent with $\alpha_2(Y,t)$.

The definition of $F(g)\ A_2(g)$ bears much resemblance with that of $N_2(g)\ A_2(g)$, if $N_2(g)$ is considered as a proximorphism of (Y,t) onto itself. The only difference is that with respect to $N_2(g)$ one may allow for different observations occuring in $A_2(g)$ to have different precision. But nevertheless it is rather obvious how to combine both steps of blurring $A_2(g)$ to obtain a statement

$$F(g)\ N_2(g)\ A_2(g). \tag{23}$$

Again, this will be consistent with $\alpha_1(X,s)$ if $N_2(g)A(g)$ is consistent with $\alpha_2(Y,t)$. If $F(g)N_2(g)A_2(g)$ is comparable with $N_1(g)A_1(g)$, we say that the quality of the application g in PT_1 is <u>reduced</u> to the quality of g in PT_2 and the quality of the proximorphism F(g). An approximative reduction of PT_1 to PT_2 in this case would thus be defined by the metatheoretic formula

$$\forall\, g \in G\colon\ F(g)N_2(g) = N_1(g). \tag{24}$$

It comprises the following postulates:

(i) Each success of PT_1 is a success of PT_2 (N_1 small implies F and N_2 small),

(ii) each failure of PT_2 is a failure of PT_1 (N_2 large implies N_1 large),

(iii) there are some successes of PT_2 which are failures of PT_1 (those $g \in G$ where N_2 is small but F and hence N_1 are large).[8]

What makes this definition provisional are certain doubts about whether the explication of success and failure in terms of logical consistency is any more than a first, simplified step. It seems that often A(g) does not conflict directly with the axiom $\alpha(X,s)$ but with some law-like hypotheses on g, like "this body is rigid" or "these measurements are statistically independent".[9] But to analyse these problems in more detail would be a subject of further research.[10]

NOTES

1. I am indepted to Alois Nitsch for preparing the computer plot.
2. The correct analogy to the previous example would be the observation that the subsets $D \subset Y \times Y \times \mathbb{R}$ and $(\phi\times\phi\times 1)\,[d]$, where $d \subset X \times X \times \mathbb{R}$, are tangential at some focus (y,y,0).
3. Cf. for instance Landau/Lifschitz (1979) § 15.
4. This is very similar to the notion of "comparison filter" introduced by Werner (1982).
5. Often s will have less dimensions than S and d' will be dense in S. If x F is always open in S, $t' \subset s'F = S$ is automatically satisfied. This will also be the case for the example in section 4.
6. Using a "t-regular fundamental system" defined by D. Mayr, this volume, would at least allow us to

distinguish between the cases $\lambda > 1$ and $\lambda \leq 1$.

7. Although my definitions are based on G. Ludwig's theory concept, I think that a great part could be translated into the Sneedean framework. The set s then should be identified with the set of (partial) models, S with the set of (partial) potential models, all axioms which are not of the form $x \in s \Longleftrightarrow a(x)$ are to be considered as "constraints" and so on. This type of translation suggests itself if, for example, the two reconstructions of classical mechanics in Sneed (1979) and Ludwig (1981) are compared. The "reduction relation ρ" then seems to be related to Ludwig's embedding map, or, in the approximate version, to our proximorphism F.
Note, however, that there is another natural approach of interpreting Sneed's "models" as models of $\mathcal{MT}_\Sigma$ (cf. Scheibe (1981)). It is interesting that from this view Ludwig's intertheoretic relation of "restriction" seems to be the due pendant to Sneed's ρ-relation.
8. Strictly speaking, the existence of those succeses is not implied by (24) alone. But otherwise $\mathcal{PT}_1$ and $\mathcal{PT}_2$ would have the same successes and failures and hence be empirically equivalent.
9. A nice example is given by R. Werner, this volume: Arbitrary correlation data are consistent with a classical statistical theory, but additional "locality" assumptions make Bell's inequalities necessary.
10. I am indepted to Wolfgang Balzer for critical remarks and to David Pearce for linguistic corrections.

Heinz-Jürgen Schmidt
Fachbereich Physik
Universität Osnabrück
Postfach 4469
4500 Osnabrück
Fed. Rep. of Germany

REFERENCES

Bourbaki, N.: 1968, 'Theory of sets', Addison-Wesley, Reading, Mass.

Ehlers, J.: 1984, 'On Limit Relations between, and Approximative Explanations of, Physical Theories', to appear in the Proceedings of the 7th International Congress of Logic, Methodology and Philosophy of Science, North Holland.

Kuipers, T.A.F.: 1983, 'The paradigma of concretization: the law of Van der Waals', Poznan Studies in the Philosophy of the Sciences and the Humanities.
Landau, L.D. and Lifschitz, E.M.: 1979, Mechanik, Akademie-Verlag, Berlin.
Ludwig, G.: 1974, 'Einführung in die Grundlagen der Theoretischen Physik, Band II, Bertelsmann, Düsseldorf.
Ludwig, G.: 1978, 'Die Grundstrukturen einer physikalischen Theorie', Springer, Berlin-Heidelberg-New York.
Ludwig, G.: 1981, 'Axiomatische Basis einer physikalischen Theorie und theoretische Begriffe', Z. f. allg. Wiss. th. XII 1, p. 55-74.
Mayr, D.: 1979, 'Zur konstruktiv-axiomatischen Charakterisierung der Riemann-Helmholtz-Lieschen Raumgeometrien und der Poincaré-Einstein-Minkowskischen Raumzeitgeometrien durch das Prinzip der Reproduzierbarkeit', Doctoral dissertation, München.
Mayr, D.: 1983, 'A Constructive-Axiomatic Approach to Physical Space and Spacetime Geometries of Constant Curvature by the Principle of Reproducibility', in D. Mayr and G. Süßmann (eds.), 'Space, Time and Mechanics', Reidel, Dordrecht.
Scheibe, E.: 1981, 'A Comparison of Two Recent Views on Theories', in A. Hartkämper and H.-J. Schmidt (eds.), 'Structure and Approximation in Physical Theories', Plenum Press, New York.
Sneed, J.D.: 1979, 'The Logical Structure of Mathematical Physics', 2nd ed., Reidel, Dordrecht.
Werner, R.: 1982, 'The Concept of Embeddings in Statistical Mechanics', Doctoral dissertation, Marburg.

A. Kamlah

A LOGICAL INVESTIGATION OF THE PHLOGISTON CASE

1. IDEALISM AND REALISM ON EXISTENTIAL HYPOTHESES

The concept of Phlogiston which dominated chemistry in the 18th century until about 1780 has been used as an example in the debate between scientific realists and idealists on scientific progress in history of science. The aim of this paper is to clarify the Phlogiston theory by formalization and axiomatization in order to draw some conclusions which are relevant for this debate. (For a philosophical analysis see Kitcher 1978.)

What is the issue of this debate, and what are the conflicting standpoints? In the last two decades scientific realism and the theory of T. S. Kuhn, which I call here "idealism", have given the most important interpretations of scientific progress. Kuhns "idealist" standpoint is well known and characterized and discussed in many text books on philosophy of science. Therefore an extensive exposition is not needed here. The most important feature of this account is the assumption of revolutionary changes in the history of science, which lead to the replacement of fundamental conceptual schemes, so called paradigms or disciplinary matrices, by others which are completely different. Each of these schemes or scientific world views deals in some way successfully with the phenomena known at the time of its domination. But we cannot expect at any time to be in possession of the best, last, and final world view, since we have to be prepared for future scientific revolutions. Thus scientific theories to a large extent are our way to deal with the world, and are more products of our minds than of nature itself. To claim that they give an objectively true account which is at least approximately independent of our historical situation would be absurd. The world is different for scientists having different disciplinary matrices. There may appear different entities, quantities, and attributes in these different worlds. We are never facing the naked world but always a world dressed into our concepts.

W. Balzer et al. (eds.), Reduction in Science, 217–238.

To scientific realists this philosophy of science seemed completely dissatisfactory and a thoroughly wrong account of human knowledge. Their attitude towards knowledge is quite opposed to the "idealist" account. Unfortunately scientific realism is not easily characterized, since it is not a homogeneous doctrine. Since we are dealing here, however, only with one of its aspects, it may be sufficient to mention only one characteristic feature, which is indeed common to most scientific realists accounts of scientific progress. One of the important claims of scientific realists is that scientific theories are true or false, and another is that the true ones really explain the phenomena. This sounds rather trivial but is in fact a disputable thesis.

If two logically conflicting theories T_1 and T_2 are empirically equivalent, such that they predict the same phenomena, one may seriously ask oneself how one of them can be held to be true and the other to be false, since there is, as we have assumed, no way simultaneously to confirm the first and to falsify the second. On the other hand we cannot hold both theories to be true since their conjunction is a logical contradiction. One way out would be to apply the predicates true and false only to observation sentences and not to theories which contain so called theoretical terms. Philosophers who tend to the conclusion that the metalinguistic predicates "true" and "false" do not apply to theories at all are called <u>instrumentalists</u>. These people only talk about empirically adequate or inadequate theories. Idealists, like T. S. Kuhn in his famous essay (1962, 1970), would not even attribute empirical adequacy to theories. They can only characterize them as being more or less successful within a certain disciplinary matrix.

One of the most important arguments of realists against instrumentalists and Kuhnian idealists - and this is their second above mentioned thesis - is that theories could not really be considered to <u>explain</u> phenomena, if they are not taken seriously, i.e. as being true or false. And any theoretical explanation rests on the assumption of the entities the theory is about. If a theory which talks about molecules is neither true nor false then there are no molecules <u>in a certain sense</u>, which I do not want to analyze here. If a theory which says that there are molecules is true, then there are indeed molecules. Thus the <u>realist</u> claims that the entities assumed by a theory explain the described phenomena, and can do this only if they are real. The instrumentalist and the idealist, however, can only let theories successfully account for some phenomena, but cannot consider them as their true explanations, and therefore they cannot believe in the real existence of the underlying entities. Particularly the idealist á la Kuhn cannot assume paradigm-independent existence of any fundamental entity in natural science.

One of the favorite examples of an explanation by the assumption of certain entities is the agreement of more than twenty mutually independent methods for the measurement of Avogadro's number. How could this be possible if matter did not really consist of atoms quite independent of the presently preferred disciplinary matrix? Otherwise we would have to assume a miracle or cosmic coincidence. Therefore realists consider as an important feature of science that it contains existential hypotheses which explain otherwise very improbable phenomena.

So far the characterization of the conflicting standpoints. We may already guess that idealists and realists will have quite different theories of scientific change. While for the idealist science may replace theories or disciplinary matrices like old clothes by new ones, realists will tend to point out that good explanations cannot simply get lost in a scientific revolution. If they are good and striking they must be quite near to the truth and important parts of them have to be retained by the new theory. Above all the underlying entities have to reappear - maybe somewhat modified - in the new theoretical account since the success of the explanation, in which they appear, counts as strong evidence for their existence.

The implications of scientific realism for conceptual change in the growth of science should now be studied in some more detail before we deal with the breakdown of the phlogiston theory which is one remarkable case of discontinuity in the ontological assumptions of a science. The argument of the realist runs as follows: If an otherwise very improbable fact is explained by the existence of certain entities, it is on the other hand very probable that these entities really exist. Therefore the realist claims at least for mature science (assuming that it yields better explanations than immature science):

"(1) Terms in mature science typically refer
(2) The laws of a theory belonging to a mature science are typically true." (Putnam 1978, p. 20)

This auxiliary thesis of contemporary scientific realism, which is called internal realism by Putnam has a certain appeal, since it makes this doctrine more relevant for the present stage in the history of science. Naturally the realist does not want to place the stuff of the world or the real entities at an inaccessible epistemological distance. A variety of realism which talks about real but unattainable and unknown entities is hardly better than any kind of idealism. Therefore it is important for the realist to defend the thesis that there are real things in the world as described by physics and that we know them already in some way. If any change in the theoretical outlook of physics implies a change in the entities involved we would never have the confidence to have found the bottom of reality. Science would be nothing but a series of errors and illusions.

Unfortunately the inference from a good explanation of some phenomena by some entities to the existence of these entities is not generally valid. A hypothesis H explains evidence E, if it makes it much more probable than it would be otherwise. The same could also be done by a rival hypothesis H' which assumes entities of a different kind. We cannot be sure that such a rival hypothesis does not exist.

There are many examples of such pairs of explanatory rival hypotheses in the history of science which have frequently been supposed to undermine the position of internal realism. One of these is the pair phlogiston chemistry versus oxygen chemistry, which was debated in the second half of the eighteenth century, and which I want to discuss here in detail.

One may add Newton's light particles versus light waves, continental bridge hypotheses like the assumption of Atlantis or Gondwana versus Wegener's continental-drift theory, and heat stuff theories versus molecular-motion theories for the explanation of caloric phenomena. Also positive versus negative charge theories of electricity are significant here, and I think that some more examples of that kind could be found if we tried to explore the history of science more systematically.

I want first to deal with the phlogiston example and later draw some general conclusions from it. We shall be led in our analysis by the following question: How is it possible to explain phenomena (to remove their apparent improbability) by erroneously assuming entities which are actually nonexistent? Is there some logical relationship between the true and the incorrect rival hypotheses? The phlogiston example will help us to come nearer to the answer, even if we may still feel unable to find a logical expression for this relation.

The logical kinship between the rival theories will turn out to be the relation of reduction. Phlogiston theory is reducible to modern chemistry. The new theory inherits the information given by the old one and reproduces its explanatory achievements.

We may gain from this discussion also some hints which will bring us nearer to answering the question whether more advanced theories usually talk about the same entities as their predecessors. A reduction of a theory T to another theory T' does not imply automatically that T and T' share common objects they are talking about. Their structural relationship seems to be more important than the common furniture of the scenery they depict. Finally, we shall deal with the issue of commensurability, the question if there is at all a language common between the old and the new theory. In the phlogiston case the answer will be clearly positive.

2. THE PHLOGISTON CASE

In the eighteenth century most chemists believed that there was a chemical element called phlogiston. This element emanates from substances in the process of combustion. When wood is burnt there is always some ash left which is less heavy than the wood was originally, and thus it seems plausible to assume that a component of the wood has gone. The assumption that there is a substance of such a kind had to face more difficulties the more facts became known about combustion, production of gases of different kinds, about chemical reactions with acids, and when chemical reactions were more and more studied quantitatively. Finally the phlogiston theory was given up, and phlogiston was replaced by oxygen which is present where phlogiston was assumed to be lacking, and which is lacking where phlogiston was assumed to be present. (For an excellent presentation see Conant 1948.)

The proponent of the oxygen theory was Lavoisier, who is commonly held to be the Newton of chemistry. Lavoisier not only replaced phlogiston by oxygen. He laid the foundations of modern chemistry. The coincidence of the introduction of new methods and of a new kind of theoretical reasoning in chemistry with the overthrow of the phlogiston theory gave this theory change a more important role in the history of chemistry than might be justified. The concept of phlogiston thus became a symbol of an outdated kind of science and its overthrow a revolutionary act. Lavoisier's wife burnt demonstratively the books of Stahl and other phlogiston theorists (Mason 1974, p. 368). Thus this overthrow became one of the most reported examples of a revolutionary innovation in science,and T. S. Kuhn has seen in the step from phlogiston theory to oxygen theory a gestalt switch which is significant for scientific revolutions. We shall see that the logical structure of this theory change does not justify such a qualification as revolutionary.

Phlogiston theory and oxygen theory can be formulated within the same language. In particular, the observational language for each theory which contains the spatial mass density distributions of the chemical compounds, is completely the same. The theories differ in the numerical values of three matrices which relate chemical compounds to their molecular weights, to the elements contained in them, and the chemical elements to their atomic weights. Using these three matrices one computes from the mass density distributions of the compounds those of the elements. These matrices and functions are the theoretical terms of the theory. One obtains the one triple of matrices from the other by a simple linear transformation of the spatial mass density distributions of the elements, thus there is also made possible a translation of the theoretical terms of one theory into those of the other. There is however one qualification to be made here. A translation of the aforementioned kind may lead to assumptions which look rather odd and artificial. This is however no argument against such a translation in principle.

The possibility of a translation from phlogiston theory into the language of oxygen theory and vice versa makes it questionable whether the existence of phlogiston or the existence of oxygen are characteristic assumptions of the two different theories. Realists and those who want to refute them usually talk as if it would be possible to speak about the existence of theoretical entities quite independently from the framework of certain theories. Thus it is an important point made by realists that whatever an electron is in Lorentz', Bohr's or Heisenberg's theory, all three theories talk about the same kind of objects, namely electrons. We shall see, however, that in the phlogiston-oxygen case it is to a certain extent quite a matter of convention to decide under which circumstances phlogiston or oxygen exists or not. Therefore arguments pro and contra realism cannot be drawn from the discussion of phlogiston, as long as it is not specified in detail what is meant by "existence".

We shall come to these insights by applying logical reconstructions of chemical theories. I am no adherent of an unbridled structuralism. One should use formal methods as little as possible but certainly as much as necessary, and I think they are necessary in our case. We shall see, when we study chemical theories with logical or mathematical methods, that in this, as in many other cases, the logical reconstruction of theories reveals features which cannot be recognized by an informal treatment. (Thus I claim my approach to be more promising than Kitcher's in his careful and thorough 1978 analysis of the phlogiston case.) Phlogiston becomes a problem in current discussion only because no one looks at the mathematical formulation of the theory, and everyone discusses the scientific revolution of chemistry in the 18th century informally and in purely ordinary-language terms. If we avoid this mistake and discuss chemistry in a mathematical form the same problem will not remain and for most applications the mathematical functions of the phlogiston theory may be redefined in terms of those of oxygen theory.

I shall use a quite modern formulation for my purpose, and I shall disregard completely the fact that chemists in the eighteenth century were not able to express their ideas in the same way. I still think that I do not miss the point by my anachronistic account. Furthermore I do not try to do justice to the historical facts. I just want to formulate a phlogiston theory which is the best I can think of and which explains as many chemical facts as possible. After all the problem at issue is not only an historical one but also a problem of logical analysis.

The theory is best reconstructed within a two level scheme consisting of an observational and a theoretical level. The observational level contains the spatial mass density distributions of the substances which are observed in nature, of the chemical compounds. I call these functions σ_k, where k is an index of the substance. The function σ_k has location and time as its arguments. We may combine these functions to a single function σ. The type of σ is given by:

$$\sigma : \mathbb{N} \times D \rightarrow \mathbb{R}$$

The set $\mathbb{N}$ of natural numbers here has the role of the set of substance indices. D is a space-time region in which the chemical processes with which one is dealing take place.

These functions however are not yet sufficient for the

formulation of the chemical theory. We need in addition to them theoretical functions, the spatial mass density distributions of the elements ρ_k. These elements cannot be directly observed, they are hypothetically assumed for the explanation of chemical reactions. In chemical reactions we observe that the initial and the final substance appear always in the same quantitative proportions, and we try to explain this fact by assuming

1. that chemical elements are neither destroyed nor created, their total quantities being constant in time
2. that chemical elements are bound in compounds in fixed proportions
3. that the sum of the densities of the elements equals the sum of the densities of the compounds (the total mass density).
4. that all densities are nonnegative

These four postulates are the axioms of our chemical theory, which is so far completely general and does not yet contain any specific chemical assumptions about separate elements, compounds, and reactions.

We now have to specify the type of the mass density distributions for elements and of some other functions which we need for the formulation of specific chemical facts. If we combine the density distributions ρ_k to one single function ρ, we shall obtain for it the following type:

$$\rho : \mathbb{N} \times D \to \mathbb{R}$$

$\mathbb{N}$ is used here as the set of numbers for the chemical elements. D is the same set as before. For this theoretical function ρ (or for the functions ρ_k) we may formulate the laws of the theory: We still need, however, for this purpose three other functions, which are lawlike terms, the molecular weights m_i, the atomic weights a_k and the numbers $n_{k,i}$ of atoms of different kinds k in the separate molecules of compunds i :

$$m : \mathbb{N} \to \mathbb{R}$$
$$a : \mathbb{N} \to \mathbb{R}$$
$$n : \mathbb{N}^2 \to \mathbb{N}$$

We are now ready for the formulation of the theory of chemistry:

A chemical system is a structure $\langle D, \sigma, \rho, m, a, n\rangle$ for which the axioms A1-a5 hold. These axioms are:

A1 $-\mathrm{div}\,\vec{j}_k = \frac{\partial \rho_k}{\partial t}$; $\vec{j}_k$ is the current density related to ρ_k.

The axiom A1 expresses the preservation of the chemical elements by a continuity equation.

A2 $$\rho_k(\vec{x},t) = \sum_i \frac{a_k}{m_i} n_{k,i}\sigma_i(\vec{x},t)$$

This axiom is the most important one. To understand it, we may consider a simple example. Let there be water with the density σ_{H_2O} and sulphuric acid with the density $\sigma_{H_2SO_4}$ present in a chemical system. If we divide these densities by their molecular weights m_i, we obtain the densities measured in mol/cm^3. We are now looking for the densities of hydrogen and of oxygen in our system. H has two atoms in each water molecule and also in each sulphuric-acid molecule. O has one atom in the former and four in the latter. Therefore the respective coefficients n are:

$$n_{H,H_2O} = 2;\ n_{H,H_2SO_4} = 2;\ n_{O,H_2O} = 1;\ n_{O,H_2SO_4} = 4$$

Multiplying the molecular densities of water and sulphuric acid by these factors and adding them up we obtain the atomic densities of hydrogen and oxygen (measured in the unit mol/cm^3). Multiplying these densities by the atomic weights a_k we finally obtain the mass densities measured in g/cm^3, symbolized by ρ_k.

The next axiom is necessary and sufficient for the derivation of assumption 3. If we sum over the index k in A2, we will obtain

$$\sum_i \rho_k(\bar{x},t) = \sum_i \sigma_i(\bar{x},t)$$

if and only if the following holds:

A3 $$m_i = \sum_k a_k n_{k,i}.$$

It says that the molecular weight of the compound is the sum of the atomic weights involved multiplied by the weight factors $n_{k,i}$. To these laws a very crucial one has to be added. All densities ρ_k/a_k are positive or zero.

A4 For all $i,\vec{x},t$: $\rho_i(\vec{x},t)/a_i \geq 0$

We shall see later that this very law restricts the translatability of oxygen theory and phlogiston theory into each other. Axiom A4 may be satisfied for one of the theories and violated for the other.

We have now to add a specification of the nontheoretical structures of our general theory. Informally it has already been explained that σ is the observable function of the theory. Thus we have:

If $\langle D,\sigma,\rho,m,a,n\rangle$ is a chemical system,
$\langle D,\sigma\rangle$ is a nontheoretical chemical system.

So far the description of the chemical theory has been completed. This theory is indeed quite a primitive fragment of a modern thermodynamical theory of chemical reactions, but historically it was just this fragment which had to be established before one could start to discuss stability conditions as a function of temperature, pressure and the densities of the elements. This does not mean however that chemists like Lavoisier were able to formulate the equations A1 - A4. But even if they did not do that, the equations express just what their theory essentially amounted to.

We have now to try to specify the phlogiston theory T_{Ph} and the oxygen theory T_O. We have first to give a list of the known chemical compounds so that every such substance will get a number. Then in the oxygen theory the coefficients m_i and a_k will have the values which are contained in any textbook of chemistry, and the $n_{k,i}$ may be taken from the usual formulations of the chemical formulas. Thus for $CaCO_3$ (which is lime) we have

$$n_{Ca,CaCO_3} = 1;\ n_{C,CaCO_3} = 1;\ n_{O,CaCO_3} = 3$$

(As above I use here symbols like Ca and $CaCO_3$ as names for the index numbers.) So there is nothing special about T_O, the oxygen theory. The reader who wants to know more about it may read a textbook of chemistry, and apply the prescription just given for writing down the $n_{k,i}$ from the chemical formulas, which he finds in these books. I have more to say about phlogiston theory T_{Ph}, which cannot be read up in modern textbooks, particularly since my phlogiston theory is an artificial one. The version of this theory which I want to treat first is an unsophisticated one insofar as it is still unable to account for the existence of what we now call molecular oxygen. Later I want to discuss briefly also a more advanced version which is compatible with the existence of this important gas. In one respect our theory will be quite sophisticated, however, it will ascribe a negative atomic weight to phlogiston.

We are now going to relate oxygen theory T_O to phlogiston theory T_{Ph}. The symbols $\bar{\sigma}$, $\bar{\rho}$, $\bar{m}$, $\bar{a}$, $\bar{n}$ of theory T_{Ph} are marked by a stroke "–". I have already mentioned that phlogiston theory assumes much phlogiston in compounds which contain little oxygen and vice versa. Thus we may substitute for

	C_2	the compound	C_2Ph_2
for	CO	" "	CPh
for	CO_2	" "	C_2

Thus we may assume that if max_C is the maximum number of oxygen atoms per carbon atom in a carbon oxide, then the number of phlogiston atoms $\bar{n}_{Ph,i}$ in the compound i of carbon is given by

$$\bar{n}_{Ph,i} = max_C - n_{O,i}$$

In general we may determine the number of phlogiston atoms in a molecule by

$$\bar{n}_{Ph,i} = \sum_{k \neq O} n_{k,i}\, max_k - n_{O,i} \qquad \text{(O in this formula means oxygen, not zero!)} \qquad (C1)$$

This is not valid however for the molecules N_2 and O_2, which need to be considered separately. It is important to make clear that this transformation is designed in a way that the axiom A1 will also be valid for phlogiston. A simple calculation shows this, which the reader may try to perform by himself. We shall now deal with two components of air. We assume that air is simply just one homogeneous stuff, namely

Ai_2, consisting in nothing else than the element Ai, called "air". When a candle is burning in a closed vessel, the process of combustion will stop after some time. We explain this in the following way: the candle evaporates while burning phlogiston which becomes bound to the air leading to phlogistized air Ai_4Ph. When the entire air in the vessel is phlogistized, the candle will stop burning. Thus we have

$$\bar{n}_{Ph,Ai_2} = 0;\ \bar{n}_{Ph,Ai_4Ph} = 1;$$
$$\bar{n}_{Ai,Ai_2} = 2;\ \bar{n}_{Ai,Ai_4Ph} = 4 \qquad \text{(C2a)}$$

The values of these coefficients are chosen in such a way that we can account for the fact that air contains in fact 4/5 nitrogen molecules and 1/5 oxygen molecules, and that the candle will stop burning if these O_2 molecules are consumed. The air in the phlogiston theory can absorb as many phlogiston atoms as we today believe that it can emit oxygen atoms. Until now we know only the coefficients $\bar{n}_{Ph,i}$ and $\bar{n}_{Ai,i}$. For all other elements we simply have:

$$\bar{n}_{k,i} = \bar{n}_{k,i};\ \bar{n}_{k,Ai_2} = 0;\ \bar{n}_{k,Ai_3Ph} = 0 \qquad \text{(C2b)}$$

i. e. all other atoms appear in the compounds in the same numbers as usual. From these coefficients we may easily obtain the atomic weights of the phlogiston theory. For all $k \neq$ Ph and $k \neq$ Ai we will have

$$a_k = a_k + \max_k a_O \qquad \text{(C3)}$$

and for Ai and Ph we have $\bar{a}_{Ai} \cong 18; \bar{a}_{Ph} \cong -16$. (C4)

Phlogiston has to have a negative atomic weight. The numbers are chosen in such a way that they will yield the same molecular weights $\bar{m}_i$ and the same densities σ_i as in the oxygen theory, except for the components of atmospheric air. From the preceding considerations we may derive:

$$\bar{\sigma}_{Ai_2} = \frac{36}{8}\sigma_{O_2};\ \bar{\sigma}_{Ai_4Ph} = \sigma_{N_2} - \frac{28}{8}\sigma_{O_2} \qquad \text{(C5)}$$

$$\bar{m}_{Ai_2} = 36;\ \bar{m}_{Ai_4Ph} = 56 \qquad \text{(C6)}$$

All other values of $\bar{\sigma}$ and $\bar{m}$ are the same as for the corresponding σ and m. By the characterization of $\bar{n}_{k,i}$ and $\bar{a}_k$ the phlogiston theory T_{Ph} is completely described.

From this description we may derive the following linear transformations between the densities $\bar{\rho}_k$ of T_{Ph} and ρ_k of T_O:

$$\bar{\rho}_k/\bar{a}_k = \rho_k/a_k \text{ for } k\neq O, \text{ Ph, N, } A_i \qquad (C7)$$

$$\bar{\rho}_{Ph}/\bar{a}_{Ph} = \sum_k \max_k \rho_k/a_k + \tfrac{1}{2}\sigma_{N_2}/m_{N_2} - \rho_O/a_O$$

$$\bar{\rho}_{Ai}/\bar{a}_{Ai} = 2\sigma_{N_2}/m_{N_2}$$

(The index O denotes oxygen.)

$$\bar{\rho}_N/\bar{a}_N = \rho_N/a_N - 2\sigma_{N_2}/m_{N_2}$$

What makes the difference between T_O and T_{Ph}? We obtain one theory from the other simply by a transformation, which is linear in the functions ρ and σ. The transformation is completely consistent with A2. It leaves A1 untouched, if we assume that atmospheric nitrogen N_2 is a completely inert gas like helium or argon. This assumption is justified for the chemical reactions of the eigtheenth century since one needs high pressures and temperatures for reactions with N_2. A3 is not affected either. Only for A4 will there be a difference. If the ρ_k are transformed into the $\bar{\rho}_k$ of the phlogiston theory T_{Ph}, the equations of axiom A4 are transformed into equations of a different kind. Therefore it is not trivial that A4 will hold for T_{Ph} if it holds for T_O. The existence of certain compounds may violate A4 for T_{Ph}. Thus there can be no pure molecular oxygen in some region of space in the theory T_{Ph}. Its density $\bar{\rho}_{Ph}/\bar{a}_{Ph}$ will inevitably become negative. So, as T_{Ph} has been constructed it is compatible with quite a lot of chemical facts. It accounts for the fact that portions of oxides of metals are heavier than the portions of metals from which they are produced, that combustion stops after some time, and many others. But it cannot account for molecular oxygen. As is well known, J. Priestley produced molecular oxygen and thus he refuted the theory T_O. He, however, did not conclude that he had discovered a new element and isolated it as a more or less pure biatomic gas. He rather concluded that there is a purer air than atmospheric air, the latter being partly loaded with phlogiston, and that he had produced pure dephlogistized air. We may easily adapt our theory to Priestley's results if we interpret atmospheric air as a mixture of pure air Ai_2 and loaded air Ai_rPh_s, which is a compound of r Ai-atoms with s Ph-atoms. (We are using r and s as adjustable parameters.) The reaction which yielded oxygen in Priestley's experiment was the decomposition of HgO, mercuric oxide.

$$2HgO \rightarrow 2Hg + O_2$$

We may replace this reaction by the following in the notation of T_{Ph}:

$$2s \cdot Hg + 2 \cdot Ai_rPh_s \rightarrow 2s \cdot HgPh + r \cdot Ai_2$$

That is: loaded air plus mercury yields mercuric oxide plus pure air. This reaction is clearly different from the former since it does not start from mercuric oxide alone but from mercuric oxide plus loaded air. This difference, however, is not so important since Priestley did not experimentally exclude any influence of atmospheric air on the reaction, and we furthermore choose the adjustable coefficients r and s in such a way that already very small portions of loaded or phlogistized air, i.e. such that s >> r, are sufficient to keep the reaction running.

The equations of T_{Ph} are not greatly modified for $T_{Ph,r,s}$; as we may call the modified phlogiston theory. We have now

$$\bar{n}_{Ph,Ai_2} = 0 \qquad \bar{n}_{Ph,Ai_rPh_s} = s$$

$$\bar{n}_{Ai,Ai_2} = 2 \qquad \bar{n}_{Ai,Ai_rPh_s} = r$$

$$\bar{a}_{Ai} = 14 + \frac{s}{r}16; \; \bar{a}_{Ph} = -16$$

$$\bar{\rho}_{Ph}/\bar{a}_{Ph} = \sum_k \max_k \rho_k/a_k + 2\frac{s}{r}\sigma_{N_2}/m_{N_2} - \rho_O/a_O$$

All other equations remain unchanged, and so are the conclusions drawn from the equations for T_{Ph}. It is still impossible for $T_{Ph,r,s}$ that there exists pure oxygen, but if it is mixed up with some nitrogen in the proportion s : r, we may explain this mixture in terms of $T_{Ph,r,s}$ as pure air Ai_2.

We may ask again as we did for T_{Ph}, the first version of the theory, what may happen to the axioms A1 - A4 if we apply the transformation of σ and ρ into $\bar{\sigma}$ and $\bar{\rho}$. Again we observe that only A4 changes its meaning, which states that densities of the elements have to be positive. This axiom is essential for hypotheses about how molecules are composed of atoms of different kinds, since the elements have to be chosen in a way which guarantees positive densities for each of them.

3. IMPLICATIONS FOR THE DISCUSSION OF THEORY CHANGE

One may ask whether atomic densities really have to be positive. In our theory there is no intrinsic reason for this assumption. I have talked several times in the previous pages about atoms or molecules. This talk was not at all necessary for the theory, however, it was just convenient. Therefore it might also be possible to drop axiom A4. On the other hand, it is quite clear that this axiom is a consequence of the idea that matter consists of atoms. A negative number of atoms makes no sense. But are we bound to the atomic hypothesis? And if we do not want to dispense with this hypothesis, we may introduce an assumption like Fermi's hypothesis of an electron sea. We fill up the space with phlogiston. At some places, where comtemporary chemists suggest that there is oxygen, we assume a reduced density of phlogiston. Thus we may, in a more sophisticated theory of phlogiston than I presented here, also account for completely pure molecular oxygen. This last solution seems to be very ad hoc, indeed.

What does follow from these considerations about the existence of phlogiston? The existence of this element is theory dependent. If we accept axiom A4, we have to reject both versions of T_{Ph} at the moment when molecular oxygen is discovered, that is, when a substance is discovered with all the characteristic physical attributes of oxygen which comes from a reaction in which an oxide is reduced and turned into a metal.

But if we accept a Fermi sea of phlogiston, we may still claim that there is such an element. The theory dependence of the existence hypothesis of phlogiston now shows, on the other hand, that nothing is gained by the claim that science talks about the same object while theories change. The theory dependence of phlogiston is neither in favour of the anti-realist nor in favour of the realist. The world is more complicated than some scientific realists want it to be.

What follows for the incommensurability thesis? It is clearly refuted for our example. There is a translation between the concepts of both theories, and as long as certain phenomena are not observed, both theories can explain the data. The assumptions which are implied by axiom A4 for T_O and for T_{Ph} or $T_{Ph,r,s}$ are different of course. But it is quite possible to say in the language of T_O that A4 for T_{Ph} or $T_{Ph,r,s}$ is simply false.

We may, however, still assume, that there is a gestalt switch leading form T_{Ph} to T_O. But this is essentially a psychological process. If we are able suddenly to interpret a vast amount of data which looked quite confusing before, this need not at all be due to the presence of a new language or symbolism applied. It may rather be the case that we suddenly understand the data by application of the same language which we have used already for a long time. This is what usually happens when Sherlock Holmes solves a criminal case. He does not apply a new theory about human nature. He simply tries to find an order in contingent facts by application of his theory of human behaviour.

In the language of our reconstruction we may thus conclude: Phlogiston theorists and oxygen theorists do not differ in their theories but only by the assumed values for their theoretical functions n, a and ρ.

This result is neither in favour of the idealists nor in favour of the realists. Against the idealists it implies commensurability, continuity of language in a theory change, and against the realists it states that a change of assumed fundamental entities of nature may be conventional, in such a way that these entities cannot be claimed to explain anything.

4. THE REDUCTION OF PHLOGISTON THEORY TO OXYGEN THEORY

As already mentioned, phlogiston theory is reduced to oxygen theory. I want to construct the reduction relation between both theories by applying Adams' logical reconstruction of reduction (E.W.Adams 1959, p. 260 ff.), which has to be slightly modified for this purpose. I use Stegmüller's terminology. The symbol "I" denotes the set of intended applications (W. Stegmüller 1973, p. 134 ff.), "M" the model set of the theory and "M_p" the set of "possible models". Assume there is a correspondence relation R the converse of which, $\breve{R}$, is a function mapping the possible models $x' \in M'_p$ of the reducing theory to the possible models $x \in M_p$ of the reduced theory, ie.

$$R \subseteq M_p \times M'_p \quad ; \quad R : M'_p \rightarrow M_p$$

Using the abbreviation

$$R|A = \{x | \bigvee y(R(x,y) \wedge y \varepsilon A)\}$$

we may express Adams' reduction relation as follows:
R reduces the theory $T = \langle M_p, M, I \rangle$ to $T' = \langle M'_p, M', I' \rangle$ iff

$$I \subseteq R|I' \quad \text{and} \quad R|M' \subseteq M$$

Unfortunately there are very few instances of this kind of relation in science. In most cases the reducing theory describes some processes which are forbidden by the laws of the reduced theory but are not intended applications of it either. Thus wave optics describes dispersion phenomena which are forbidden by geometrical optics. Therefore we may consider geometrical optics as a theory for devices of large dimensions, and thus light deflections in narrow slits and grids no longer belong to the set of intended applications of geometrical optics. We may reduce geometrical optics as a theory of large-scale phenomena to wave optics if we introduce a restriction set E (applying an idea of G. Ludwig, 1972, p. 78 ff., in a somewhat modified way):

$$x \varepsilon E \leftrightarrow x \text{ is a large scale optical process}$$

and replace Adams' reduction relation by the following one:

$$I \cap E \subseteq R|I' \quad \text{and} \quad E \cap R|M' \subseteq M$$

Thus we may say that geometrical optics is only reduced to wave optics after it is restricted to large scale phenomena, and that the restriction set E contains only the possible models representing large scale devices. For any physicist it will be immediately clear that in nearly all interesting cases of reduction we have to deal with a restriction set E. If we reduce Newton's mechanics to Einstein's, E will contain only the possible models for slow mechanical processes. If we reduce ideal gas theory to van der Waal's theory of gases, E will contain only gases of high temperature and low density. The list of examples for restriction sets E may be prolonged nearly indefinitely.

I do not claim that relation R3 is already a completely

satisfactory reconstruction of reduction in physics. It does not account for the fact that in most cases we can only obtain an approximation of the reduced theory by logical derivation. In the case of phlogiston theory, however, no such approximation is involved. Therefore relation R3 is sufficient for our present purpose.

The restricted reduction relation (R3) now has to be applied to the phlogiston- oxygen case. For the sake of simplicity, only the simple phlogiston theory T_{Ph} and not the sophisticated theory $T_{Ph,r,s}$ should be considered. For $T_{Ph,r,s}$ the situation is a bit more complicated but not much different in principle. We have to construct the correspondence relation R and the restriction set E for our two theories.

The sets of possible models are: $M_p = \langle \bar{D},\bar{\sigma},\bar{\rho},\bar{m},\bar{a},\bar{n}\rangle$

$$M'_p = \langle D,\sigma,\rho,m,a,n\rangle$$

$\breve{R}$ is a function: $\breve{R}$: $M'_p \rightarrow M_p$.

If $\langle\langle \bar{D},\bar{\sigma},\bar{\rho},\bar{m},\bar{a},\bar{n}\rangle, \langle D,\sigma,\rho,m,a,n\rangle\rangle \varepsilon R$ then $\bar{D} = D$, $\bar{m} = m$, $\bar{\sigma} = \sigma$ except for the indices N_2, O_2, and Ai_2, Ai_4O, for which C5 and C6 hold (see section 2). The matrix $\bar{n}$ is related to n by C1 and C2, the vector $\bar{a}$ is related to a by C3 and C4, the density function vector $\bar{\rho}$ is related to ρ by C7 (see again section 2).

Next we have to specify the restriction set E. It contains all possible models $x \varepsilon M_p$ except for those which describe processes

1. in which $\sigma_{N_2} \geq 4\sigma_{O_2}$
2. in which atmospheric nitrogen reacts with other chemical substances

If one applies these two terms R and E to the theories T_O and T_{Ph} one may easily verify that the reduction relation which was defined by (R1) and (R3) holds between them.

The fact that phlogiston theory is reducible at least by a restricted reduction to T_O is much more important than any possible claims about the entities underlying these theories and their persistence in theory changes. Why this is so we have to discuss in the last section.

5. THE STRUCTURAL RELATIONSHIP BETWEEN RIVAL EXISTENTIAL HYPOTHESES

What can we learn from the phlogiston case for the general problem of theory change and the persistence of fundamental entities? A general reflexion on the concept of explanation has already shown that to any explanation by an existential hypothesis there may be a rival.

We may understand better the relation between two competing explanations if we apply the Bayes formula to them. An explanation of evidence E by hypothesis H can be understood in terms of personal probabilities as follows ($p(E)$ being the probability for the expectation of E):

H is a good explanation for E
<u>iff</u> H and E are true and
$p(E/H) >> p(E)$

This explication of "weak statistical explanation" may be too broad for our purpose, but nonetheless sufficient. Can we conclude that a theory which, if it were true would explain E very well, is made probable by E?

We may apply the Bayes formula. Let H_i be a set of possible hypotheses, and E the empirical evidence. Then we have

$$p(H_k/E) = \frac{p(H_k)\cdot p(E/H_k)}{\sum_i p(H_i)\cdot p(E/H_i)}$$

$p(H_k/E)$ on the left side of the equation is the probability of H_k given E, $p(E/H_k)$ the known probability of E given H_k, and $p(H_k)$ is the apriori probability of H_k. If we look at the formula, we see that much depends on the a priori probability $p(H_k)$. If the a priori probability of H is very low, we have not gained much by a good explanation of E by H. Let us look at an example:

In the desert of Peru one has found gigantic drawings on the ground, which may very well be viewed from a plane or a helicopter, and which could easily been explained as signs addressed to extraterrestrial intelligent beings. The appearance of these drawings would be much less improbable if they had indeed this function. This does not, however, make it very likely that extraterrestrial intelligent beings exist. The reason is the low value of $p(H)$, which yields a low value of $p(H/E)$ even if $p(E/H)$ is not much smaller than 1.

We face another situation if there are two rival hypotheses H and H' with equal a priori probability, which explain E equally well, i. e. which yield p(E/H) = p(E/H'). In such a case also p(H/E) and p(H'/E) are equal. From this observation we may conclude that only if H has a much higher a priori probability than its rival - it may be the much simpler theory for example -, and if H explains E much better, can we conclude that p(H/E) is comparable to 1. This is, however, no guarantee that we can be practically sure that H is true. The function p(H/E) is a conditional probability and if new evidence E' is found which cannot be explained by H, the value of $p(H/E \wedge E')$ may be much lower and our wonderful theory H may dwindle away, and with it its underlying entities.

This happened for example to the hypothetical primeval continental bridges, Atlantis between Europe and America, and Gondwana between India and Madagascar.

These now-submerged continents were assumed by paleontologists to explain the otherwise very improbable striking similarities between the fauna and flora of countries separated by large oceans.

There is however an alternative hypothesis now generally preferred namely Alfred Wegener's continental drift hypothesis. If South America has drifted away from Africa and South India from Madagascar or vice versa, whilst these areas were attached in former paleontological epochs, the relationship between the plants and animals in them may be equally well accounted for.

The continental drift hypothesis H_W is today preferred for two reasons: 1. $p(H_W)$ has been enhanced by physical considerations. A continental drift does not seem impossible any more. 2. New empirical facts E' have been discovered which are explained by H_W but not by H_B, the continental bridge hypothesis. Thus $p(H_W/E \wedge E')$ is now much larger than $p(H_B/E \wedge E')$.

So far this all seems to be clear from conventional considerations in terms of inductive logic. There is no reason to assume that mankind has already found the true and final furniture of nature. What happened to continental bridges may also happen to the presently assumed elementary particles.

But the examples considered in this paper, above all the abandonment of phlogiston, show that this is not all what has to be discussed if one deals with conflicting existential hypotheses. In most of these cases there is a structural relationship between them. In any of the considered cases a certain pattern which appears in reality has to be explained: the distribution of light in optical instruments, of electrical charge or heat in rigid bodies, of chemical substances in space, of animal species in a country. Such a pattern can equally well be explained by a mould as by a model from which it is a copy. We may equally well assume positively or negatively charged particles, phlogiston or oxygen, positrons or holes in a Fermi sea of electrons, to explain the same phenomena. So there exists in all these cases a map or translation which tranforms the state descriptions of one theory into those of the other. This map constitutes a reduction relation for the old and the new theory as we have seen in section 4 for the phlogiston case. Such a map exists also for the particle and the wave theory of light, even if their state descriptions are not in the mould-model relation. Thus we see that what are retained in theory change are not in the first place the fundamental entities, the furniture of nature,but rather their structure. The things in the world are only the medium or the channel of the information which contains the structure. Information may be realized in a different way and the medium or language of realization is less important for explanations than the message itself. Therefore it is the information or structure that is inherited in theory change, not the stuff in which it is imprinted. The continental brigde versus continental drift example is a bit different from those of physics and chemistry. But also in this case it is not the structure - which is the species' distribution in geological epochs - which is exchanged, but rather the channel for its spreading.

My last considerations may be rather poetic or "philosophical" for a hard-boiled structuralist, and should only be considered the first step to an investigation leading to a sober logical reconstruction of reduction which embraces also the inductive relations between the older and the more advanced theory. The difficulties in such an enterprise should not lead to a denial of the explicandum.

Andreas Kamlah
Fachbereich: Kultur- und Geowissenschaften
der Universität Osnabrück
Postfach 4469, D-4500 Osnabrück

REFERENCES

Adams, E. W.: 1959, 'The Foundations of Rigid Body Mechanics and the Derivation of its Laws from those of Particle Mechanics', in L. Henkin, P. Suppes, A. Tarski (eds.): The Axiomatic Method, Amsterdam: North Holland Publ. Comp.
Conant, J. B.: 1948, 'The Overthrow of the Phlogiston Theory: The Chemical Revolution of 1777 - 1789', in J.B.Conant, L.K.Nash (eds.): Harvard Studies in Experimental Science, vol. 1, Cambridge (Mass.): Harvard Univ. Pr.
Kitcher, P.: 1978, 'Theories, Theorists and Theoretical Change', The Philosophical Review, 87, 519 - 547.
Kuhn, T. S.: 1962, The Structure of Scientific Revolutions, Chicago/London: Univ. of Chicago Pr.
Kuhn, T. S.: 1970, 2nd edition of 1962, with a "Postscript".
Mason, S. F.: 1974, Geschichte der Naturwissenschaft, (2. Aufl.), Stuttgart: Kröner.
Stegmüller, W.: 1973, Probleme und Resultate der Wissenschaftstheorie und analytischen Philosophie, vol 2, Theorie und Erfahrung, zweiter Halbband, Berlin/Heidelberg/New York: Springer.

Theo A.F. Kuipers

UTILISTIC REDUCTION IN SOCIOLOGY: THE CASE OF COLLECTIVE GOODS

1. INTRODUCTION

One aim of methodological individualism in sociology is to explain social regularities on the basis of some regularity in the behaviour of individuals. An additional aim may be to explain such an individual regularity in turn on the basis of the assumption that the individuals intend to maximize expected utility, or at least that they behave as if they intend to do so. Explanations of social regularities exemplifying both features will here be called utilistic reduction.

In this paper we shall present a detailed account of one such utilistic reduction, viz. the explanation of Mancur Olson's hypothesis 'the larger the group, the farther it will fall short of providing an optimal ammount of a collective good' (Olson (1965), p.35)[1].

Olson's own explanation, indicated in the title of his book 'The Logic of Collective Action', has already been streamlined and simplified by S. Lindenberg (1982). However, Lindenberg's account still has many shortcomings, as we have pointed out extensively in our (1983c). Compared with the expositions of Olson and Lindenberg, our account in (1983c) brings to the fore a number of hidden elements and necessary refinements, and it is shown that the resulting explanation has exactly the structure of so-called heterogeneous reduction.

In this paper we will report the main findings of our (1983c); but in several respects the analysis will also be improved.

Section 2 spells out the three fundamental steps of the reduction of Olson's hypothesis.

In Section 3 we will discuss three additional issues, of increasing generality. First, we will show that the expectations of the individuals with respect to the transformation process of collective behaviour into social effects are largely correct, and hence rational, in the light of the objective transformation rules. Nevertheless - and this is well-known in the 'Olson-literature'- the utilistic be-

W. Balzer et al. (eds.), Reduction in Science, 239–267.

haviour of the individuals based on these expectations leads to rather irrational consequences. Second, the treated utilistic reduction is essentially an as-if explanation. We will argue that a new explication of intentional explanations, presented in our (1983b), provides the adequate means to transform any as-if-utilistic reduction into a genuine intentional-utilistic reduction. Third, and finally, in any type of reduction in sociology (i.e. utilistic or not) transformation rules will be necessary. We will discuss a number of types of such rules, and possibilities to explain them.

In the remainder of this section we shall clarify and illustrate the notion of heterogeneous reduction.

When we talk in this paper about reduction we mean throughout deductive micro-reduction, as opposed to something like approximative reduction or to forms of reduction not involving different ontological levels. In the light of these restrictions a theory (or law, or set of laws) T' is said to be reduced to T if and only if 1) T is of a 'lower' or 'more fundamental' ontological level than T' and 2) T' is derivable from T, possibly with the aid of transformation rules relating some concepts of the two theories. We speak of heterogeneous reduction if (non-analytic) transformation rules are necessary to accomplish the deduction, otherwise we speak of homogeneous reduction [2].

Homogeneous reduction turns out to be decomposable into two steps. In the first step, the individual step, some individual law is derived on the basis of the reducing theory and special assumptions. An individual law is here understood as any uniform feature in the behaviour of the individuals of the reducing theory. In the second homogeneous step, the aggregation step, this individual law is aggregated in some way or other, for instance, with or without using statistical assumptions. This aggregation leads to some aggregated law, which completes a homogeneous reduction.

In the case of heterogeneous reduction we also start with these two homogeneous steps. In an additional, third step, the transformation step, transformation rules are applied to the aggregated law, in order to derive the law intended to be reduced.

We will illustrate heterogeneous reduction, and hence also homogeneous reduction, by a global sketch of the kinetic reduction of the ideal gas law, i.e. the reduction of the ideal gas law (PV = RT) to the kinetic theory of gases.

Individual step. From the kinetic gas theory, in particular Newton's laws of motion, and the assumption that the molecules collide elastically with the wall, one derives the individual law that the momentum-exchange m_e in a collision between a molecule and the wall is given by:

$$m_e = 2mv_w$$

where m is the mass of the molecule and v_w the velocity of the molecule in the direction perpendicular to the wall.

Aggregation step[3]. Through aggregation, using some statistical assumptions, we can derive from this individual law the 'perfect gas law' (for one mole of gas):

$$\text{PGL} \quad PV = (2/3)N\bar{u}$$

Here V indicates the volume of the gas, P its pressure, $\bar{u}$ the mean kinetic energy of the molecules and N Avogadro's number of molecules in a mole. With some qualifications, which are irrelevant here, we can say that PGL uses only 'aggregated' quantities and hence is an aggregated law.

Transformation step. The usual form of the transformation rule is:

$$\bar{u} = (3/2)kT = (3/2)(R/N)T$$

where T indicates the (absolute) temperature of the gas, k Boltzmann's constant (= R/N) and R the ideal gas constant. It is easy to check that with this rule it is possible to derive from PGL the ideal gas law:

$$\text{IGL} \quad PV = RT$$

In Kuipers (1982) we have given a detailed analysis of the transformation step [4], showing that 1) the realistic restriction to 'asymptotic' behaviour can be made, 2) the apparent symmetry (PGL can also be derived from IGL and the transformation rule) can be removed, 3) an additional transformation rule is required for pressure, and, finally, 4) both transformation rules can be formulated in such a way that they express so-called ontological identities.

2. THE REDUCTION OF OLSON'S HYPOTHESIS

2.1. Preparations

Let us now turn to Olson's hypothesis. First we need to make a number of terminological remarks. Point of departure are groups of people defined in some way. Somebody belonging to a group is called a member of that group. A good is called a collective good for a group if all members of the group will benefit from it, whether or not one participates in the pursuit of that good. One very common kind of participation is to be a (contribution paying) member of the organization (e.g. a labour union) pursuing the collective good. However, we will speak of participants without any specific form of participation in the pursuit of the collective good in mind. The number of participants divided by the total number of members of the group (the size of the group) will be called the degree of participation.

If the collective good is not a matter of degree but an 'all or nothing' object (e.g. a bridge), the crucial notion will be the chance that the collective good will be realized, the chance of realization for short. On the other hand, if the collective good is something that can be realized in principle at any level, the crucial notion will be the level at which the collective good is realized, the level of realization. In the first case we will speak of the chance-variant, in the second case of the level-variant. Mixed variants are of course also possible. Due to the unclear notion of 'chance' in the present context, the chance-variant is more vague than the level-variant. Nevertheless, we will present the analysis in the chance-variant.

Now we are in a position to reformulate Olson's hypothesis as follows:

OH The larger the group the smaller the chance of realization.

The basic idea of the whole reduction is the decomposition of OH into the following two hypotheses:

OH-I The larger the group the lower the degree of participation.

OH-II The lower the degree of participation the smaller the chance of realization.

In the homogeneous part of the reduction, OH-I is then derived as an aggregated regularity on the basis of an individual regularity. This individual regularity says roughly that an individual stops participating if 'his' group reaches a certain size, and this is explained as the consequence of maximizing expected utility.

In the heterogeneous step OH is derived from OH-I with OH-II as a transformation rule, transforming 'degree of participation' into 'chance of realization'.
Although the aggregation and transformation step may sound rather trivial in this sketch, it will turn out that all three steps require careful treatment in order to get a rigorous deductive account.

2.2. Individual step

The individual regularity to be derived in the first step will be explained by a number of hypotheses about the considerations individuals might use in choosing between participation and non-participation. Although the presentation will suggest that the individuals are consciously deliberating, this is not essential. That is, our aim and claim in this section is only an 'as-if explanation'. As announced in the introduction, we will discuss how to transform such an as-if-explanation in Section 3.

Since any set of individuals could constitute, in principle, a real group, having or not (yet) having a collective good, we will call every set of individuals a group. In many cases, however, it will be clear that the following implicit assumption is made:
'suppose that this set constitutes a real group'.

To begin with, the following abbreviations should be interpreted in this light:

X	a (fixed type of) collective good
I	the universe of all individuals (under consideration)
i	a variable for individuals, i.e. $i \in I$
G	a variable for sets of individuals, groups, i.e. $G \subseteq I$
P(i,G)	i participates in the pursuit of X for G
N(i,G)	not-P(i,G)
P(G)	$\{i \in G \ / \ P(i,G)\}$, the subset of participants of G
N(G)	$\{i \in G \ / \ N(i,G)\}$ = G-P(G)
n(G)=n	the size of G
m(G)=m	the size of P(G), $o \leq m \leq n$
m/n	the degree of participation in G
EU-A(i)	i's (subjective) expected utility of (i's) action A

As a general hypothesis we will assume the classical principle of utility theory: individuals maximize their expected utility in the choice between participation and non-participation (the P/N-choice, for short):

MH <u>Maximization hypothesis</u>
For all G and all i in G
P(i,G) iff EU-P(i,G) $\geq$ EU-N(i,G)

The equation in MH will be called the <u>participation equation</u>. Note that MH contains the harmless but convenient assumption that individuals participate when their expected utilities of the two actions are equal.

To be able to apply MH we have of course to introduce <u>specification hypotheses</u> for the utilities and expectations of the individuals. We start with the following

<u>Utility hypotheses</u>

U1 For all G and all i in G
i's utility of X for G is positive and independent of G; it is represented by $U_X(i)$.

U2 For all G and all i in G
i's subjective costs, i.e. negative utility, of participation in the pursuit of X for G are positive and independent of G; it is represented by $C_P(i)$.

Of course we assume that no other utility than those specified in U1/2 (i.e. U1 and U2) are involved. Note that it is not difficult to think of situations where one or more aspects of U1/2 do not seem very reasonable. The most delicate assumption is surely contained in U2, viz. the fact that the subjective costs of participation are assumed to be independent of the group and hence of the size of the group. It may well be, for example, that the (financial) contribution decreases with increasing size, and this is likely to decrease the subjective costs as well.

Turning to the expectations of the individuals we will first formulate the relevant hypotheses for an arbitrary individual i. These hypotheses will also be called the <u>subjective transformation rules</u>. Lateron we will assume, in addition, <u>objective transformation rules</u>, i.e. rules factually governing the transformation process of collective behaviour into social phenomena. We start with:

E1 $R_i = R_i(n,m)$
i.e. i's estimate R_i of the chance of realization of X for G depends only on the size of G and P(G), i.e. n and m, and so we can denote it by $R_i(n,m)$.

Let $b_i(n,m)$ denote $R_i(n,m+1) - R_i(n,m)$, $m < n$. For obvious reasons, $b_i(n,m)$ will be called the <u>marginal effect</u> according to i. Our next assumption is:

E* $b_i(n,m) = b_i(n)$
i.e. $b_i(n,m)$ does not depend on m, and hence can be represented as $b_i(n)$.

For convenience we introduce the conjunction

E*1 E1 and E*

The two remaining assumptions are dependent on E*:

E*2 $b_i(n) > o$
i.e. the marginal effect, according to i, is always positive

E*3 $b_i(n) > b_i(n+1)$
i.e. the marginal effect decreases with increasing size of the group according to i

Of these assumptions, E*2 is quite plausible and E*3 will turn out to be crucial.

We combine and generalize these assumptions for all individuals:

E*H <u>Expectation hypothesis</u>
For all i, R_i satisfies E*1/2/3, i.e. E*1, E*2 and E*3.

In order to formulate and prove the intended individual regularity we have to define a technical notion.

<u>Def. 1</u> i has <u>switch (group) size</u> S(i) if
for all G, for which i in G, holds
$P(i,G)$ iff $n(G) < S(i)$
(and hence $N(i,G)$ iff $n(G) \geq S(i)$)

Now it is possible to prove the following theorem:

<u>Th.1</u> MH, U1/2, E*H imply together:
SL <u>Switch Law</u>: all i have a (specific) switch group size, i.e. for all i there is a natural number S(i) such that S(i) is the switch group size of i.

Proof An individual i in G is supposed to argue as follows according to MH, U1/2 and E1: if there are z participants, besides myself, then the utilities I can expect are respectively

$EU\text{-}P(i,G) = R_i(n,z+1).\ U_X(i) - C_P(i)$,

$EU\text{-}N(i,G) = R_i(n,z).\ U_X(i)$

The resulting participation equation reduces, on the basis of E*, to

$b_i(n) \geq C_P(i)/U_X(i)$,

in which the hypothetical number z has disappeared [5]. According to U1/2 and E*2 both sides of this equation are positive. Moreover, the quotient on the right side is positive. But, according to E*3, $b_i(n)$ decreases with increasing n. Hence, there will exist, as a rule, a (finite) switch group size, viz. the smallest number n for which $b_i(n) < C_P(i)/U_X(i)$. For the case that $b_i(n)$ does not 'pass' the quotient, despite E*3, we define of course $S(i) = \infty$.

Q.e.d.

It is interesting to note that the notion of the switch size is a purely behavioural concept, i.e. it does not refer to utilities and expectations. Consequently, the switch law SL is a purely behavioural regularity, i.e. a regularity in individual behaviour. In what follows it will also become clear that the role of the utilistic hypotheses (MH, U1/2, E*H) in the utilistic reduction of Olson's hypothesis is restricted to the presented explanation of SL. These facts have two interesting consequences. 1) If a rival explanation of SL can be given, on the basis of different utilistic hypotheses or on the basis of non-utilistic hypotheses, the rest of the story could remain the same. 2) As we have said before, we only claim that the preceding story gives at least an as-if- explanation. To transform the whole resulting as-if-utilistic reduction into a genuine intentional-utilistic reduction it will be sufficient to transform the given as-if explanation of SL into an intentional explanation of SL.

2.3. Aggregation step

Although OH-I may already sound plausible in the light of the switch law, an exact derivation of it is still problematic. For, consider the following more explicit formulation of OH-I (leaving out here and in what follows the obvious quantifiers):

OH-I If G_2 is larger than $G_1 (n_2 > n_1)$ then the degree of participation in G_2 is lower than that in $G_1 (m_2/n_2 < m_1/n_1)$.

Now, SL does not allow the derivation of OH-I, for, on the one hand, OH-I is, as it stands, a claim about any two groups of different size, whereas, on the other hand, SL does not imply any restriction on the switch sizes of different individuals.

Consequently, in order to derive 'something like OH-I' from SL, the derivation should deal, in some way or another, with the _same_ individuals. If we succeed in this, this 'something like OH-I' may well be called an _explication_ of OH-I, for it will be a revision of OH-I which captures, in a precise way, what we can defensibly mean by, and what is nevertheless 'close to', the initial formulation.

For these purposes we introduce the notions of _fusion_ and _division_ of groups, again hinting at, but formally not requiring, real groups.

Def. 2 G_3 is a _fusion_ of G_1 and G_2 ($G_3 = G_1 \oplus G_2$) if $G_1 \cap G_2 = \emptyset$ and $G_1 \cup G_2 = G_3$. G_1 and G_2 are then said to constitute a _division_ of G_3.

Now, if G_3 is a fusion of G_1 and G_2, we can argue, on the basis of the switch law, as follows. There may be individuals who were participants in G_1 or G_2, for whom the larger size of G_3 reaches (and probably passes) their switch size. Hence, these individuals become non-participants in G_3. However, the converse cannot occur, i.e. SL excludes that there may have been non-participants in G_1 or G_2 who become participants in G_3.

In sum, it is possible to prove

_Th.2_ SL implies:

LF _Law of Fusion_: if $G_3 = G_1 \oplus G_2$ then $P(G_3) \subseteq P(G_1) \cup P(G_2)$

One might claim that LF itself is already an explication of OH-I. But note that LF is a purely qualitative statement, whereas OH-I is quantitative, though only comparative. It turns out that a (mathematical) consequence of LF, hence also a consequence of SL, is in this respect closer to OH-I, viz.

Oh-Ie If $G_3 = G_1 \oplus G_2$ then $m_3 \leq m_1 + m_2$

That this is 'something like OH-I' becomes more clear from an equivalent 'mean-value-formulation' (the equivalence being due to the analytical fact that $n_3 = n_1 + n_2$ if $G_3 = G_1 \oplus G_2$), viz.

$$\text{If } G_3 = G_1 \oplus G_2 \text{ then } \frac{m_3}{n_3} \leq \frac{n_1}{n_1+n_2} \cdot \frac{m_1}{n_1} + \frac{n_2}{n_1+n_2} \cdot \frac{m_2}{n_2}$$

which reads: the degree of participation in a fused group is not higher than the weighted mean value of the degrees of participation in the fusing groups. Of course, the weakening of 'lower than' to 'not higher than' is the price of soundness.

Although the mean-value-formulation 'sounds' more like OH-I than the 'number-formulation', the former is just a clarifying alternative to the latter. However, for the explication of OH instead, we will be forced into such a mean-value-formulation (in terms of chances of realization), because there an equivalent number-formulation will not be available.

It will of course depend on the nature of groups and of the collective good whether the idea of fusion and division is purely theoretical or has some practical applications. If a group is defined as the set of all those individuals in a society who share interest in the collective good under consideration, the composition of this group only changes through 'natural' mutations. Hence, in this case of unique groups, fusion and division are purely hypothetical. If, however, groups are defined in another way, i.e. independent of the collective good, as for instance families, or, more general, communities and if the collective good is only related to each group seperately, then actual fusion and division becomes possible. Hence, for such non-unique groups experimental testing of the law of fusion and hence of OH-Ie is possible.

Given the structure of LF and OH-Ie it is reasonable to call them regularities in collective behaviour, or collective regularities for short. Given the way in which they could be derived from SL it is also reasonable to call them aggregated regularities. Moreover, it is also clear that the joint derivation of OH-Ie in the individual and the aggregation step satisfies all requirements for homogeneous reduction as described in the introduction.

2.4. Transformation step

The last step concerns the transformation of collective

behaviour into social phenomena. In the individual step we made assumptions about the expectations individuals had concerning this transformation. In contrast to the there postulated subjective transformation rules, E*1/2/3, our concern here are objective transformation rules.

Let us first restate the verbal version of the main transformation rule:

OH-II The lower the degree of participation the smaller the chance of realization.

The following rule seems to be more or less implicit in OH-II and corresponds to E.1:

TR1 $R = R(n,m)$
i.e. R, the objective chance of realization of X for G, depends only on the size of G and P(G), i.e. n and m, and hence can be represented as R(n,m).

The objective marginal effect of participation b(n,m) is of course defined as $R(n,m+1) - R(n,m)$, $m < n$. Although we will not assume that the marginal effect depends only on n, we will assume that it is always positive (compare E*2):

TR2 $b(n,m) > o$

It is easy to check that TR2 is equivalent to

TR'2 if $o \leq m < m' \leq n$ then $R(n,m) < R(n,m')$

and this is rather close to a formal version of OH-II, assuming TR1. Hence, let us take TR1 and TR2 as a provisional explication of OH-II and let us see how far we can come, on the basis of them and OH-Ie, in deriving OH:

OH The larger the group the smaller the chance of realization.

We refer here to the explication OH-Ie of OH-I in fusion terms, which we had to design in order to be able to derive it from the switch law, for we want to retain of course this partial explanation in the inclusive explanation of OH.

To achieve this, it is clear that we will have to explicate OH in fusion terms as well. In the case of OH-I there were two equivalent formulations, a number- and a mean-value-formulation. However, since the chance of realization is not a matter of natural numbers, we are now forced into a mean-value-formulation:

OHe If $G_3 = G_1 \oplus G_2$ then

$$R(n_3,m_3) \leq \frac{n_1}{n_1+n_2} \cdot R(n_1,m_1) + \frac{n_2}{n_1+n_2} \cdot R(n_2,m_2)$$

Note that TR1 is used in the transition from OH to OHe.

It is easy to see that OH-Ie and TR1/2 do not yet imply OHe, but only:

If $G_3 = G_1 \oplus G_2$ then $R(n_3,m_3) \leq R(n_1+n_2,m_1+m_2)$

To derive OHe from this we need an additional transformation rule, of which the formulation in fusion terms reads:

TR3 $R(n_1+n_2,m_1+m_2) \leq \frac{n_1}{n_1+n_2} \cdot R(n_1,m_1) + \frac{n_2}{n_1+n_2} \cdot R(n_2,m_2)$

Putting

OH-IIe = TR1 & TR2 & TR3 = TR1/2/3

we have indicated the proof of

Th.3 OH-Ie and OH-IIe imply OHe

It has not been noted by Olson and Lindenberg that something like TR3 is a hidden assumption in the claim that (something like) OH can be explained.

It is important to realize that the need for TR3 has nothing to do with the fusion-formulations of OH-I and OH: the fusion-formulations of the latter two forces only a fusion-version of a straightforward idea, viz. that the chance of realization may not increase with increasing size of the group, when the degree of participation remains constant (despite OH-I!), or formally:

If $n < n'$ and $m/n = m'/n'$ then $R(n',m') \leq R(n,m)$.

In our (1983c), Section 3.5, we have illustrated the relevance of TR3 by presenting a possible counterexample to it, viz. in terms of fixed costs for the realization of a collective good for a group, i.e. costs which are independent of the size of the group. If there are such fixed costs, they need to be payed twice for two groups, but only once for a fused group. This creates the possibility that the chances for the collective good increase with fusion despite a possible decline of the degree of participation.

Although we have already hinted a little at the relation between the (objective) transformation rules TR1/2/3 and the subjective transformation rules E*1/2/3, the precise relation will be studied in Section 3.1.

In Section 3.3, where we will discuss types of (objective) transformation rules in general, we will also briefly comment on the nature of TR1/2/3. But it will already be clear by now that they are non-analytic, and hence that the transformation step is a truly heterogeneous step, from which it follows that the total explanation of OH (i.e. OHe) is indeed an example of heterogeneous (utilistic) reduction, as claimed in the introduction.

2.5. Survey

The deductive pattern of the treated example can be represented by

$$\text{TS}\ \frac{\text{AS}\ \dfrac{\text{IS}\ \dfrac{\text{MH} + \text{U1/2} + \text{E*H}}{\text{SL}}}{\text{OH-Ie}} \qquad \text{OH-IIe (=TR1/2/3)}}{\text{OHe}}$$

where IS stands for individual step, AS for aggregation step and TS for transformation step.

Before we lay down the general structure of utilistic reduction we will first state a division of regularities (or laws, or rules) which we have already suggested from time to time. Of course the formulations should be interpreted with careful flexibility.

Individual regularities or, more precisely, *regularities in individual behaviour* (such as the switch law): those and those circumstances lead to that and that individual behaviour.

Collective regularities or, more precisely, *regularities in collective behaviour* (such as OH-I): those and those circumstances lead to that and that collective behaviour.

Transformation rules (such as OH-II): that and that collective behaviour leads to that and that social phenomenon.

Social regularities (such as OH): those and those circumstances lead to that and that social phenomenon.

In these general terms we get as a general pattern:

UTILISTIC REDUCTION OF A SOCIAL REGULARITY		
IS	maximization hypothesis	specification hypotheses
AS	individual regularity	
TS	collective regularity	transformation rules
	social regularity	

If all transformation rules used in a utilistic reduction are analytic it is a case of homogeneous reduction. As soon as there is at least one empirical transformation rule it is a heterogeneous reduction.

3. RATIONALITY, INTENTIONAL INTERPRETATION AND TYPES OF TRANSFORMATION RULES

In this section we will deal with three topics of increasing generality. First, we will evaluate the rationality of the individualsin the case of the utilistic reduction of Olson's hypothesis concerning collective goods. Second we will show how any <u>as-if</u>-utilistic reduction can be transformed into an <u>intentional</u>-utilistic reduction. Finally, we will give a short survey of possible transformation rules for any type of reduction in sociology.

3.1. Rational expectations and irrational consequences

We will first investigate the relation between the objective and the subjective transformation rules, i.e. between TR1/2/3 of Section 2.3 and E*1/2/3 of Section 2.1. As we have seen, both rules play a crucial role in the reduction. If the postulated beliefs about the transformation process of the individuals, were incompatible with, or highly different from, the objective transformation rules, the reduction story would only have peculiar applications. But, if the beliefs of the individuals are largely correct in the light of the objective transformation rules, i.e. if their expectations are largely rational, there can be many natural situations where all reduction conditions are, roughly, satisfied.

Let us first note that the two sets of rules are compatible. That is, there is at least one (formal) example for the objective chance of realization R which satisfies TR1/2/3 as well as E*1/2/3. It is easy to verify that $R(n,m) = m/n$ is such an example.

Our main question, however, is the following: how much is true of E*1/2/3 if TR1/2/3 are true? In the light of the fact that TR1 trivially implies E1, this question reduces to: how much is true of E*, E*2 and E*3 if $R(n,m)$ satisfies TR2 and TR3?

Let us first formulate weaker versions of E*2 and E*3, viz.

E2 $b_i(n,m) > o$

E3 $$\frac{S_i(n,n)}{n} > \frac{S_i(n+1,n+1)}{n+1}$$

where $S_i(n,m)$ is defined as $R_i(n,m) - R_i(n,o)$, i.e. $\sum_{z=o}^{m-1} b_i(n,z)$.

It is easily checked that E*2 and E*3 are in fact the conjuntion of E2 and E3, respectively, with E* (i.e. $b_i(n,m) = b_i(n)$), in the same way as E*1 is, by definition, the conjunction of E1 and E*.

It is clear that TR1/2/3 do not imply E*. On the other hand, it is not only a trivial fact that they do imply E1, but also that they imply E2. Hence, the remaining question is: do TR1/2/3 imply E3? This is an important question, for E*3 is, as we have seen, a crucial element in the reduction, and it is clear that E3 grasps the main point of E*3 in this respect, viz. the <u>average</u> marginal influence decreases with increasing size of the group. The answer to this question is 'yes' and hence we have in total the following theorem.

<u>Th.4</u> TR1/2/3 imply E1/2/3

<u>Proof</u> E1 and E2 are trivially implied, as already indicated. The general validity of TR3 obviously requires, substituting $R(n,m) = R(n,o) + S(n,m)$,

(1) $$S(n_1+n_2,m_1+m_2) \leq \frac{n_1}{n_1+n_2} \cdot S(n_1,m_1) + \frac{n_2}{n_1+n_2} \cdot S(n_2,m_2)$$

and a similar condition for $R(n,o)$.
It is not difficult to check that

(2) $$\frac{n'}{m'} S(n',m') \leq \frac{n}{m} S(n,m) \text{ for } n' \geq n,\ m' \geq m$$

is not only a sufficient but also a necessary condition for (1).
Substitution of $m' = n' = n+1$ and $m = n$ in (2) leads to

(3) $S(n+1,n+1) \leq S(n,n)$

From TR2 trivially follows that

(4) $S(n,n) > o$

Finally, (3) and (4) trivially imply E3 (for S). Q.e.d.

We summarize Th.4 provisionally in the slogan: the expectations of the individuals are largely rational. This point has not been made by Olson and Lindenberg. As to the remaining 'possible distance' between TR1/2/3 and E*1/2/3 the following points are relevant. If R satisfies, in addition to TR1/2/3, the independence condition corresponding to E*, i.e.

TR* $b(n,m) = b(n)$

it follows from Th.4 that R has all the properties E*1/2/3. Nevertheless, there need not be quantitative correspondence between $R_i(n,m)$ and $R(n,m)$ in this case.
However this may be, if TR* holds in addition to TR1/2/3, the individuals base their behaviour on qualitatively correct insights in the transformation process.

If TR* is approximately true the behaviour of the individuals is still based on more or less correct insights. If, however, TR* is fundamentally false (whereas TR1/2/3 are true) the assumption E* is no longer rational. Inspection of the proof of Th.1 shows that the decision to participate should then no longer be made independent of what the others are going to do. Hence, the above mentioned slogan, i.e. 'the expectations are largely rational', should be qualified with 'provided the objective marginal effect roughly depends only on the size of the group'. To be sure, the qualification will be true in many cases.

Despite the rational expectations of the individuals, their behaviour leads to rather irrational consequences. This irrational aspect of Olson's logic of collective action has been discussed extensively in the literature. It is also known under the heading 'the tragedy of the commons'. Taylor (1976) shows that the individuals are subject to (a generalized form of) the so-called prisoner's dilemma, with non-participation as the dominant strategy.

Here we will only show these irrational consequences for an extreme case. Let us first assume, in general, that the individuals have perfect knowledge about the transformation process concerning X, i.e.

PK For all i, $R_i(n,m) = R(n,m)$

Note that PK presupposes TR1 as well as E1. Let us further assume,

TR* $b(n,m) = b(n)$

as well as TR2 and TR3. From PK and Th.4 it follows that E*1/2/3 are all satisfied. Let us also assume U1/2.

Consider now a group G the size n of which ia larger than, or equal to, the switch size S(i) for all i in G. If they all behave according to MH, i.e. maximize their expected utility, nobody will participate, i.e. $P(G) = \emptyset$.

Inspection of the proof of Th.1 shows that all these assumptions imply:

(5) For all i in G, $b(n) < C_P(i) / U_X(i)$

Every individual i in G can now objectively expect the utility:

(6) $R(n,o)\ U_X(i)$

If everybody would, despite (5), participate in the pursuit of X, i.e. not maximize expected utility, but 'behave socially', each individual could then objectively expect the utility:

(7) $R(n,n)\ U_X(i) - C_P(i)$

This expected utility is, for all i in G, larger than that according to (6) if, using $nb(n) = R(n,n) - R(n,o)$,

(8) For all i in G, $nb(n)\ U_X(i) > C_P(i)$

It is important to note that the conjunction of (5) and (8), i.e.

(9) For all i in G, $b(n)\ U_X(i) < C_P(i) < nb(n)\ U_X(i)$

is a real possibility, despite its extreme character.

In the case of (9), the conclusion is straightforward: 'social behaviour' by everybody leads to a larger expected utility for everybody than 'maximizing expected utility behaviour' by everybody. It is intuitively clear that this paradoxical phenomenon will not only occur in the extreme case (9), but that it will also occur in some partial sense in less extreme cases.

From the fact that the foregoing story was based on PK it follows already that the irrational consequences of maximizing expected utility behaviour cannot be due to the lack of rationality of the expectations of the individuals. In the first half of this subsection we have seen that the expectations may be called rational, even without perfect knowledge, i.e. without quantitative correspondence. Also in these cases it will be possible to derive similar irrational consequences. Hence, under rather general conditions the source of the described kind of irrationality cannot be the expectations. It is of course also absurd to assume that the source could be the utility assignments U1/2, for they function as primitives in the diagnosis of irrationality. Hence, the remaining candidate, viz. the principle of maximizing expected utility itself, is indeed the source of irrationality.

3.2. Intentional interpretation

In this section we will deal with the intentional interpretation of an as-if-utilistic reduction. To be precise, we will show how the actions of the individuals can be explained intentionally in terms of the conscious pursuit of maximal utility or, for short, how they can be explained as intentional-utilistic.

Of course we do not claim that such an intentional interpretation is always realistic. The only thing which matters is that such an interpretation will be realistic in some cases and our problem is what we precisely mean by that in such cases. We will present our analysis in terms of the reduction of Olson's hypothesis, but it will be clear that the content of the example does not play a substantial role.

In the preceding paragraphs we have alluded to an intuitive notion of intentional explanation of actions which needs further explication. Now one might expect that this explication will be of a nomological nature, given the 'nomological approach' of reduction in this paper. In particular, one might expect that the current nomological explication of intentional explanations (NI-explanations) will provide the adequate intentional interpretation.

However, in the literature many different objections have been made to NI-explanations. We will present one such objection in some detail. Suppose P(i,G), i.e. i participates in the pursuit of X for G. For the whole of this section we will implicitly assume that i is a member of G, if i and G occur as free variables. For completeness' sake we will first specify the nomological *as-if* explanation, with the maximization hypothesis MH in the role of nomological premise.

$$\frac{EU - P(i,G) \geq EU - N(i,G) \quad MH}{P(i,G)}$$

This is an *as if* explanation because i's (conscious) wish of maximal utility does not play a role.

In the NI-explanation of P(i,G) the following *desiderative* premise is explicitly stated

WMU(i,G): i wishes (wished) maximal utility in the P/N-choice

Moreover, the *epistemic* premise

DU-P(i,G) EU-P(i,G) $\geq$ EU-N(i,G)

is interpreted as the (conscious) belief of i that his expected utility of participation 'dominates' that of non-participation. For convenience, not-DU-P(i,G) will be equated with DU-N(i,G). Finally, both premises are taken into account in the *new* nomological premise:

CPMU For all G and all i in G,
if WMU(i,G) then
P(i,G) iff DU-P(i,G)

CPMU states roughly that everybody is always consistent (C) in his pursuit (P) of maximum utility (MU). The resulting NI-explanation is clearly a deductive argument

$$\frac{WMU(i,G) \quad DU\text{-}P(i,G) \quad CPMU}{P(i,G)}$$

The main objection to this NI-explanation concerns the role of the nomological premise CPMU. It belongs to the nomological view that CPMU should be interpreted as an empirical (i.e. falsifiable) premise [6] and that the explanation is only adequate if there are good reasons to assume that all required premises are true, hence also CPMU. Suppose now that there have been observed one or more convincing counterexamples to CPMU, i.e. 'inconsistent individuals'. Hence, CPMU is false. Now it follows that P(i,G) cannot be explained intentionally, according to the NI-explanation, in terms of the pursuit of maximal utility, even if the particular individual i has nothing to do with these counterexamples.

The obvious first reply to this objection is that CPMU should be statistically reformulated, roughly as follows: most individuals are mostly consistent. This reformulation would enable then a 'statistical intentional explanation' for now it is derivable that P(i,G) is probable.

A broader formulation of the objection excludes also this reply: the nomological or statistical premise is <u>irrelevant</u>, for, what does it matter for the intentional explanation of P(i,G) whether or not (almost) everybody is (almost) always consistent? The only consistency-question which can matter for this is whether or not i has been consistent in the particular case.

We take it that this objection has made it clear that there is something wrong with the NI-explication of our intuitions about intentional explanations. So let us make a new start by asking: What conditions seem to be intuitively necessary in order to allow us to say that P(i,G) is <u>intentionally interpretable</u>, i.e. can be explained intentionally, <u>in terms of the pursuit of maximal utility</u>, without making any assumption in advance concerning the precise structure of an intentional explanation? The following conditions seem plausible in any case: i wished maximal utility, i believed that participation was the best alternative in this respect, and i has been consistent. Note that the first two conditions imply that the third reduces to the fact that i participates; for, in the described circumstances he would have been inconsistent if and only if he had not participated.

The three mentioned conditions do not only seem necessary in order to justify the claim that P(i,G) is intentionally interpretable, but they also seem sufficient: it is at least difficult to think of other conditions we could or would like to add.

If this is so and if we continue to assume that an intentional explanation can be reconstrued as a deductive argument then it is plausible to assume that the three conditions function in this argument as the premises of the intentional explanation of P(i,G). Due to the already mentioned logical connection between the three, this set of premises reduces to WMU(i,G), DU-P(i,G) and P(i,G). Because this set includes P(i,G), it is evident that P(i,G) cannot be the conclusion of the underlying argument. For this would not only be trivial but it would also make the desiderative and epistemic premises redundant, and this is absurd.

Hence, our conclusion is that we have to look for a new explication of intentional explanations in which not only the desiderative and epistemic premises function truly as premises, but also the action-statement itself. In particular, we have to look for another, suitable conclusion.

In our (1983b) we have presented and defended extensively a general explication of everyday intentional explanation which satisfies these requirements. Here we will describe the general idea only briefly [7]. It will turn out that the present case of intentional-utilistic explanations forms a special case in two respects.

An intentional explanation of an action refers, by definition, to an intention, i.e. a conscious goal, of the actor. The crucial idea of the new explication is that the conclusion of the underlying argument of an intentional explanation is not that the actor has performed a certain action, but that the performed action was intentional, i.e. more or less consciously directed to a goal. In a complex premise, the explanation in the narrow sense, this goal is specified. Schematically, the argument reads as follows:

i aimed at goal g with action a
i performed a intentionally

The transition from the premise to the conclusion is, under any reasonable interpretation of the conclusion, an application of the rule of inference called Existential Generalization. The only thing we need to assume for this claim is that 'i performed a intentionally' can be explicitly defined as 'there is a goal which was aimed at by i with a'. In the present context we will speak of Intentional Generalization (IG) and of IG-explanations. The reverse transition, from the conclusion to the premise, is called Intentional Specification (IS), which is of course not a deductive but a synthetic step.

An IG-explanation will be called true or false (adequate or inadequate) if the premise is true or false, respectively.

It will be clear that it is possible that there is more than one true IG-explanation for one action.

In order to compare the present IG-explication with the current nomological (NI-)explication we will decompose the premise of an IG-explanation, i.e. i aimed at goal g with action a, into three components:

IG.1 i wished goal g
IG.2 i believed action a to be useful for g
IG.3 i performed a
―――――――――――――――
i performed a intentionally

With some temporal qualifications, which do not matter here, the conjunction of these components is equivalent to the original complex premise. The corresponding NI-explanation is the following:

NI.1 i wished goal g (i.e. the same as IG.1)
NI.2 i believed action a to be necessary for g
NI.3 some generalization of the conditional
if NI.1 and NI.2 then i performs/performed a
―――――――――――――――
i performed a

It is easy to see that the three NI-premises together imply the three IG-premises, hence they imply the 'IG-conclusion'. Hence, in the light of the first implication there is a well-defined sense in which an NI-explanation is stronger than the corresponding IG-explanation.

This greater strength, however, is precisely responsible for the 'explicative objections' that have been raised against NI-explanations, especially by adherents of the so-called 'Practical Syllogism':

- the nomological premise (NI.3) is irrelevant,
- we do not claim by an everyday intentional explanation that the action could have been predicted,
- that the actor has performed a certain action is a presupposition when we start to look for an intentional explanation,
- frequently, alternative actions will have been possible, in which case the epistemic premise (NI.2) is much too strong.

All these objections do not apply to IG-explanations, as is easy to verify. Here we will only remark briefly a point which is of equal importance. The IG-explication rests on an unproblematic logical rule of inference. In this respect it is also highly superior to the Practical-Syllogism-explication (see note 6), which needs to assume a *magical* meaning postulate.

A straightforward example of a (non-utilistic) IG-explanation can be given for *participation*:

$U_X(i) > o$	i wishes the collective good X for G
$b_i(n) > o$	according to i, participation (in the pursuit of X) is useful for the production of X
P(i,G)	i participates in the pursuit of X for G
(Conj.)	i aims, with his participation, at X for G
(Int.Gen.)	i participates intentionally

It is of course impossible to construe an analogous IG-explanation for non-participation (unless we assume that the individual believes that non-participation is useful). However, it is possible to give intentional explanations for both actions, on a different level, viz. in terms of the goal 'maximal utility'.

The IG-explanation of P(i,G) in terms of maximal utility obviously is:

WMU(i,G)	i wishes maximal utility
DU-P(i,G)	i believes that participation is necessary and sufficient (hence useful) for maximal utility
P(i,G)	i participates in the pursuit of X for G
(Conj.)	i aims, with his participation, at maximal utility
(Int.Gen.)	i participates intentionally

Note that the premises are exactly those which we had derived earlier as the presumable premises. It is clear that an analogous explanation can be given for N(i,G). This type of explanation will be called *intentional-utilistic* (IU-)*explanation*.

IU-explanations are in two respects special cases of IG-explanations. In the first place, all IU-explanations refer to the same goal, viz. maximal utility. In the second place, the epistemic premise specifies that the relevant action is not only useful, but even necessary and sufficient.

We will call an action *IU-interpretable* if all premises of the corresponding IU-explanation are true. This makes it easy to formulate the answer to our original question *what* an intentional interpretation of maximization behaviour would look like, viz. the actions of the individuals are IU-interpretable. At the same time we get the answer to the question *when* an intentional interpretation of a utilistic reduction is realistic, viz. when the actions of the individuals are IU-interpretable, in which case we will speak of intentional-utilistic (IU-)reduction.

Our analysis should of course be such that the maximization hypothesis (MH) follows from the hypothesis that the actions of the individuals are IU-interpretable and this provides a nice touchstone for the analysis. The suggested intentional-utilistic interpretability hypothesis comes down to the following:

IUIH For all G and all i in G,
if P(i,G) then WMU(i,G) and DU-P(i,G), and
if N(i,G) then WMU(i,G) and DU-N(i,G)

It is easy to check that the following theorem is provable,

Th.5 IUIH iff WMU and MH

where WMU indicates that everybody wishes maximal utility in the P/N-choice, i.e.

WMU For all G and all i in G: WMU(i,G)

Besides the fact that IUIH implies MH, as was to be expected from an adequate account, it is also clear from Th.5 what IUIH adds to MH, viz. WMU. In words, utility maximization behaviour is intentionally interpretable if the individuals (consciously) wish maximal utility.

It is easy to see that the presented intentional interpretation of the maximization hypothesis designed for the utilistic reduction of Olson's hypothesis can be generalized to any context where a similar maximization hypothesis is useful in the reduction of social regularities.

Moreover, the scheme for utilistic reduction at the end of Section 2 can of course be extended to a scheme for intentional-utilistic reduction by adding on top the (deductive) 'neutralizing step' (NS):

$$\text{NS}\ \frac{\text{intentional-utilistic interpretability hypothesis}}{\text{maximization hypothesis}}$$

As has been said already, an intentional interpretation of utilistic reduction will not be realistic in every context. In some contexts, however, it will be realistic, viz. in situations where people are more or less forced to choose consciously between alternative actions. If the maximization hypothesis leads to the explanation of regularities in such contexts, then an intentional interpretation seems unavoidable. But there are of course also many contexts where there is no pressure for a conscious choice at all. If utilistic reduction is nevertheless possible in such a case, intentional interpretation may be wrong. How else to explain the maximization hypothesis in that case, we leave as an open question.

With respect to Olson's hypothesis itself, it is clear that there will be collective goods for which an intentional interpretation is realistic as well as goods for which this is not the case. However this may be, in Section 2 we have seen that utilistic reduction of this hypothesis on the basis of the maximization hypothesis is perfectly possible. In this subsection we have seen that an intentional interpretation <u>can</u> be given, that fits nicely with our intuitions concerning intentional explanation.

3.3. <u>Types of transformation rules</u>

In the context of sociology, transformation rules relate by definition collective behaviour with social phenomena. They will be indispensable ingredients in any type of heterogeneous reduction in sociology (and economics), whether or not the reduction is based on some form of utility theory.

In this final section we will discuss some types of transformation rules, including possible ways of explaining them. The following diagram represents the types to be considered:

Types of transformation rules	
analytic	
	causal
empirical	social
	institutional

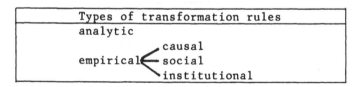

With the global division into analytic and empirical transformation rules we subscribe to Lindenberg (1977). A transformation rule is <u>analytic</u>, and hence does not require explanation, if it follows from a definitional link between social phenomena and collective behaviour. For example, the degree of (internal) democracy can be defined as the degree of participation in decision processes. If all transformation rules in a case of reduction are analytic, this reduction is by definition homogeneous. Note that, in the context of collective goods, it is difficult to think of goods which are linked in this analytic way with collective behaviour.

Heterogeneous reduction presupposes at least one empirical, i.e. non-analytic, transformation rule, of which we will distinguish three types, called causal, social and institutional rules. These distinctions are made along the lines of the role of intervening subjects, i.e. subjects acting 'between' collective behaviour and social phenomena.

In the case of causal transformation rules there are no intervening subjects at all: the link between collective behaviour and social phenomena is purely a matter of causal processes. A typical example is: a decreasing number of smokers leads to a decreasing number of cases of lung cancer. Of course, the explanation of causal transformation rules is outside the scope of sociology.

Social transformation rules are by definition links following from actions of intervening subjects, which need not only be real human beings but also other types of acting subjects, like firms, cities, etc.. In most cases the transformation rules for collective goods will be of this type.

Except for the extreme case of institutional transformation rules (see below), social transformation rules presuppose that the collective behaviour creates circumstances in which the intervening subjects have some freedom of action. If these subjects maximize their expected utility in this 'space of action', it is of course possible that the apparent social regularity between their circumstances and the social phenomenon is again utilistically reducible, in which case new transformation rules will turn up.

For most social transformation rules not all these assumptions will be realistic. But this does not exclude explanations of such rules which perfectly fit into the program of methodological individualism, viz. explanations assuming intentional, but not necessarily utilistic, actions of the intervening subjects. In our ((1983c) Section 3.5) we have sketched two examples of such explanations for the transformation rules concerning collective goods.

We define institutional transformation rules as extreme cases of social transformation rules, viz. rules where the behaviour of the intervening subjects has been institutionalized. A typical example is the rule, in parliamentary democracy, that a bill gets the status of law if and only if it gets a majority of votes in parliament. Though there are intervening subjects involved, e.g. the president's signature, their role is fixed. Lindenberg (1977) seems to have thought only of this type of empirical transformation rules.

In many cases it will be obvious what institutional transformation rules are involved. In principle, however, the assessment of institutional rules is a matter of empirical research, think e.g. of historical or anthropological research. Suppose that the existence of such a rule has been extablished. Now, the quest for an explanation of that rule is in a sense meaningless. Of course, it makes sense to ask 'external' questions, like why (and how) the rule has been brought about, why it is maintained and why it is abolished or replaced. But it does not make sense to look for a 'deeper' mechanism which produces the institutional rule.

For this reason, institutional rules are very similar to the transformation rules in the reduction of the ideal gas law. If carefully formulated, the latter are, like 'water is H_2O', examples of ontological indentities: universal relations which need empirical justification but no (causal) explanation [8]. Institutional rules differ only from ontological identities in that they are no universal connections, but artificial connections. This difference not only provides the reason why the indicated external questions make sense, it leaves after all some room for a distinction between sociology and physics [9].

Department of Philosophy
Westersingel 19
9718 CA Groningen
The Netherlands

NOTES

1) For empirical evidence for this hypothesis, see Olson (1965) and (1982)
2) The terms 'homogeneous' and 'heterogeneous' reduction were introduced by Nagel ((1961),Ch.11)
3) In our (1983a) we have argued that (advanced) textbooks in physics, though giving a correct account of the aggregation step for the ideal gas law, make a mental error with respect to the wall-attraction in the aggregation step leading to the more realistic law of Van der Waal's. It is also shown that this error can be avoided, but only by a complete reform of the aggregation step.
4) Our (1982) is in many respects a critical evaluation of Nagel's (1961) account of this example in the light of Causey's (1977) ontological approach to reduction.
5) It is precisely this aspect which gives the whole analysis

a <u>static</u> character: the individuals do not need to bother about the behaviour of the others. By changing the assumptions such that the (subjective) marginal effect depends, as a rule, on the number of participants the analysis would achieve a <u>dynamic</u> character. Taylor's (1976) 'supergames' might be useful for such an analysis. Moreover, important qualifications of the regularities seem unavoidable.

6) This is the crucial difference between the nomological explication and the 'semantic' explication of Von Wright in terms of the so-called Practical Syllogism. In the latter account of CPMU is (implicitly) assumed to be a meaning postulate.

7) There are also some differences in presentation. In particular in our (1983b), intentional-utilistic explanations are presented as intentional explanations of <u>choices</u> between alternative actions, to distinguish them from 'direct' intentional explanations of the chosen <u>action</u>. Unfortunately, this, in our opinion very adequate, distinction leads to many complicated formulations, which are certainly not suitable for a summary.

8) See note 4. Causey's distinction (Causey (1977)) between causal connections and identities is entirely based on whether or not an explanation is required. Like causal connections, identities need empirical justification. This distinguishes identities from analytic connections which need neither explanation, nor justification.

9) I like to thank the Netherlands Institute of Advanced Study (NIAS) at Wassenaar for the possibility of this research. I also thank Wolfgang Balzer, Carl Doerbecker (†), Henk Flap, Bert Hamminga and Henk Zandvoort for their comments on the draft of my (1983c), which forms the basis of this paper.

REFERENCES

Causey, R.L.: 1977, <u>Unity of Science</u>, Synthese Library 109, Reidel, Dordrecht.

Kuipers, T.A.F.: 1982, 'The reduction of phenomenological to to kinetic thermostatics', <u>Philosophy of Science</u>, 49.1,' 107-119.

Kuipers, T.A.F.: 1983a, 'The paradigm of concretization: the law of Van der Waals', to appear in <u>Poznan Studies in the Philosophy of the Sciences and the Humanities</u>.

Kuipers, T.A.F.: 1983b, 'The logic of intentional explanation', to appear in: <u>The Logic of Discourse and the Logic of</u>

Scientific Discovery (Eds. J. Hintikka and F. Vandamme).

Kuipers, T.A.F.: 1983c, 'Utilistische reductie in de sociologie', manuscript, partly to appear in Mens en Maatschappij, under the title 'Olson, Lindenberg en reductie in de sociologie'.

Lindenberg, S.: 1977, 'Individuelle Effekte, kollektive Phänomene und das Problem der Transformation', H. Eichner, W. Habermehl (Hrg.), Probleme der Erklärung sozialen Verhaltens, Meisenheim, Hain, 64-84.

Lindenberg, S.: 1982, 'De onvolledigheid van algemene hypothesen: een pleidooi voor verklarende modellen', Mens en Maatschappij, 57.4, 373-391.

Nagel, E.: 1961, The Structure of Science, Routledge, London.

Olson, M.: 1965, The Logic of Collective Action, Harvard UP, Cambridge.

Olson, M.: 1982, The Rise and Decline of Nations, Yale UP, New Haven.

Taylor, M.: 1976, Anarchy and Cooperation, Wiley London.

[illegible] *Scientific Discovery* (Eds. T. Nickles and [illegible]), [illegible]
[illegible] *Nagarjuna* [illegible] *Mahayana* [illegible]
[illegible] *Problem der Transformation* [illegible]
[illegible] *Mathematical* [illegible] 7, 4, 3 [illegible] 19[illegible]
Nagel, E. 1961 *The Structure of Science*, Routledge, London.
Olson, M. 1965 *The Logic of Collective Action*, Harvard University Press.
Olson, M. 1982 *The Rise and Decline of Nations*, Yale University Press.
[illegible] M. [illegible] *Personal Knowledge*, [illegible] London.

David Pearce and Michele Tucci

INTERTHEORY RELATIONS IN GROWTH ECONOMICS: SRAFFA AND WICKSELL

This essay is devoted to a formal reconstruction of two major rival theories of growth or capital economics belonging respectively to the classical and neoclassical schools: they are due to Piero Sraffa and Knut Wicksell. Contrary to the conventional view, we argue that these two theories are closely related and, in fact, in good agreement. More precisely, we show that in a well-defined sense Wicksell's theory can be reduced to Sraffa's; and we discuss some of the implications of this for the methodology of economics.

1. INTRODUCTION

The appearance, in 1960, of Sraffa's influential book ushered in a new wave of criticism of the theory-structure of neoclassical economics. Much of this criticism centred on what was alleged to be an incorrect representation of the concept of capital, and it was adapted in various ways and directed simultaneously at each of the main branches of neoclassical theory. Especially under fire were the prominent theories of Walras and Wicksell.[1]

Wicksell's theory, in particular, was already criticised from the neo-Ricardian standpoint in the work of Garegnani (1960). Subsequently, many non-Sraffian economists joined in the controversy (see, e.g. Burmeister and Dobell (1970), Burmeister (1976)) and the debate is still active today. For recent counter-arguments, see Burmeister (1980) and also Laise and Tucci (1980), (1981), (1982). Elsewhere we have examined some aspects of the dispute by studying the relationship of Sraffa's theory to some theories of Walrasian type (Pearce and Tucci (1982b), (1984)). Our task in the present paper will be to analyse the connections between Sraffa and Wicksell.

We shall approach the problem from what is still these days an unconventional angle. We shall not enter the debate from the traditional perspective of the theoretical economist, nor study in depth the individual arguments produced in the controversy. Instead, we shall propose a formal reconstruction

W. Balzer et al. (eds.), Reduction in Science, 269–293.

of the theories in question, so that the relations between them can be determined from a logical viewpoint. The following words of introduction are designed to encourage both the philosopher of science unversed in economic theory and the economist unfamiliar with logical metascience to read further.

The economist dipping into this paper will encounter familiar theories in a rather unusual setting. The systems of Wicksell and Sraffa are commonly recognised to be rival theories representing very different and conflicting accounts of the economic world. Here those theories are logically reconstructed and compared within a model-theoretic framework, and the resulting analysis sweeps out a wide area of common ground belonging to each theory.

In the philosophy of science, the use of model theory, and techniques of logical reconstruction generally, has a well-established background and history, especially in the domain of physics and the natural sciences. Attempts to extend the use of formal frameworks to the social sciences are of rather more recent origin. Nevertheless, we can see no real obstacle against their use and, in fact, mathematical economics seems well-suited to the process of formalisation.

In the social sciences, the problem of how to deal with different, and possibly rival, theories does not yet have a widely-accepted solution, especially so in the case of economics. At a very intuitive level, it could be said that theory-criticism in economics is composed of a variety of informal and formal factors that, taken in combination, furnish a general picture of the way in which a given economic school attacks and rejects features and assumptions of theories belonging to a rival school. Clearly, the methodology governing the process of acceptance and rejection of theories in economics is somewhat vague and obscure. And empirical evidence is rarely a relevant factor here, because there is so little general consensus among economists about how 'facts of the world' should be brought to bear on the appraisal of abstract theories.

In the light of these difficulties, this work intends to develop a new direction for handling intertheory relations in economics; in this context they are relevant to the dispute between classical and neoclassical theories. Our method of reconstruction is intended to depict each of the theories from an independent conceptual basis, and is designed to be as faithful as possible to the original, or most common, form of the theories, so as not to prejudice substantially the question of what relation if any exists between them. The theories are then further analysed in order to throw into sharper relief their shared, as well as distinguishing, features. The key methods employed here are those of translation, in the syntactical sense, and definable relations between models, in the semantical sense.

As many economists would be doubtless quick to observe, such methods cannot generally be considered to provide an exhaustive comparison of rival economic schools. Any formal reconstruction of an economic theory is bound to omit significant philosophical, ideological and political factors which constitute the 'breeding ground' of each school or research tradition, and set heuristic guidelines in the search for acceptable theories. The received opinion is that economic theories possess certain 'qualities' that can hardly be conveyed in a strictly formal manner, either in the usual representation, or in a rational reconstruction of the kind given here. On the other hand, the claim that there is no link whatsoever between the formal structure of a theory and its broader interpretation would probably be ranked as extremist. Most economists seem to agree that within any given school the formal and informal ingredients are finely interwoven, so that an economic line of thought may be accurately characterised by a suitable mixture of the two. At this juncture the results of the present paper can be seen to be pertinent, for we are able to indicate a clear sense in which Sraffa's and Wicksell's theories are, from a model-theoretic standpoint, 'virtually' the same. More precisely: under suitable assumptions, statements belonging to the latter theory may be translated into statements belonging to the former in such a way that true sentences of the one theory correspond to true sentences of the other. Similarly, any inconsistencies or anomalies in one of the theories would, on the strength of this translation, give rise to corresponding inconsistencies or anomalies in the other theory. Further, it should be remarked that the translation here is not to be regarded as a conceptual curiosity or a mathematical 'quirk', but is intended to be faithful to the actual meanings of concepts appearing in the two different languages.

At this point the economist who sides with the conventional reading of the dispute between classical and neoclassical theories is faced with an awkward problem: How is it possible that the Sraffa theory, so long and vigorously opposed to theories of Wicksell's kind, shares with the latter a significant 'basic' structure ? Is it conceivable for economic theories to be antagonistic and, formally, so closely related at the same time ? Or, in other words, can those non-formal 'qualities', underlying rival economic theories, make all the difference ?

It would be beyond the scope of the present analysis to provide answers to such questions. But we would like to stress here that, whilst it is clearly important to embed the formalism of mathematical economics into a richer context where broader interpretative issues can be dealt with, one might view with some suspicion any claim to the effect that contrasting and conflicting aspects of rival theories are in no way reflected by some distinctive feature or other of the mathematical formalism.

Turning to the philosophy of science at large, it is remarkable that economics has so rarely been seen as a legitimate source of material on which to apply and test logical and methodological concepts. Problems of explanation, reduction and progress in economics have received scarce attention from philosophers of science in general. And, with one or two exceptions, there are hardly any case studies focusing on the formal reconstruction of economic theories.[2] This is somewhat surprising in view of the fact that theoretical economics is a mature and mathematically sophisticated discipline for which the tools of axiomatic and logical analysis would appear to be particularly appropriate; and because economics is an active arena of debate, controversy and theoretical dispute between rival schools.

One aim of the present paper is thus to draw attention to the fact that economics possesses interesting and non-trivial examples of reduction, and to show how some of the techniques and concepts usually adopted in the metatheory of the natural sciences may be fruitfully applied in branches of the social sciences as well. Actually, our approach below will be to reconstruct economic theories in a framework slightly more abstract and general than the usual one. However, though more general, our concept of reduction will retain (and extend) several of the basic ideas of the standard account. Our framework is discussed and defended at greater length in Pearce and Rantala (1983) and elsewhere in this volume. A short resumé of the essentials appears in the next section.

2. THEORIES AND INTERTHEORY RELATIONS

Mathematical economists traditionally construe their theories in an informal, or perhaps semi-formal, manner. They deal, by and large, with economic models defined by systems of mathematical equations, inequalities, and so forth. Different models are compared by studying the kinds of 'solutions' they admit once their exogenous parameters have been fixed. And it is often claimed, in intertheory debates, that a given model is 'logically inconsistent', or that one theory is merely a special case of another, more general theory, where the weight of preference tends to fall in favour of the latter. This kind of criticism has been influential in shaping the development of economic theory, even if it does not always meet the standards of rigour prescribed by methodologists of science for the 'ideal' case.

One way of coming closer to those standards is through a logical reconstruction of economic theories. The aim of this is not simply to increase the level of formalisation, but to achieve an enhanced grasp of the conceptual structure of the theory. One tries to make precise the logical form of the theory, make explicit certain assumptions that may be 'hidden' in the usual presentation, and clarify the exact relations holding between rival theories. Thus, if a theory is axiomatised, or if its logical structure is otherwise laid bare, one may analyse by formal methods its metatheoretical properties; for instance to check whether it stands in a relation of consistency, contradiction, entailment, or whatever, to some other chosen theory.

The price to pay for this higher level of formalisation is the need for a more complex assortment of logical concepts and devices. At the same time, it would be rash to suppose that all of the ingredients, formal or informal, that make up an actual economic theory could be readily captured in a logical reconstruction. Even to attempt this would require a considerable increase in the complexity of the framework, at the risk of a questionable gain in clarity. The formal reconstruction should therefore be viewed as a sort of technical unit with which to work.

However, once economic theories have been assigned a suitable canonical description, they may be studied with the same degree of exactness afforded to axiomatic theories in mathematics and physics. The chief problem is thus to find a canonical format that is adequate to the philosophical task at hand, in this case the analysis of intertheoretic relations.

To accomplish this, let us start by observing that an economic theory can generally be conceived as a structural entity; for instance as a tuple $< X_1,\ldots,X_n >$, where the X_i are variables or parameters representing specific economic quantities, such as 'price', 'wage', 'rate of profit', and so forth. Some of these 'variables' are taken to be exogenous, in the sense that they assume specific mathematical 'values' which are determined from the 'outside' - i.e. independently of the theory - though the theory may restrict the range of admissible values by means of appropriate constraints. Typical exogenous variables will be those representing technical components of the production process. The remaining, endogenous variables are determined entirely form 'within' the theory, in the sense that their values are restricted by the relevant equations, and other functional relations, which the theory asserts to hold between the $X_1,\ldots,X_n$. 'Production price' will typically be taken as endogenous to the theory. The tuple $< X_1,\ldots,X_n >$, satisfying the mathematical conditions imposed by the theory, is often referred to as an economic *model*. From the logical point of view, however, this terminology is slightly misleading, for it is clear that what is intended by a 'model' here is really a *class* of mathematical structures, because different assignments of values to some of the variables may give rise to different admissible values for the others. For example, a change in the commodity set or in the production techniques may induce a corresponding change in the assignment of prices. Similarly, when one speaks, say, of a neoclassical production function f, one generally has in mind a class of mathematical functions, since the domain and range of f, or even its shape, may be allowed to alter within the constraints of the theory in question. To avoid confusion, it is thus preferable to distinguish between the variable or *functor* $\underset{\sim}{f}$ and the corresponding *functions* f which $\underset{\sim}{f}$ may denote. This is nothing other than the familiar distinction in logic between a term (e.g. a predicate or name) and its extension (e.g. a relation or individual). In other words, we can regard a suitable list of 'variables', together with the mathematical relations specified by the laws of the theory, as giving rise to a collection of structures or *models* in the usual logical sense.

Developing this line of thought a little farther, suggests that, formally, an economic theory may be characterised by the following four components. First, a vocabulary or *similarity type* τ which fixes the shape of the models under consideration. Here, τ is just a set of *symbols* plus *sorts*, the latter determining which kinds of objects the symbols refer to. Naturally, each exogenous and endogenous variable has a name in τ, as do the various sets and operations required in the underlying mathematics of the theory, e.g. in real and complex analysis. Sometimes it is useful to indicate which elements of τ are exogenous and which are endogenous, but we do not formally introduce this distinction into our theory-representation, because it is not always rigidly adhered to in economics: in a given theory some variables may change category according to context. Next, we can depict the 'conceptual structure' of the theory by means of a collection N of structures of type τ (in symbols: $N \subseteq Str(\tau)$); they represent, so to speak, all the possible worlds or models admitted by the theory. Naturally, each member of N is a structure comprising a domain for each sort in τ and a relation, operation or individual (from the appropriate domains) corresponding to each symbol in τ. Third, we consider a class M of <u>models</u>, or structures in N that additionally satisfy the basic laws, equations, and so forth, of the theory. And, finally, certain kinds of <u>transformations</u> belonging to the theory may be represented by (classes of) relations or maps between elements of N; for instance, transformations of scale in the production process, technological transformations, or even time transformations, when a theory is viewed dynamically, can be thus characterised.

In this way, an economic theory can be taken to be a structure

$$T = \langle \tau, N, M, R \rangle$$

where τ is a similarity type, $N \subseteq Str(\tau)$, $M \subseteq N$, and R is a collection of suitable relations between structures in N. In what follows, however, we shall omit this last component of a theory T, since, in the present case study, we shall not be concerned with any characteristic transformations as such.

We assume in general that N and M (and, where appropriate, R) are set-theoretically definable entities, and, in addition, that N and M are logically definable collections. That is to say, there is some logic L and sets of axioms Σ, Θ (sets of L-sentences in the type τ), such that N and M are respectively the classes of all models of Σ and Θ. In the usual terminology, N and M are thus <u>L-elementary</u> (or <u>generalised elementary</u>) <u>classes</u> for some logic L. Any such logic will be called <u>adequate for</u> T. To gain maximum flexibility, we use the term 'logic' here in a quite general sense, as it occurs in abstract logic. However, for the present applications we require only $L_{\omega_1\omega}$, i.e. the extension of first order logic which admits countable conjunctions and disjunctions of formulas; and no special familiarity with this logic will be needed to follow the arguments set out below. As usual, we assume that for any logic L and type τ admitted by L there are associated: a collection $Str_L(\tau)$ of structures, a collection $Sent_L(\tau)$ of sentences, and a truth relation $\models_L \subseteq Str_L(\tau) \times Sent_L(\tau)$.

Naturally, we do not suppose that all of the properties

of an economic theory are reproduced in the above description. What we hope to capture is the basic structure of the theory. The picture may then be further refined to cope with additional features and specific applications of the theory, according to context. Nevertheless, this characterisation seems to be adequate for a formal rendering of intertheoretic relations, and thus we shall adopt it as our 'technical unit' for the purposes of comparing the theories of Sraffa and Wicksell.

In this framework a number of intertheory relations of a reductive kind may be defined. Here we cannot enter into a detailed discussion, but we refer instead to Pearce and Rantala (1983c) and (this volume). For the present, let $T = \langle \tau, N, M \rangle$ and $T' = \langle \tau', N', M' \rangle$ be theories, and let L,L' respectively be logics adequate for T and T'. Then, by a <u>reduction of T to T', relative to</u> $\langle L, L' \rangle$, we mean a pair $\langle F, I \rangle$ of maps such that:

(2.1) (i) $F: K' \to M$ is <u>onto</u> (where K' is a non-empty subclass of M');

(ii) $I: Sent_L(\tau) \to Sent_{L'}(\tau')$; such that

(iii) K' is definable in $L'(\tau')$, and for all $\mathrm{M} \in K'$ and $\phi \in Sent_L(\tau)$:

$$F(\mathrm{M}) \models_L \phi \iff \mathrm{M} \models_{L'} I(\phi).$$

In other words, a reduction of T to T' amounts to a structural or semantic correlation F which takes models of T' onto models of T, and a compatible syntactical translation I which maps L-sentences in the language (type) of T into L'-sentences in the language (type) of T'. The domain and range of F are definable in the appropriate logics, and the translation preserves truth (relative to F). It follows from this that any sentence ϕ of $L(\tau)$ is true in T (in the sense that it holds in all models in M) if its translation is true in T' (i.e. $\forall \mathrm{M}' \in M'$: $\mathrm{M}' \models I(\phi)$). Consequently, (2.1) fulfils the chief syntactic and semantic properties required of a reduction relation between theories.

3. THE THEORIES OF SRAFFA AND WICKSELL

The Sraffa theory we shall depict here is a system of prices, wages and rates of profit for a finite number of industries producing one commodity each. For simplicity we assume that only one 'consumer good' is produced in the economy, the remaining goods being considered as intermediate products. We also make the assumption of constant returns to scale; though we shall not formalise this condition explicitly, since it is not an intrinsic feature of Sraffa systems. Its role comes into play when the Sraffa and Wicksell theories are compared.

Mathematically, this type of Sraffa model is usually defined by a set of technical matrices and labour vectors representing the different technological possibilities of production. One 'technology' is then selected as optimal with respect to wages and prices for a given, fixed rate of profit. The level of wage and rate of profit is assumed uniform throughout the productive sector; and we suppose that an infinite set of 'technologies' is available.

This mathematical representation can be slightly modified so as to obtain a structure that is also a model in the logical sense. The modification involved is inessential from the economic point of view: it simply allows us to view a Sraffa model as a structure of a certain many-sorted similarity type. This structure also contains a model of the underlying mathematical apparatus of the theory; and we take this to be a suitable model of analysis that is 'strong' enough to include the natural numbers, real numbers, and all the machinery required for defining infinite dimensional Euclidean spaces and the usual operations of matrix algebra, e.g. matrix addition and multiplication, the taking of inverses, etc. In the sequel, we leave this purely mathematical part as implicit; for an explicit treatment of models of analysis the reader is referred to Pearce and Rantala (1983b).

The productive part of the economic structure can be described by means of functions from the set of technologies into Euclidean spaces, and the economic variables of the model are taken to be real-valued functions on the technology and commodity sets. Technically it is expedient to work with an economy possessing a (countably) infinite set of industries only finitely many of which are engaged in production; we shall sometimes refer to these as the non-null industries, and they alone may have non-zero labour and production coefficients.

Let $\mathcal{M}$ be a three-sorted structure of the form

$$\mathcal{M} = \langle \mathcal{M}_0; C, T, B, l, w, p, c, r, g, t^*, n \rangle$$

where $\mathcal{M}_0$ is a model of analysis; C is a finite, non-empty set; T is an infinite set; $B: T \to \mathbb{R}^{\omega} \times \mathbb{R}^{\omega}$; $l: T \to (\mathbb{R}^+)^{\omega}$; $w: T \to \mathbb{R}$;[3] $p: T \times C \to \mathbb{R}^+$; $c \in \mathbb{R}^n$; $r \in \mathbb{R}^+$; $g \in C$; $t^* \in T$; $n \in \mathbb{N}^+$.

Let τ' be the type of $\mathcal{M}$; and, for any $t \in T$, denote by $B_{ij}(t)$ the value of $B(t)$ at coordinates (i,j), and by $l_i(t)$

the ith component of $l(t)$. Further, we denote by $\overline{B}_k$ and $\overline{l}_k$ respectively the projections of $B[T]$ and $l[T]$ to their first k components.

Now, let N_0' be the class of all models $\mathcal{M}$ as above of type τ' such that:

(3.1) $n = |C|$;

(3.2) For all $t \in T$, and for each $i,j > n$,

$B_{ij}(t) = l_i(t) = 0$;

(3.3) $\overline{B}_n \times \overline{l}_n$ is a compact subset of $\mathbb{R}^{n^2+n}$ with some continuity properties;

(3.4)
$$c\begin{pmatrix}0\\1\\1\\\vdots\end{pmatrix} = 0 \text{ and } c\begin{pmatrix}1\\0\\0\\\vdots\end{pmatrix} > 0 .$$

And let N' be the least class of models of type τ' including N_0' and closed under isomorphism.

The intended interpretations of the various components of the model $\mathcal{M}$ above are as follows: C is a set of commodities, one of which, g, is a single consumer good. T is a set of technologies, and B is a technical coefficient function such that, for each technology t, $B_{ij}(t)$ denotes the quantity of the jth commodity used in the production of the ith commodity. Here, C is taken as an ordered set containing n elements (3.1), and without loss of generality we take g to be the first element. l is a labour coefficient function specifying for each industry and technology the amount of labour employed. p is a price function, and w denotes constant wage per unit of labour. Finally, c represents the level of net production, set at a positive real value for commodity g - taken as the first component of the vector - and zero elsewhere (3.4); r is a uniform rate of profit, and t* is the optimal technology. The remaining conditions ensure that no economic activity occurs in the 'null' industries, and that the technical and labour functions satisfy some desirable mathematical properties (3.3).[4]

Now let M_0' be the class of all models in N' satisfying:

(3.5) For all $t \in T$, $B(t)p_t(1+r) + l(t)w(t) = p_t$;

(3.6) $cp_t = 1$;

(3.7) $w(t^*) = \max_{t \in T} w(t)$;

where p_t is the vector formed by evaluating p(t,x) for each x in C. The class of models of the Sraffa theory is then defined as the least class M' of structures in N' including M'_0 and being closed under isomorphism. Thus, the theory of Sraffa may be characterised by the triple

$$ST = < \tau', N', M'>$$

and it is clear that we may take the infinitary logic $L_{\omega_1\omega}$ as a logic adequate for ST in the sense indicated earlier (see also the discussion in Pearce and Rantala (1983b)).

The basic axioms of the Sraffa theory are contained in (3.5)-(3.7). The first condition sets production prices equal to the sum of all costs incurred in the production process. The second fixes a numeraire of prices, and (3.7) defines t* as an optimal technology, i.e. one that maximises wage at a rate of profit that is exogenously determined. [5] These three equations can be jointly satisfied, that is to say:

(3.8) The class M' is non-empty.

For, the fact that there are models satisfying (3.5) and (3.6) follows for the existence of solutions for the matrix equations. Such solutions have been widely studied in the literature, and they amount to continuous functions of B and 1. By compactness (3.3) and the Weierstrass theorem it follows that a technology t*, optimal in the sense of (3.7), can always be found. Actually, (3.7) does not generally define a unique t*, but the optimal technologies are all 'equivalent' in that they yield the same values for all endogenous functions. This justifies our selection of a single representative from the equivalence class. [6]

We now turn to the characterisation of a Wicksell theory without land. [7] This theory provides an economic description of the production of a single consumer commodity over a finite period of time. It also incorporates the value of capital employed in production, expressed as a coefficient of units of the final commodity produced. [8] Wicksell's production function describes the possible distributions of labour over time in order to achieve a certain level of production. And once again the theory embodies an optimisation process that selects that distribution which minimises production costs and maximises profit, given a prior, fixed amount of capital.

Mathematically, it is expedient to represent the temporal dimension of the theory by a (countably) infinite sequence of equal time intervals. Since the total period of production is finite, we suppose that only finitely many intervals, say the first k, enjoy economic activity; these may be called non-stagnant. Consequently, the production function will assign zeros to all remaining time intervals.

Let us now consider two-sorted structures of the form:

$$M = < M_0; \{G\}, f, A, a_0, \ldots, a_{k-1}, K, i, \overline{w}, y, \overline{p}, k >,$$

where M_0 is a model of analysis; $\{G\}$ is a singleton; $f \subseteq \mathbb{R}^{\omega} \times \mathbb{R}$; $A \in \mathbb{R}^+$; $a_0, \ldots, a_k \in \mathbb{R}$; $K \in \mathbb{R}^+$; $i, \overline{w} \in \mathbb{R}$; $y, \overline{p}: \{G\} \to \mathbb{R}^+$; $k \in \mathbb{N}$. And let τ be the type of M. We now define N_0 to be the class of all models of type τ satisfying:

(3.9) For all $\vec{x} = < x_0, x_1, x_2, \ldots >$, $\vec{x}' = < x_0', x_1', x_2', \ldots >$, z, and $m \geqq k$, $((f(\vec{x}, z) \Rightarrow x_m = 0) \;\&$ $\& \; (f(\vec{x}, z) \;\&\; f(\vec{x}', z) \Rightarrow \vec{x} = \vec{x}'))$;

(3.10) f, viewed as a partial function from $\mathbb{R}^{\omega}$ into $\mathbb{R}$, is C^2, i.e. differentiable with continuous first partial derivatives;

(3.11) $\overline{p}\,(G) = 1$.

And we take N to be the least class of models of type τ including N_0 and closed under isomorphism.

In the above structure, G represents a single consumer good. f is a production function indicating the total quantity z of G produced by any given assignment $< x_0, x_1, x_2, \ldots >$ of labour over time. k indexes the non-stagnant intervals, thus only $x_0, \ldots, x_{k-1}$ may have positive values (3.9). $a_0, \ldots, a_{k-1}$ denote the actual quantities (or optimal quantities) of labour employed in each interval, and A is the total amount of all labour. K represents the total value of capital, i denotes the constant rate of interest (profit), and $\overline{w}$ the constant unitary wage. Finally, the production price $\overline{p}$ is set equal to one (3.11), and y is the total amount of G produced.[9]

Let M_0 be the class of all models in N satisfying:

(3.12) $$f(a_0, \ldots, a_{k-1}, 0, 0, \ldots) = y;$$

(3.13) $$\partial f_j \big]_{a_0, \ldots, a_{k-1}, 0, 0, \ldots} = \overline{w}(1+i)^j \quad (j=0,k-1)$$

where, for each j, ∂f_j is the first partial derivative of f with respect to x_j;

(3.14) $$\sum_{j=0,k-1} a_j = A;$$

(3.15) $$\sum_{j=1,k-1} a_j \overline{w}(1+i)^j = K.$$

Then the class of Wicksell models can be defined as the least class M of structures in N including M_0 and closed under isomorphism. Here, (3.12) and (3.14) express the obvious relations between the various components. (3.13) specifies $a_0,\ldots,a_{k-1}$ as an optimal assignment of labour over the non-stagnant time intervals, and (3.15) expresses the total capital expenditure as the sum of all costs incurred.

Wicksell's theory may now be characterised by:

$$WT = < \tau,N,M > ,$$

and once again we can take $L_{\omega_1\omega}$ as a logic adequate for the theory. As in the case of Sraffa, we have an analogous existence theorem for WT:

(3.16) The class M is non-empty.

The existence of 'solutions' for the equations that define M depends on the shape of the production function f. However, f is customarily taken to satisfy the standard neoclassical property of strict convexity of isoquants. In any case, it can be shown that the class of functions f giving rise to solutions is large, and includes the vast majority of cases considered to be relevant in neoclassical economic theory. Studies of some properties of f that yield solutions with monotonically decreasing relationships between K and i can be found in Laise and Tucci (1980), (1981) and (1982).

4. THE REDUCTION OF WICKSELL TO SRAFFA

In this section we shall define a structural correlation F from a subclass H of Sraffa models onto the class of Wicksell models, and a translation I from sentences of type τ into sentences of type τ', such that relative to a suitable logic L <F,I> is a reduction of WT to ST. Furthermore, we shall indicate how F can be extended to a mapping with domain M' such that, for all models $\mathcal{M}$ in M'-H, F($\mathcal{M}$) is 'almost' a model in M.

We start by defining H to be the class of all Sraffa models in which B(t) is a pseudo-triangular matrix, for all $t \in T$. That is to say,

$$B(t) = \begin{bmatrix} 0 & b_{12}\cdots & b_{1n} & 0 & 0 & 0 & \cdots \\ & 0 & b_{2n} & \cdot & & & \\ & & 0 & \cdot & & & \\ 0 & & 0 & \cdot & & & \\ & & & \cdot & & & \end{bmatrix}$$

where all entries in the matrix on or below the diagonal are zero. Now, under the assumption of constant returns to scale, the expression for the quantity of labour incorporated in c units of g is given by:

$$(4.1)\quad c(I-B(t))^{-1}l(t) = cl(t) + cB(t)l(t) +\ldots+ cB(t)^{n}l(t)+\ldots$$

where I is the unit matrix. Thus, for any $M \in H$ we can define a partial function f_M on $\mathbb{R}^{\omega}$ by setting

$$(4.2)\qquad f_M(x_0,\ldots,x_n,\ldots) = 1 \iff$$

$$\iff \exists t \in T\ (x_0=ul(t)\ \&\ x_1=uB(t)l(t)\ \&..\&\ x_{n-1}=uB(t)^{n-1}l(t)\&..)$$

where $u \in R^{\omega}$ is the (row) vector $<1,0,0,0,\ldots>$. Notice that, since B is pseudo-triangular, $B(t)^m = 0$ for $m \geq n$ if n is the number of non-null industries in M; so terms in the series on the right hand side of (4.1) become zero after the nth.

Now we suppose that B has sufficient continuity properties that the resulting function f_M is C^2.[10] And, by constant returns to scale, f_M will be linear homogeneous: so for any positive real α and tuple $\vec{x} \in \mathbb{R}^{\omega}$, we can set $f_M(\alpha\vec{x}) = \alpha$ whenever $f_M(\vec{x}) = 1$. In this manner the domain of f_M comprises all real multiples of tuples $\vec{x}$ obtained from (4.2).

Let $M = < M_0;C,T,B,l,w,p,c,r,g,t^*,n > \in H$. We define a mapping F from H into structures of type τ by

$F(M) = < M_0;\{G\},f,A,a_0,\ldots,a_{k-1},K,i,\bar{w},y,\bar{p},k >$, where

$G = g$

$k = n$

$f = f_M$

$a_0 = cl(t^*),\quad a_1= cB(t^*)l(t^*),\ldots,\quad a_{k-1}= cB(t^*)^{n-1}l(t^*)$

$A = \sum_{j=1,k-1} a_j$

$i = r$

$\bar{w} = w(t^*)$

$\bar{p}(G) = 1$

$y = f(a_0,\ldots,a_{k-1},0,0,\ldots)$

$K = \sum_{j=1,k-1} a_j\bar{w}(1 + i)^j.$

Lemma (4.3) For all $\mathcal{M} \in H$, $F(\mathcal{M}) \in M$.

Proof. Clearly, the only completely non-trivial task is to show that $F(\mathcal{M})$ satisfies axiom (3.13); the remaining axioms are immediately satisfied from the definition of F. The argument is as follows. From the Sraffa optimality condition (3.7) it can be shown (Garegnani (1975), Lippi (1979)) that r and $w(t^*)$ must obey:

$$(4.4)\quad (I-(1+r)B(t^*))^{-1}l(t^*)w(t^*) = \min_{t \in T} (I-(1+r)B(t)l(t)w(t^*)).$$

In effect this means that there is agreement between the macroeconomic optimality criterion (3.7) ("given r, maximise w in the set of technologies" or "given w, maximise r") and the microeconomic optimality condition, viz. "given w and r, minimise production costs in the set of technologies". Now, once $f_{\mathcal{M}}$ is given, (4.4) can be turned into

$$(4.5)\quad \min\ x_0\bar{w} + x_1\bar{w}(1+i) + \dots + x_{n-1}\bar{w}(1+i)^{n-1}$$

$$\text{sub}\quad f_{\mathcal{M}}(x_0, x_1, \dots, x_{n-1}, \dots) = 1.$$ [11]

And from (4.5) we can obtain a system of equations based on the partial derivatives of $f_{\mathcal{M}}$ at each argument by employing the usual Lagrange multiplier technique for constrained minimisation, keeping in mind that $f_{\mathcal{M}}$ is a zero homogeneous function. Therefore, substituting a_j for x_j $(j=0,n-1)$ and setting $\bar{p}(G) = 1$, (3.13) is satisfied. □

Next, it is clear that by varying the set of technologies and the number of non-null industries, and by choosing B and l appropriately, the method sketched above allows us to obtain every Wicksell model from some Sraffa model obeying the pseudo-triangularity condition. Thus we have:

Lemma (4.6) For all $\mathcal{N} \in M$ there is an $\mathcal{M} \in H$ such that $F(\mathcal{M}) = \mathcal{N}$.

Combining (4.3) and (4.6) we obtain the required result,

Theorem (4.7) WT is reducible to ST.

Proof. We have already demonstrated the existence of a structural correlation F between the two theories that satisfies the first condition of reduction, (2.1). It only remains to show that there is some logic L in which the relevant classes M, H and M' can be defined and some translation I between L-sentences such that (2.2) and (2.3) hold. Clearly, for L we can again choose $L_{\omega_1\omega}$ which is obviously strong enough to define H (in the type τ')! A moment's reflection also shows that this logic suffices for characterising definitions of each symbol of τ by formulas in the type τ' of ST. The appropriate definitions can simply be read off from the conditions that define F(M) from M in each case. Thus it is a straightforward matter to specify a recursive mapping I of all L-formulas of type τ into L-formulas of type τ' which satisfies

$$\forall M \in H, \forall \varphi \in Sent_L(\tau) \quad (M \models_L I(\varphi) \iff F(M) \models_L \varphi)$$

and so meets the remaining requirement for reduction (2.3). Once again, the only completely non-trivial part of the translation concerns the definition of Wicksell's production function f. Actually, rather than regarding f as a function with countably many argument places we can more easily treat it as a unary function from the vector space $\mathbb{R}^\omega$ into the reals. Then, if 'α', 'v' and 't' denote variables ranging over scalar reals, vectors (in $\mathbb{R}^\omega$) and technologies, respectively, and if Pr_i ($i<\omega$) is the projection function that collapses any vector to its ith component, f can be defined by

$$\forall\alpha\forall v \; (f(\alpha v) = \alpha \iff \exists t \; (Pr_1(v) = ul(t) \;\&\; Pr_2(v)=uB(t)l(t) \;\&\ldots\&\; Pr_n(v)=uB(t)^{n-1}l(t) \;\&\ldots));$$

where the defining formula is a countable conjunction, i.e. an expression of $L_{\omega_1\omega}(\tau')$. From this we conclude that there is a reduction, relative to $L_{\omega_1\omega}$, of WT to ST.[12] □

The model-theoretic analysis of Wicksell and Sraffa can be pushed a stage further. Suppose we take an arbitrary Sraffa model $M \in M'$ for which B(t), instead of being required to be pseudo-triangular for each $t \in T$, satisfies only the usual condition of <u>vitality</u>.[13] As before, the quantity of labour per c units of commodity is expressed by (4.1), i.e.

$$c(I-B(t))^{-1}l(t) = cl(t) + cB(t)l(t) +\ldots+ cB(t)^n l(t) +\ldots$$

where now this series is infinite: terms do not generally vanish to zero after a fixed, finite number. However, it can be shown that this series <u>converges to zero</u>, i.e. if we substitute x_0 for the first term, x_1 for the second, and so on, then $x_n \to 0$ as $n \to \infty$. Now, we can define from M a structure of type τ in the same way as before, using the definition of f_M in (4.2). The resulting structure, $F(M)$, is not strictly a Wicksell model, since the number k of non-stagnant time intervals is now infinite, and (3.9) is not satisfied. But it is a routine matter to verify that the basic axioms of WT, (3.12)-(3.15), are true in $F(M)$ (making the obvious changes to the summations, etc.), because the Wicksell optimality conditions hold in the Sraffa model, and solutions for $a_0, a_1, \ldots$ are obtainable as in the previous case. Accordingly we may call $F(M)$ a limit model of Wicksell; notice that the convergence of (4.1) means that all the relevant sums in the limit model - quantity of labour, capital, net product, etc. - are finite.

How can we interpret the situation just described ? One approach would be to drop the requirement on admissible structures for WT that only finitely many time intervals are economically significant. In this case the limit models could be regarded as genuine Wicksell models, where the theory is redescribed by replacing the original model classes N and M by appropriate (larger) classes, N_1 and M_1, say. Then the reduction of WT to ST would be characterised by extending the original map F to a function from M' into M_1 and applying essentially the same translation I as before. This would not amount to a very drastic alteration of the theory, since the basic Wicksellian relations between labour, production, capital, wage, and so forth, remain the same. What might be objected is that this move constitutes a conceptual change in the theory which makes it less historically faithful to Wicksell's writings.

If this objection is taken seriously, one may proceed by retaining the earlier formulation of WT and by noting instead that any limit model $F(M)$ as defined above is approximately a Wicksell model, in the sense that one can find models in M that are arbitrarily 'close' to $F(M)$ by choosing larger and larger finite values of k. In this context, the proximity of two models means simply that the values of the exogenous and endogenous components are 'almost' the same. So, a model $M' \in M$ is 'close to' $F(M)$ if $a_j^{F(M)} = a_j^{M'}$ $(j<k)$ and $a_j^{F(M)} \approx a_j^{M'}$ $(j \geq k)$; $A^{F(M)} \approx A^{M'}$; $K^{F(M)} \approx K^{M'}$; $y^{F(M)} \approx y^{M'}$; etc. To make this idea precise one could define a suitable topology on M and then show that each limit model is in fact the limit of a convergent sequence of Wicksell models. Under this interpretation one would say that the reduction of WT to ST is 'composed' of two structural cor-

relations: an exact correspondence between H and M, and a limiting or approximate correspondence between M'-H and M.[14]

5. CONCLUDING REMARKS

Let us now try to asses something of the significance of the preceding results for the wider context of economic methodology.

First, we should deal with a question that inevitably arises when one scientific theory T is purported to be reducible to another theory T': Can the direction of the reduction be reversed; that is, can T' also be reduced to T ? In our framework, this question has a clear and precise meaning. A reduction from ST to WT would have to include a model-theoretic mapping F' from some subclass of M onto M' and a 'compatible' translation I', say, from τ'-sentences into sentences of type τ. Prima facie it cannot be excluded that some such pair <F',I'> exists, but there are several formal and informal considerations that make it highly unlikely in practice. And, there are quite solid grounds for supposing that such a relation, if unearthed, would have to involve an unorthodox reconstrual of the theories.

What can be easily observed is that the present reduction <F,I> of Wicksell to Sraffa cannot straightforwardly be inverted to obtain a reduction in the other direction: F' could not simply be an inverse of F, nor I' an inverse of I. The reason is that, as far as the productive side of the economy is concerned, Sraffa's conceptual framework is the more expressive of the two. It includes a more complete description of the employment of labour and commodities in the production process. In terms of structures, this is reflected by the fact that in general a single Wicksell model corresponds via F to a class of Sraffa models.[15] And of course syntactically it means that Sraffa's technical and labour functions, B and l, cannot be explicitly defined in terms of Wicksell's production function f; hence F and I are not bijective mappings. On the other hand, we have already seen that, from the logical point of view, the additional complexity of WT on the properly economic constraints is no obstacle to translation; for, 'capital' is formally defined by the axioms of WT, and so may be replaced by terms that have a direct translation into the language of Sraffa.

If we agree that there is no plausible means to reduce ST to WT, we are left with the following question: What does this apparently asymmetrical relation between the theories imply on a broader philosophical and methodological front ? We cannot attempt anything like a complete answer in the present paper, but we shall close with a couple of related remarks that seem worth bearing in mind. The first observation relates to what in our view is not implied by the formal results, the second to what can be legitimately concluded from the logical analysis of Sraffa and Wicksell.

The conclusion that should not be too hastily drawn is that the reducing theory is somehow more fundamental or more progressive than the reduced theory. There is of course a natural sense in which the word 'reduction' is linked to the concept of scientific progress. The association typically arises in one of two kinds of situation. In one case a theory is superceded by a better theory dealing with essentially the same domain of phenomena. It may happen, and some philosophers would insist that it always happens, that the earlier theory is reducible, at least approximately, to its successor. If so, the process of scientific change exhibits an obvious measure of continuity; yet the supplanting theory is held to be more progressive than its predecessor in that it corrects mistakes in the latter, it solves more problems, or it achieves a more complete explanation of the 'facts'. On the other hand, some alleged instances of reduction involve the explanation of an accepted scientific law or theory by means of some more comprehensive or fundamental law or theory. In this case the domains of the two theories may differ, as when macro-laws concerning large scale effects are explained by micro-laws dealing, for example, with the behaviour of atoms, molecules, genes, and so forth. The domain of the reducing theory then underlies, so to speak, the domain of the reduced theory; and again the former is held to be more progressive in virtue of its yielding a 'deeper' kind of explanation. These reduction types, with their consequent emphasis on the superiority of the reducing theory, are a commonplace in the physical and perhaps biological sciences. Physics is especially well-placed in this respect, because it is rich in widely-accepted (if occasionally contested) cases of reduction, and because the concept of progress in physics is rather well-developed notwithstanding the presence of many conflicting accounts of progress. The same is not always true of other scientific disciplines. For example, the question of progress hardly ever arises in connection with instances of reduction in mathematics; and economics owns neither a large supply of accepted cases of reduction, nor a well-articulated theory of economic progress.

What mathematical economics does offer is a rich source of potential, and for the most part unexplored, instances of reduction. It is too soon as yet to judge whether the pattern of reduction in economics resembles that of physics, say, when seen from an epistemic, ontological or methodological point of view. (Of course, experience drawn from other disciplines may lead us to ask interesting questions; e.g. Can macroeconomic laws be reduced to microeconomic laws ?) In fact, on the evidence of the present case study, reduction in economics might well have a quite distinctive flavour. With Sraffa and Wicksell we are clearly dealing with rival theories, neither of which has superceded or replaced the other, and neither of which can claim to provide, from a conceptual vantage point, a more primary or fundamental foundation for economic analysis. These and other factors besides suggest that here we are dealing with an example that does not tightly fit the paradigmatic pattern of reduction in the physical sciences; and, in particular, it would be rash to infer any premature value judgements regarding the superiority of the reducing theory over the reduced.

This is not to say that no interesting conclusions follow from the reduction of Wicksell's theory to Sraffa's. What is striking about this case from a methodological standpoint is that it shows how much basic agreement there is between two theories that are not only conceptually different but even belong to rival economic research traditions with all their accompanying political and ideological differences. To gauge the measure of similarity existing between ST and WT, let us run through the implications of the model-theoretic connections in more detail.

Economists often talk about the 'consistency' or 'inconsistency' of their theories. They do not usually have in mind precisely the logical concept of consistency, but the latter is no doubt a part of their meaning. Indeed, the present analysis provides some clues about the logical consistency of ST and WT. First, there is the standard 'semantic' claim of consistency: that each theory possesses a model, expressed by (3.8) and (3.16). Secondly, the reduction of WT to ST brings additional evidence of their mutual consistency in the syntactical sense, for the reduced theory cannot be formally inconsistent - in the sense of entailing a contradiction - unless the reducing theory is too. It should also be clear that the two theories are mutually compatible in that they share common models; allowing for translation. Moreover, each Wicksell model is definable in a suitable sense from a Sraffa model, which tells us that there is no question of one theory yielding economically viable 'solutions' where the other does not; and this property seems to be close to the economists' idea of relative consistency. Finally, let us consider the purely deductive relation between the theories, which amounts to the fact that, under suitable assumptions, the laws of WT are derivable from those of ST; this ensures agreement between them over a wide range of economic 'phenomena'.

Suppose that the class M' of Sraffa models is axiomatised by an $L(\tau')$-sentence θ' and the class M of Wicksell models by an $L(\tau)$-sentence θ.[16] Now, the reduction of WT to ST implies that there exist additional assumptions, axiomatised say by σ (defining the class of models H), such that for all $L(\tau)$-sentences ϕ:

$$\theta \models \phi \quad \Leftrightarrow \quad \theta', \sigma \models I(\phi)$$

So, in particular, the translation of any law or consequence of WT is entailed by a suitable extension of ST. And, if the limit models are recognised as genuine Wicksell models, then the deductive relation is still stronger: in this case translations of statements true in WT follow directly from ST with no additional assumptions, i.e.

$$\forall \phi, \qquad \theta \models \phi \quad \Rightarrow \quad \theta' \models I(\phi).$$

And, of course, the same property holds of statements that are anomalous for the theories: any sentence contradicting Wicksell's theory has a translation which is false also for Sraffa.

In the face of this, it is hard to sustain the conventional belief that these two theories are fundamentally at odds in their description of economic 'reality', or to provide any clear grounds for favouring one of them at the expense of the other. Since each theory has a distinctive 'approach' to the subject - ST is more comprehensive on the productive side of the economy, whilst WT assigns capital an explicit role - it cannot be excluded that in some applications, where additional assumptions are brought to bear, one system will be more adequate or illuminating, in some contexts, than the other. But as foundational models for theoretical economics, the two theories are conceptually comparable and in essential agreement under that comparison. In short, the logical study of Sraffa and Wicksell yields no support for those who wish to contrast and differentiate these two strands of the classical and neoclassical approaches to growth economics.

NOTES

1. A host of leading economists participated actively in the debate, including Burmeister (1976), Dobb (1970), Garegnani (1960), Hahn (1982), Robinson (1971) and Samuelson (1962), to mention only a few.
2. For some recent work in this area, see e.g. the various contributions to Balzer, Spohn and Stegmüller (1982).
3. Throughout, we denote by $\mathbb{R}$ ($\mathbb{R}^+$) the nonnegative (resp. postive) reals, and by $\mathbb{N}$ ($\mathbb{N}^+$) the nonnegative (resp. positive) natural numbers.
4. In Sraffa's theory, B, l, r and c would commonly be construed as exogenous terms, w and p as endogenous.
5. Axioms (3.5) and (3.6) are drawn from Chapter 1, and (3.7) from Chapter 3, of Sraffa (1960).
6. Cf. Tucci (1976) and (1979).
7. Cf. Wicksell (1934). This book contains both the theory reconstructed here as well as the more complex theory, with land, that is usually associated with Wicksell. As far as the comparison with Sraffa is concerned, our choice is not restrictive, however. In principle one could incorporate the component of land into ST and then compare this theory to the more complex Wicksell theory. The results of such an analysis should resemble in all essential features the results obtained below.
8. Wicksell's concept of 'capital' is clearly not the same as 'money' in the usual sense; he refers to K as "merchandise credit".
9. In Wicksell's theory, f, K, A and $\bar{p}$ are usually taken as exogenous terms, $a_0,\ldots,a_{k-1}$, $\bar{w}$, i and y as endogenous.
10. We leave it as an exercise for the eager mathematically inclined reader to express the exact continuity properties that B[T] and l[T] should possess in order for f_M to be C^2 for any model .
11. This passage relies on standard textbook material such as can be found in Henderson and Quandt (1971).
12. If we think of the definition of f as being given in the form $\forall v(f(v) \leftrightarrow \exists t\ \phi(v,t))$, then clearly ϕ can be taken as a countable conjunction of first order formulas. However, it is clear that _in each model in_ H f is defined by a _finite_ formula, since terms in ϕ are all of the form '0=0', after a finite stage. This suggests that with some additional notational complexity we could rework the definition of f in a first order language and therefore define a translation between first order formulas of τ and τ'. We shall not pursue this possibility here, however.

13. For the condition of vitality, see e.g. Pearce and Tucci (1982a).
14. For the method of handling approximate reduction relations by means of converging sequences of models in a topological space, see e.g. Mayr (1981) and the contributions of Ludwig, Mayr, Scheibe and Schmidt in the present volume. However, the method of nonstandard analysis employed by Pearce and Rantala (1983b) and elsewhere in this volume can also be applied to the example at hand. This would yield a third and intuitively appealing perspective on the limits models. To use this method here we would assume that the mathematical basis of WT could be constituted by nonstandard analysis, so that M_0 could be taken as either a standard or nonstandard model of analysis. Now, the essential point about a limit model F(M) is that in it the sequence $a_0, a_1, \ldots$ converges to zero. So, starting with any standard model $M \in M'$-H, we could define F(M) to be a nonstandard Wicksell model in which k is a nonstandard infinite natural number and a_j is infinitesimal for all infinite j. F(M) would then be approximated, in a suitable sense, by a (standard) Wicksell model formed by taking the standard parts of each individual and relation in F(M); for a general description, see Pearce and Rantala (1983b) and (this volume). This approach requires no significant conceptual alteration to WT, since the limit process is completely characterised in the underlying mathematics. At the same time it retains the regular form of reduction as a mapping from M' to M.
15. There may exist a translation between two theories even if the structural correlation is defined by a (many-many) relation instead of a (many-one) function. However, here a translation from the language of ST into that of WT is unlikely to be forthcoming because for any $M \in M$, models in the class F^{-1}(M) are not even elementarily equivalent, i.e. in agreement on all first order sentences.
16. We shall not write out in full here the axioms θ and θ' since we are concerned only with the deductive form of the connection between ST and WT.

David Pearce
Institut für Philosophie
Freie Universität Berlin
Habelschwerdter Allee 30
1000 Berlin 33, BRD.

Michele Tucci
Istituto di Economia Politica
Facoltà di Economia e Commercio
Università di Roma
Via del Castro Laurenziano
Rome, Italy.

REFERENCES

Balzer, W., Spohn, W. and Stegmüller, W. (eds.): 1982, Philosophy of Economics, Springer-Verlag, Berlin-Heidelberg-New York.

Burmeister, E.: 1976, 'Real Wicksell Effect and Regular Economics', in M. Brown, K. Sato and P. Zaremka (eds.), Essays in Modern Capital Theory, North Holand, Amsterdam.

Burmeister, E.: 1980, Capital Theory and Dynamics, CUP, Cambridge.

Burmeister, E. and Dobell, A: 1970, Mathematical Theories of Economic Growth, MacMillan, London.

Dobb, M.: 1970, 'The Sraffa System and Critique of the Neo-Classical Theory of Distribution', De Economist, 4, 347-62.

Garegnani, P.: 1960, Il capitale nelle teorie della distribuzione, Giuffre, Milan.

Garegnani, P.: 1975, 'Beni capitali eterogenei, la funzione della produzione e la teoria della distribuzione; Nota matematica', in G. Lunghini (ed.), Produzione, capitale e distribuzione, ISEDI, Milan.

Hahn, F.: 1982, 'The Neo-Ricardians', Cambridge Journal of Economics, 6.

Henderson, J. and Quandt, R.: 1971, Microeconomic Theory, 2nd Ed., McGraw-Hill Kogakusha, Tokyo.

Laise, D. and Tucci, M.: 1980, 'Note sulla teoria wicksellliana del capitale', Rivista di Politica Economica.

Laise, D. and Tucci, M.: 1981, 'La teoria del capitale di Wicksell. Una replica', Rivista di Politica Economica.

Laise, D. and Tucci, M.: 1982, 'La teoria del capitale di Wicksell. Una ulteriore replica', Rivista di Politica Economica.

Lippi, M.: 1979, I prezzi di produzione, Il Mulino, Bologna.

Mayr, D.: 1981, 'Investigations of the Concept of Reduction, II', Erkenntnis, 16, 109-29.

Pearce, D. and Rantala, V.: 1983, 'New Foundations for Metascience', Synthese, 56, 1-26.

Pearce, D. and Rantala, V.: 1983b, 'A Logical Study of the Correspondence Relation', Journal of Philosophical Logic.

Pearce, D. and Rantala, V.: 1983c, 'Logical Aspects of Scientific Reduction', in H. Czermak and P. Weingartner (eds.), Epistemology and Philosophy of Science, Hölder-Pichler-Tempsky, Vienna.

Pearce, D. and Tucci, M.: 1982a, 'On the Logical Structure of some Value Systems of Classical Economics: Marx and Sraffa', Theory and Decision, ,

Pearce, D. and Tucci, M.: 1982b, 'A General Net Structure for Theoretical Economics', in Balzer et al. (1982).
Pearce, D. and Tucci, M.: 1984, 'Intertheory Relations in Growth Economics: Walras and Sraffa', forthcoming.
Robinson, J.: 1971, 'The Measure of Capital: the End of the Controversy', Economic Journal, 81, 597-602.
Samuelson, P. A.: 1962, 'Parable and Realism in Capital Theory: the Surrogate Production Function', Review of Economic Studies, 29, 193-206.
Sraffa, P.: 1960, Production of Commodities by means of Commodities, CUP, Cambridge.
Tucci, M.: 1976, 'A Mathematical Formalization of a Sraffa Model', Metroeconomica, 28, 16-24.
Tucci, M.: 1979, 'La determinazione di techniche ottimali in un modello di Sraffa', Istituto Matematico: G. Castelnuovo, University of Rome.
Wicksell, K.: 1934, Lectures on Political Economics, Vol. 1: General Theory, Routledge, London.

L. Hamminga

POSSIBLE APPROACHES TO REDUCTION IN ECONOMIC THEORY

"Idea vera debet cum suo ideato convenire"

Spinoza, Ethica, Pars I, Ax. VI

1. INTRODUCTION

To what extent can the nature of theory development in economics cast light on the philosophical problem of reduction? Nowadays, "reduction" has a rather technical connotation, referring to ever more complicated definitions of this concept that arose in the logical-philosophical reduction programme largely triggered off by Chapter 11 of Nagel's famous "Structure of Science" (Nagel (1961)). This programme developed its successive specifications of the concept of reduction partly inspired by problems in comparing different physical and biological theories, partly by analysis of the logical nature of the concept of reduction itself.

Entering economics on the technical level to which this discussion has been raised, one hardly knows how to start and what to look for. As I have suggested elsewhere [1], many times theoretical economics develops very methodically and regularly, but to describe these strategies of theory development one needs a set of philosophical concepts and axioms different from the ones to which we are used in the philosophy of science: theoretical economics, I claim to have argued succesfully, typically does not aim at "theories" - whatever that may mean - but at proving so called <u>interesting theorems</u>. I have called such proofs <u>propositions of economic theory</u> (PET's), and they can be written as

$$V\ (\ FEA\ ,\ EI\ ,\ S\ \Rightarrow IT)$$

where an interesting theorem (IT) is proven from FEA-conditions (foundations of economic analysis), which are the basic conditions of a research programme (such as the

W. Balzer et al. (eds.), Reduction in Science, 295–318.

Ricardian, the Neo-Classical, etc.) as a whole, and EI-conditions (explanatory ideal), which serve the application of the FEA to a certain topic (like consumer behaviour, labour market, international trade, etc.). V specifies the number of objects (such as goods, productive factors, countries or other individuals) introduced by the equations and variables. The special conditions S are needed in order to render the total set of FEA, EI and special conditions sufficient for IT in V.

The lowest level routine activity in theoretical economics ("normal science") is centred around one such theorem. In their publications participants present new sets of special conditions S from which they have been able to derive IT, with the help of the condition sets FEA and EI (which are not subject of discussion on this level), in some field (e.g. two goods, two factors, two countries). Such a new S has to be weaker (weakening of conditions result) or incomparable (alternative conditions result) with the old sets S from which IT has already been derived in the literature. A third strategy in a normal programme is deriving some of the special conditions that are in the literature already from other, new, special conditions (conditions for conditions result). The fourth and last strategy is field extension, followed by a proof of the theorem in the extended field.

Some sets of normal programmes belong together in the sense that their interesting theorems belong to the same topic. For instance, the neoclassical application programme to international trade contains normal programmes around numerous interesting theorems, among which the Heckscher-Ohlin theorem, the Stolper-Samuelson theorem, the factor price equalization theorem and the Rybczynski theorem are perhaps the best known. Such an application programme is characterized by an FEA-EI-couple (in our example the explanatory ideal consists of the conditions that the production function for any good is the same the world over, and that countries differ with respect to their factor endowments; this is known as the "factor proportions analysis").

Finally, sets of application programmes dealing with different topics belong together in one research programme if they apply the same FEA. There is nothing deep in the notion of a "foundation of economic analysis". The question what to identify as an FEA, properly allowed to be referred to as "meta-aggregation problem", is not too interesting in itself and should be solved by maximizing convenience in writing the history of economics. There are at least two perfect examples of FEA's: the labour value foundation of analysis and the neoclassical foundation of analysis.

Given this systematics, what are we talking about when we use phrases like "economic $theory_1$ reduces to economic $theory_2$"? I have been arguing ((1983), p. 99-100) that economics, using the notion of a "theory", sometimes refer to the set FEA, sometimes to the condition sets FEA and EI together, sometimes to all sets of conditions used in deriving the theorem (FEA, EI and S), and sometimes to the interesting theorem itself. The problem of understanding the phrase is only aggravated once we realize that the modern, technical notions of "reduction" in philosophy often require a theory to be endowed with complicated entities having whole webs of interrelationships that are mapped to other theories by a host of microtubuli called intertheoretic links, relators, and what not.

The best thing to do, given this situation, is a conceptual retreat [2] in order to be able to make a running jump to how economists actually compare the different "rival" results of their intellectual endeavours.

2. REDUCTION COVERING INTERESTING THEOREMS

Such a retreat towards a more general notion of reduction can be guided by the original intention contained in the concept of reduction, to be formulated thus: T_1 reduces to T_2 iff all important jobs that we have T_1 for, can actually be done also by T_2 (whereas the reverse - if you hate symmetry in reduction - is not the case). In other words: T_1 reduces to T_2 if we do not loose anything important if we "drop" T_1 and "maintain" T_2.

Now, discussions about "dropping" and "maintaining" intellectual constructions should nog be looked for in the normal programmes of theoretical economics. The usual results there are PET's, the mathematical truth of which does only very seldomly [3] require discussion after publication. So the kind of problem situation we look for is the extra-ordinary one where economists start to shout at each other like this: "... if it has been worth while to substitute the interdependence theory, as developed by such authorities as Walras, Menger, Jevons, Marshall, Clark, Fisher, Pareto and Cassel, for the classical labour value theory, there is every reason for giving up the labour value analysis in the treatment of international trade problems as well" [4], that is, when economists discuss whether the energy of the scientific community of economists should be directed to application programmes of one, rather than of another FEA. So this obviously is the proper problem situation to study if we are to make some empirical observations concerning how economists actually reduce one "theory" to another.

As I have argued [5], the reason for economists to step over to the neoclassical - "interdependence theory" - application programmes were 1) the derivation of interesting theorems with the help of the neoclassical FEA in areas where the labour value FEA did not yield very straightforward, unambiguous,outspoken results, 2) the belief that more such results could, with the help of the neoclassical FEA be achieved in the future, and 3) the belief that the neoclassical FEA would, in the future, also yield interesting theorems covering the areas in which interesting theorems had already been proven from the old FEA.

Now, to make our running jump, I propose to use the notion reduction at first sight in such a way that it is defined by the third requirement we find economists to make in the problem situation where the question is: to which FEA should we devote the efforts of our scientific community?:

DEF (1) FEA_1 reduces at first sight to FEA_2 iff for all interesting theorems derivable from FEA_1 (with the help of some suitable EI_1 and S_1), there exist suitable sets EI_2 and S_2 such that these theorems have a counterpart, derivable from FEA_2.

This definition, I conjecture, reflects how economists actually understand the situation in which all important jobs that we had T_1 for, can also be done by T_2, or, in which we do not loose anything important if we "drop" T_1 and "maintain" T_2. Some specification of the meaning of the terms occurring in the definition, however, is necessary for the conjecture to deserve its name.

First, factual theory development in economics can only be accounted for if we insert the notion of a counterpart of a theorem: it seems that it is quite frequently not required that exactly the same theorem is derivable, but that a theorem is derivable about roughly the same features of economic reality; sometimes it is tolerated - or even celebrated - if a new FEA enables the derivation of a theorem that says something startlingly different about it.

To illustrate: from the viewpoint of the classical FEA, every theorem on comparative cost differences causing international trade should be phrased in terms of production function differences among countries. These very same features could be - and in fact were - phrased in terms of differences of factor proportions in the Ohlin-Samuelson application programme to international trade, based on the neoclassical FEA. The theorems were not the same, therefore, but some had the same non-theoretical interpretation as their classical "counterparts". As an example of theorems saying something startlingly different about the same features of economic reality we can take the classical and neoclassical theorems on the relationship between labour input ratio's of industries and the exchange ratio's of the goods produced by the industries. Classical theories result in linear dependency of these ratio's as the typical case (or even as the result of an equilibrating process), neoclassical theories treat linear dependency as an implausible, non-typical limiting case.

It is hard to give a general, logically unambiguous description of what is felt to be a "counterpart of a theorem", but it may be a consolation that this contrasts with the ease in which working economists discern some theorem as the counterpart of another.

Next, we have something to say about the suitability of some sets EI and S to some FEA. Without any restriction pertaining to suitability we could reach the state in which every FEA reduces to every other one quite easily by simply including all interesting theorems in EI and S and subsequently derive them proudly from this set of conditions containing them. "Suitability", therefore, has an important role in defining the economic puzzle: it forms a bar to "ad hoc" strategies of derivation. Again, a general, logically unambiguous description is hard to achieve. Some necessary conditions for the suitability of EI - and special conditions to some FEA, however, are that the EI - and S-conditions should be restrictions on the functions and "data" introduced by the FEA, and that they should not contain terms that are neither defined on these functions nor used in the nontheoretical formulation of the theorem to be derived. Thus, in the neoclassical research programme, EI- and S-conditions are about signs of derivatives (of various orders) of production and utility functions, sets of admitted factor endowment configurations, restrictions on combinations of such derivatives and configurations (such as "elasticities"), and so forth. As for the negative requirement: if the theorem is about "tariffs", EI- and S-conditions may contain this term, and terms defined upon it (like: "demand exerted with tariff income by the government"), but they are not allowed to include new primitive terms. More meta-economic research into the exact nature of the undeniably operative principles of "suitability" seems interesting indeed.

The third thing to note about this definition is that one can never be sure (in a logical or mathematical sense) that some FEA reduces at first sight to another, though whether or not it does, is entirely a formal matter. It is tempting to state this in Platonic terms: only the gods see all the possibilities of derivation, with the help of all possible and suitable EI's and S's. Human economists have, at any time, beliefs about what the gods see, they have these beliefs on the basis of what they see, and they need these beliefs to decide in which programmes to participate. This is another reason why reducibility of theories never stops being a controversial issue in economics. The more specified versions of the concept that I will develop below, will retain this feature. It explains why some application programmes, for instance the neoclassical application programme to international trade, were taken up, and attracted participants before there were PET's proving all the neo-classical interesting theorems for which there were classical counterparts. The neoclassical application programme to international trade started (i.e. succeeded in getting participants) after 1933, but it was not until 1949, by means of the Metzler results on the Stolper-Samuelson theorem [6]), that it "caught up" with the classical programme on the issue of how the gains from international trade are actually divided among countries. It also explains why there always remained some economists at work on labour value theorems. Up to the present day, articles appear on the "transformation problem" and the interpretation of the relationship between "value" and "price"-systems in classical theories. On the other hand, there are groups of economists still occupied with what is called the "neoricardian" programme, who operate on the basis of a reconstructed version of the Ricardian analysis, drop its labour value form but maintain Ricardo's specific articulation of the concept of "capital".

No one can be sure which of these (and many other) groups of theoretical economists really have the future, all struggle for the reduction of other programme's FEA's to their own FEA, all aim at being able to proudly display - as special cases of their own approaches - all interesting features of rival theories, and more.

3. REDUCTION COVERING OPERATIONALLY MEANINGFUL THEOREMS ONLY

On the first two pages of his "Foundations of Economic Analysis" (1947), P.A. Samuelson states: " .... only the smallest fraction of economic writings, theoretical and applied, has been concerned with the derivation of operationally meaningful theorems. In part at least this has been the result of the bad methodological preconceptions that economic laws deduced from a priori assumptions possessed rigor and validity independently of any empirical human behaviour .... The economist has consoled himself for his barren results with the thought that he was forging tools which would eventually yield fruit. The promise is always in the future; we are like highly trained athletes who never run a race, and in consequence grow stale" [7]. Samuelson's introductory definition of an operationally meaningful theorem was "simply a hypothesis about empirical data that could conceivably be refuted, if only under ideal conditions" [8], and one can hardly mention a book that had more influence on the post war requirements of logical form in communicating economic ideas.

The metatheory summarized in section 1 suggests the following interpretation of the impact of Samuelson's "Foundations". Samuelson, addressing himself to economists mainly - though not only - of the neoclassical creed, gave a more specific formulation of the requirements a theorem must meet to be an interesting theorem. And this formulation will allow us here to jump to a more specified definition of reduction at second sight.

We can restate Samuelson's exposition of the "problem situation in which the [economist] finds himself" [9] by presuming him to be endowed with a system of (n+m) variables affecting each other's values, most generally to be written down as

$$f_i(x_1, \ldots, x_{n+m}) = 0 \quad (i=1, \ldots, n) \tag{1}$$

If m=0 we can simply solve this set of equations, if independent. In economic theories, however, $m > 0$, and we have to choose a set of n endogenous variables assumed to form the system analysed, where the other m variables are assumed to be exogenous, to the shocks of which the system is exposed, as a result of which the solution values of the endogenous variables ($x_1, \ldots, x_n$) change. (Solution values are often called equilibrium values because economists usually assume that, after a shock, it takes time for the endogenous variables to follow an adjustment path to the new solution values). Thus we assume a chain of causation, where changes of exogenous variables "cause" changes in endogenous variables. Of course, a necessary condition for being able to derive anything useful from (1) about such a causal process is that the f_i be such that solutions of ($x_1, \ldots, x_n$) exist, and most preferably, that they are unique, both preferably for every part of the ranges of the exogenous variables ($x_{n+1}, \ldots, x_{n+m}$). The search for FEA-, EI- and special restrictions on f_i, $i=1, \ldots, n$ such that the requirements of existence and uniqueness of solutions are met is therefore important, though it leads to an "intermediate good" only: such existence and uniqueness theorems are themselves not operationally meaningful, but they reveal that the FEA, EI and S conditions have reached a degree of specification such as to allow the derivation of such operationally meaningful theorems by adding some more special conditions on slopes and curvatures of the functions used.

What are these operationally meaningful theorems like? Labelling the exogenous variables $(x_{n+1}, \ldots, x_{n+m}) := (\alpha_1, \ldots, \alpha_m)$ we can rewrite

$$f_i(x_1, \ldots, x_n, \alpha_1, \ldots, \alpha_m)=0 \quad (i=1, \ldots, n) \tag{2}$$

An operationally meaningful theorem has the form either of

$$\frac{\partial x_i}{\partial \alpha_j} > 0 \quad \text{or of} \quad \frac{\partial x_i}{\partial \alpha_j} < 0 \quad \text{or of} \quad \frac{\partial x_i}{\partial \alpha_j} = 0 \tag{3}$$

An operationally meaningful theorem is, therefore, a qualitative statement about the direction of the displacement of the solution value for some endogenous variable (x_i), occurring if the system receives a shock of a certain direction from some exogenous variable (α_j). Examples: if tax is on gross sales and tax increases, output is reduced; if tax is a percentage of profits, a change of its rate does not affect output; if a price percentage tariff is imposed upon goods imported in a country and we raise the tariff, this will raise the remuneration of the scarce productive factors in that country.

It was the purpose of Samuelson's "Foundations" to show the standard modes of deriving such kinds of theorems. One important method consists of deriving systems like (2) themselves as solutions of maximization problems, reflecting the maximization of profit and utility as it forms part of the neoclassical FEA, another was requiring some dynamic version of (1) to converge to its solution value. These additional requirements both yield special conditions (on the slopes and curvatures of functions), which are used to derive operationally meaningful theorems.

A more specified definition of reduction is suggested by this procedure, described by Samuelson and followed by a host of theoretical economists:

DEF (2) FEA_1 reduces at second sight to FEA_2 iff for all operationally meaningful theorems derivable from FEA_1 (with the help of some suitable EI_1 and S_1), there exist suitable sets EI_2 and S_2 such that these theorems can also be derived from FEA_2.

Methodological principles reflecting this definition seem to be actually used in practice, that is, in practical arguing in technical contexts. Samuelson's own reasons to accept the neoclassical FEA-EI couple for international trade were: "So long as we stick to Ricardian or Taussigian simple arithmetic comparative cost examples, involving only one labour factor of production, we must assume as axiomatic, and unexplained, differences in labour effectiveness in different regions" [10]. The comparative cost theorems of Ricardo's international trade theory, however, can all be derived from neoclassical conditions as well. The main new achievement of the neoclassical FEA on the level of operationally meaningful theorems was that it allowed the derivation of comparative cost differences from factor endowment differences, where the Ricardian FEA simply had to assume these cost differences.

4. REDUCTION COVERING EMPIRICAL CONTENT

What is the reason for considering the operationally meaningful theorems to be the things that really matter in judging what the FEA, or basic theory [11] has to say? For Samuelson, this is the conceivable refutability, "if only under ideal conditions", of operationally meaningful theorems. This can be illustrated as follows: suppose we have an economic theory, where m=n=2:

$$f_1(x_1, x_2, \alpha_1, \alpha_2)=0 \qquad (4)$$
$$f_2(x_1, x_2, \alpha_1, \alpha_2)=0$$

There are, in that case, four partial derivatives about which we could try to prove an operationally meaningful theorem. These are

$$\frac{\partial x_1}{\partial \alpha_1} \quad \frac{\partial x_2}{\partial \alpha_1}$$
$$\frac{\partial x_1}{\partial \alpha_2} \quad \frac{\partial x_2}{\partial \alpha_2} \qquad (5)$$

These potential operationally meaningful theorems are thought to be conceivably refutable because we can, in some neighbourhood, rewrite (4) in difference from

$$\Delta x_1 = \frac{\partial x_1}{\partial \alpha_1}\Delta\alpha_1 + \frac{\partial x_1}{\partial \alpha_2}\Delta\alpha_2$$
$$\qquad (6)$$
$$\Delta x_2 = \frac{\partial x_2}{\partial \alpha_1}\Delta\alpha_1 + \frac{\partial x_2}{\partial \alpha_2}\Delta\alpha_2$$

There are 3^2=9 possible exogenous "shocks", to be called α-displacements, on the system, represented by table 1.

	$\Delta\alpha_1$	$\Delta\alpha_2$
1	+	+
2	0	+
3	−	+
4	+	0
5	0	0
6	−	0
7	+	−
8	0	−
9	−	−

Table 1: All possible α-displacements for m=n=2

It can easily be seen that, if some signs of (5) are known, some x-displacements are excluded, given the occurence of some α-displacement. So, for instance, if $\partial x_1/\partial\alpha_1 > 0$, then α-displacement 4 of table 1 excludes $\Delta x_1 < 0$. How to deal with this problem in a more general way? For that purpose, table 2 displays the set R of all possible displacements relevant to the theory; there are, of course, $3^4=81$.

This "universe" $R = \{r_1, r_2, \ldots, r_{81}\}$ of 81 possible variable displacements allows us to proceed in a straightforward "falsificationist" way [12]: Let us define the empirical content E(FEA) of an FEA as the set of all variable displacements r which are "ruled out" by means of the set FEA, with the help of suitable sets EI and S.

Suppose, for instance, some FEA, EI and S allow the derivation of the operationally meaningful theorem $\partial x_1/\partial\alpha_1 > 0$. This "rules out" $r_{29}, r_{30}, r_{32}, r_{33}, r_{35}, r_{36}, r_{46}, r_{47}, r_{49}, r_{50}, r_{52}, r_{53}$.
These rows, a subset of R, clearly belong to the empirical content of the FEA as defined above: the rows are, to put it in straightforward Popperian terms, "potential falsifiers" of the FEA: if they are "observed", they would constitute an instance of empirical falsification.

Now how to determine the complete class of potential falsifiers, i.e. the empirical content of some FEA-EI-S conbination?

First: if the FEA does not allow the derivation of any operationally meaningful theorem in (5), there still is a set $E_0 = \{r_{37}, r_{38}, r_{39}, r_{40}, r_{42}, r_{43}, r_{44}, r_{45}\}$ which is ruled out. That is, we have Samuelson's conceivable refutability without any operationally meaningful theorem derived, merely by demanding our functional system not to shift! We should require that, as D.F. Gordon already pointed out in his (1955), if we want to adhere to refutability: the allowance of unspecified shifts of functions would be the end of any falsifiability [13].

Secondly, the sets of rows excluded by $\partial x_1/\partial\alpha_1 > 0$, to be denoted by $E(\partial x_1/\partial\alpha_1 > 0)$, and the set $E(\partial x_1/\partial\alpha_2 > 0)$, are depicted self evidently in table 2, column (7) and (8).

Thirdly, if we have signs of the partial derivatives of one endogenous variable to all (in our case: 2) exogenous variables, we can jump to sectors of table 2 where all displacements of exogenous variables (α-displacements) are non-zero. This yields 12 newly excluded rows in column (9).

Table 2: All possible variable value displacements for m=n=2

(1) r	(2) Δx_1	(3) Δx_2	(4) $\Delta\alpha_1$	(5) $\Delta\alpha_2$	(6) E_o	(7) $E(\partial x_1/\partial\alpha_1>0)$	(8) $E(\partial x_1/\partial\alpha_2>0)$	(9) $E(\partial x_1/\partial\alpha_1>0$ <u>and</u> $\partial x_1/\partial\alpha_2>0)$	(10) $E(\partial x_2/\partial\alpha_1>0)$	(11) $E(\partial x_2/\partial\alpha_2>0)$	(12) $E(\partial x_2/\partial\alpha_1>0$ <u>and</u> $\partial x_2/\partial\alpha_2>0)$	(13) $E(\partial x_i/\partial\alpha_j>0$, $i=1,2$; $j=1,2)$	(14) <u>A priori</u> probability
1	+	+	+	+									1/36
2	0	+	+	+				x				x	1/36
3	−	+	+	+				x				x	1/36
4	+	0	+	+							x	x	1/36
5	0	0	+	+				x			x	x	1/36
6	−	0	+	+				x			x	x	1/36
7	+	−	+	+							x	x	1/36
8	0	−	+	+				x			x	x	1/36
9	−	−	+	+				x			x	x	1/36
10	+	+	0	+									0
11	0	+	0	+			x	x				x	0
12	−	+	0	+			x	x				x	0
13	+	0	0	+						x	x	x	0
14	0	0	0	+			x	x		x	x	x	0
15	−	0	0	+			x	x		x	x	x	0
16	+	−	0	+						x	x	x	0
17	0	−	0	+			x	x		x	x	x	0
18	−	−	0	+			x	x		x	x	x	0
19	+	+	−	+									1/36
20	0	+	−	+									1/36
21	−	+	−	+									1/36
22	+	0	−	+									1/36
23	0	0	−	+									1/36
24	−	0	−	+									1/36
25	+	−	−	+									1/36
26	0	−	−	+									1/36
27	−	−	−	+									1/36

(1)	(2)	(3)	(4)	(5)	(6)	(7)	(8)	(9)	(10)	(11)	(12)	(13)	(14)
	[continued]												
r	Δx_1	Δx_2	$\Delta\alpha_1$	$\Delta\alpha_2$	E^o	$E(\partial x_1/\partial\alpha_1>0)$	$E(\partial x_1/\partial\alpha_2>0)$	$E(\partial x_1/\partial\alpha_1>0$ <u>and</u> $\partial x_1/\partial\alpha_2>0)$	$E(\partial x_2/\partial\alpha_1>0)$	$E(\partial x_2/\partial\alpha_2>0)$	$E(\partial x_2/\partial\alpha_1>0$ <u>and</u> $\partial x_2/\partial\alpha_2>0)$	$E(\partial x_i/\partial\alpha_j>0,\ i=1,2\ ;\ j=1,2)$	<u>A priori</u> probability
28	+	+	+	0									0
29	0	+	+	0		x		x				x	0
30	–	+	+	0		x		x				x	0
31	+	0	+	0					x		x	x	0
32	0	0	+	0		x		x	x		x	x	0
33	–	0	+	0		x		x	x		x	x	0
34	+	–	+	0					x		x	x	0
35	0	–	+	0		x		x	x		x	x	0
36	–	–	+	0		x		x	x		x	x	0
37	+	+	0	0	x	x	x	x	x	x	x	x	0
38	0	+	0	0	x	x	x	x	x	x	x	x	0
39	–	+	0	0	x	x	x	x	x	x	x	x	0
40	+	0	0	0	x	x	x	x	x	x	x	x	0
41	0	0	0	0									0
42	–	0	0	0	x	x	x	x	x	x	x	x	0
43	+	–	0	0	x	x	x	x	x	x	x	x	0
44	0	–	0	0	x	x	x	x	x	x	x	x	0
45	–	–	0	0	x	x	x	x	x	x	x	x	0
46	+	+	–	0		x		x	x		x	x	0
47	0	+	–	0		x		x	x		x	x	0
48	–	+	–	0					x		x	x	0
49	+	0	–	0		x		x	x		x	x	0
50	0	0	–	0		x		x	x		x	x	0
51	–	0	–	0					x		x	x	0
52	+	–	–	0		x		x				x	0
53	0	–	–	0		x		x				x	0
54	–	–	–	0									0

(1)	(2)	(3)	(4)	(5)	(6)	(7)	(8)	(9)	(10)	(11)	(12)	(13)	(14)
	[continued]												
r	Δx_1	Δx_2	$\Delta \alpha_1$	$\Delta \alpha_2$	E^o	$E(\partial x_1/\partial \alpha_1 > 0)$	$E(\partial x_1/\partial \alpha_2 > 0)$	$E(\partial x_1/\partial \alpha_1 > 0$ <u>and</u> $\partial x_1/\partial \alpha_2 > 0)$	$E(\partial x_2/\partial \alpha_1 > 0)$	$E(\partial x_2/\partial \alpha_2 > 0)$	$E(\partial x_2/\partial \alpha_1 > 0$ <u>and</u> $\partial x_2/\partial \alpha_2 > 0)$	$E(\partial x_i/\partial \alpha_j > 0,\ i=1,2\ ;\ j=1,2)$	A <u>priori</u> probability
55	+	+	+	−									1/36
56	0	+	+	−									1/36
57	−	+	+	−									1/36
58	+	0	+	−									1/36
59	0	0	+	−									1/36
60	−	0	+	−									1/36
61	+	−	+	−									1/36
62	0	−	+	−									1/36
63	−	−	+	−									1/36
64	+	+	0	−			x	x		x	x	x	0
65	0	+	0	−			x	x		x	x	x	0
66	−	+	0	−						x	x	x	0
67	+	0	0	−			x	x		x	x	x	0
68	0	0	0	−			x	x		x	x	x	0
69	−	0	0	−						x	x	x	0
70	+	−	0	−			x	x				x	0
71	0	−	0	−			x	x				x	0
72	−	−	0	−									0
73	+	+	−	−				x			x	x	1/36
74	0	+	−	−				x			x	x	1/36
75	−	+	−	−							x	x	1/36
76	+	0	−	−				x			x	x	1/36
77	0	0	−	−				x			x	x	1/36
78	−	0	−	−							x	x	1/36
79	+	−	−	−				x				x	1/36
80	0	−	−	−				x				x	1/36
81	−	−	−	−									1/36
TOTALS					8	20	20	44	20	20	44	56	1
EMPIRICAL FORCE					$\frac{8}{81}$	$\frac{20}{81}$	$\frac{20}{81}$	$\frac{44}{81}$	$\frac{20}{81}$	$\frac{20}{81}$	$\frac{44}{81}$	$\frac{56}{81}$	
PRACT. TESTAB.					0	0	0	$\frac{12}{36}$	0	0	$\frac{12}{36}$	$\frac{16}{36}$	

The columns (10) to (12) read the same way. Thus, if the FEA allows - in the case m=n=2 - the derivation of positive signs for all four partial derivatives we end up with 56 of the 81 rows excluded [14].

Let us, as a specification of Popper's notions, define the empirical force of a comparative static FEA as the ratio between the number of rows excluded and the total number of rows. With 56 of the 81 rows excluded, the empirical force would be about 69%. This definition seems to be in harmony with our intuition of "force" of a theory: a theory is said to be "strong" if a great many conclusions can be drawn from it. Strong theories exclude a lot and consequently run high risks of refutation. This also is exactly what Popper wanted to make clear to us (adding that such high risks of refutation are highly recommendable and the hallmark of scientific rationality).

Let us assign zero a priori probabilities to the actual occurence of zero α-displacements [15], and spread the a priori probability equally over the remaining 36 rows, as is done in column (13). Now, the practical testability is defined as the weighted sum of the excluded rows, where the weights are their respective a priori probabilities. In the case treated (where the empirical force was 69%), the practical testability of four positively signed partial derivatives together can be calculated to be 16/36 or about 44%. Again, our intuitions seem to confirm the appropriateness of this definition: FEA's that allow the exclusion of a lot of rows with zero α-displacements ("partial equilibrium analysis") seem to run lower risks for refutation than FEA's that allow the exclusion of a lot of rows with non-zero α-displacements (such as the FEA-EI-couple used to derive the Heckscher-Ohlin theorem).

Indeed, this approach opens up a field for deriving a host of meta-economic theorems about the empirical force and practical testability of the diverse FEA-, EI- and special conditions we have in economics, if not of generalised theorems concerning systems with m endogenous and n exogenous variables, where m and n are arbitrary and R has 3^{m+n} elements. Let us, however, first harvest the relevant concept of reduction: as a first approximation we could state [16] that FEA_1 reduces at third sight to FEA_2 iff (there are suitable sets EI_2 and S_2 such that) $E(FEA_1) \subset E(FEA_2)$, where "$\subset$" stands for the inclusion relation between sets.

There is a problem left, however: this version of the definition only applies to FEA's covering exactly the same variables. Only then can we identify the respective rows excluded out of the very same set R of "all possible displacements relevant to the theory".

To solve this problem, suppose the variables relevant to FEA_1 are $(x_1, x_2, \alpha_1, \alpha_2)$. Table (2) contains all possible displacements relevant to FEA_1. Suppose, next, that FEA_2 contains, besides the variables of FEA_1, one additional variable α_3. To identify $E(FEA_2)$, we need a table of 3.81=243 rows, where to every row of table (2) there correspond three rows in the extended table: one for $\Delta\alpha_3 > 0$, one for $\Delta\alpha_3 = 0$, and one for $\Delta\alpha_3 < 0$. Now, the empirical content of FEA_1 relative to the variable set $(x_1, x_2, \alpha_1, \alpha_2, \alpha_3)$, to be denoted by $E(FEA_1 \mid (x_1, x_2, \alpha_1, \alpha_2, \alpha_3))$ contains exactly three times the number of rows as compared to the empirical content of FEA_1 relative to the variable set $(x_1, x_2, \alpha_1, \alpha_2)$, shortly: $E(FEA_1 \mid (x_1, x_2, \alpha_1, \alpha_2))$. Moreover, quite unambiguously, three rows in $E(FEA_1 \mid (x_1, x_2, \alpha_1, \alpha_2, \alpha_3))$ correspond to every one row in $E(FEA_1 \mid (x_1, x_2, \alpha_1, \alpha_2))$. Thus, the proof of the invariance of empirical force and practical testability under addition of variables is trivial (multiplication of numerators and denominators with the same number).

Let $V(FEA_i)$ denote the variable set of FEA_i, $\cup$ the union of sets and $\subset$ inclusion. Then we can now define: [16]

DEF (3) FEA_1 reduces at third sight to FEA_2 iff (there are suitable sets EI_2 and S_2 such that)

$$E(FEA_1 \mid V(FEA_1) \cup V(FEA_2)) \subset E(FEA_2 \mid V(FEA_1) \cup V(FEA_2))$$

These E-calculations can without problems be performed for all comparative static theories. It can be calculated, for instance, that, in Samuelson's famous treatment of the "Keynesian system" (Samuelson (1947), p. 276 ff.), "Case 1" (p. 278-280) results in an empirical force of $^{454}/_{729}$ or about 62% and a practical testability of $^{72}/_{216}$ or about 34%. [17]

5. RELEVANCE

If the harvest of this contribution would simply consist of three more definitions, one could justly ask whether we had come any further. The use of these definitions is of course the making of claims and expressing beliefs on the "logic of the problem situation in which the scientist finds himself" [18], most preferably claims and beliefs that can be tested against the empirical facts of theory development and reasoning in economics.

With respect to reduction at first sight, I have, in section 2, already claimed that examples abound in theoretical economics. Moreover, this claim is unambiguously supported by results of the case study into the neoclassical theory of international trade [19].

However, it is logically the weakest of the three concepts of reduction and its general applicability is, of course, partly due to the rather unconstrained notion of an interesting theorem, being any theorem that comes under the scrutiny of normal programmes in economics.

It was exactly this admissiveness by economists that prompted Samuelson to restrict the aims of fruitful economics to operationally meaningful theorems, clearly a proper subset of all theorems generally considered to be "interesting". To assess Samuelson's influence on the post war requirements for communicating economic ideas is undeniably trivial: text books on mathematical economics almost invariably treat Samuelson's principles for qualitative economics; the bulk of post war PET construction actually dealt with qualitative ("comparative statical") theorems, and the meta-economical problem of exploring the limits of the general possibilities of deriving qualitative ("operationally meaningful") theorems has become a research programme in itself [20]. Nevertheless, not yet enough research has been done to conclude that reduction at second sight is a usual strategy for comparing rival FEA's.

Though reduction at third sight seems nothing but a natural extension and specification of logical means to clarify a problem situation already present at the level of reduction at second sight, we run into applicability problems: the procedures depicted, though very clearly and easily performable, have to my knowledge, never been performed in the practice of theoretical economic research. The natural place to find such performance would have been the above mentioned meta-economical research programme, but this programme seems to restrict its aims to generalizing and weakening the conditions on functional systems that allow the inference of signs to partial derivatives of endogenous variables to exogenous variables. The programme seems to satisfy the metatheory stated in section 1, though the interesting theorems here are theorems on the derivability of qualitative theorems. But what, by those means is subjected to mathematical scrutiny is not: how qualitative theorems lead to strategies of empirical refutation, but: which strategies of derivation lead to qualitative theorems. Therefore, the conclusion is unescapable and outright that section 4 of this paper started exactly where economists find it appropriate to stop!

This, clearly, calls for either conceding that the metatheory we have introduced is in fact questionable, or for giving an explanation for this fact of economic theorizing.

Let me make an attempt to do the latter. The first explanation for the irrelevance of the empirical content of qualitative theorems is that comparative statics in empirical respect has an overwhelmingly more powerful competitor in the linear regression analysis of first order differences, as D.F. Gordon pointed out as early as 1955 21). Reconsider (6) as:

$$\begin{aligned} \Delta x_1 &= c_{11}\Delta\alpha_1 + c_{12}\Delta\alpha_2 + u_1 \\ \Delta x_2 &+ c_{21}\Delta\alpha_1 + c_{22}\Delta\alpha_2 + u_2 \end{aligned} \qquad (7)$$

where u_1 and u_2 are disturbance terms for individual observations. The only case in which qualitative theorems are more powerful than linear regression on differences is the case where we have only one comparative observation (Δx_1, Δx_2, $\Delta\alpha_1$, $\Delta\alpha_2$) that is, if we have only two observed values for the variables. Then, in the linear regression analysis, the coefficients c_{11}, c_{12}, c_{21}, c_{22} cannot yet be calculated, and consequently (7) does not yet exclude any set of variable value differences to be observed. However, if we dispose of a qualitative theorem attributing a sign to som c_{ij}, it is conceivable that one single comparative observation (Δx_1, Δx_2, $\Delta\alpha_1$, $\Delta\alpha_2$) is indeed a refuting instance.

Suppose we have not one single, but two comparative observations. For a qualitative system, this provides us merely with a second testing instance. For the linear regression system (7), this means we have reached the stage where c_{11}, c_{12}, c_{21}, c_{22} can be solved, using the observations as knowns in the equations and assuming $u_1=u_2=0$: the two substitution instances of (Δx_1, Δx_2, $\Delta\alpha_1$, $\Delta\alpha_2$) yield 4 equations, to be solved quantitatively for our 4 unknowns. This means that a linear system of four first order differences in 2 equations yields quantitative information if we dispose of more than one comparative observation 22).

The conclusion of this must be that the derivation of

qualitative theorems cannot be argued for in terms of empirical content, because we always have more than one comparative observation and can hence obtain far more "operational" (though of course not necessarily true) quantitative information [23].

A second explanation is, that theoretical economists, coming to their interesting conclusions from the opposite direction, that is, not coming "up" from data on time series but coming "down" from conditions originating from the foundations of analysis and explanatory ideals they adhere to, consider the qualitative theorems to reflect fundamental tendencies, which are explicitly allowed sometimes to be offset by subsidiary factors they did not incorporate into their systems for reasons of mathematical manageability and intuitive plausibility. These subsidiary factors are then said sometimes to cause "shifts" of the functions.

If theoretical economists argue for the plausibility of some theorem to hold in the real world, they never do so by pointing at observed time series generating sets (Δx_1, Δx_2, $\Delta\alpha_1$, $\Delta\alpha_2$) that happen not to refute the theorem. As I have shown elsewhere [24], they instead point at the "plausibility" or "generality" of the conditions from which they have been able to derive the theorem.

In recent years, falsificationist ideas, not any more so dominant in present day philosophy of science have, in the methodology of economics been thought to imply a prohibition of irrefutable theories. We must conclude here that this would lead to a rejection of the larger and better reputed part of theoretical economics. A part, moreover, that had many famous and eloquent spokesmen on the level of methodology, such as John Stuart Mill, who wrote: "Economics reasons, and as we contend, must necessarily reason, from assumptions, not from facts .... The conclusions of Political Economy .... are only true .... in the abstract; that is, they are true under certain suppositions, in which none but general causes, - causes common to the whole class of cases under consideration - are taken into account" [25], and John Maynard Keynes who stated: "Economics is a science of thinking in terms of models joined to the art of choosing models which are relevant to the contemporary world. It is compelled to do this, because, unlike the typical natural science, the material to which it is applied is, in too many respects, not homogeneous through time. The object of a model is to segregate the semipermanent or relatively constant factors from those which are transitory and fluctuating so as to develop a logical way of thinking about the latter, and of understanding the time sequences to which they give rise in particular cases" [26].

It is not the task for empirical methodology to judge whether a normative rejection of the object of its enquiry as well as of the authoritative justification given for its nature by those who are generally acknowledged to know how to do the job, is itself justified. For empirical methodology, the conclusion of this paper must be as follows: our three definitions of reduction have "magnified" our metatheoretical observations in economics in successively increasing ratios. As it seems, reduction at third sight leads to an over-magnifying observational apparatus for studying theory development in economics. The more global tools of reduction at first and second sight seem more appropriate for the field: those, who wish to take empirical evidence seriously perhaps should not engage in studying avalanches with microscopes.

The discovery to have done so should, faced with the choice either to criticize the objects of research for not behaving according to some metatheoretical hypothesis, or to improve upon this hypothesis on the basis of the knowledge that led to its falsification, lead to the choice for the latter.

NOTES

The author is a member of the Philosophy Department of the Katholieke Hogeschool Tilburg, The Netherlands. Thanks are due to the Netherlands Institute for Advanced Study for an agreeable research environment, to Drs. A.M. Tingloo-de Jong for correcting the English, to Jacqueline Suos-Wayers for accurate typing under pressure, and to Dr. W. Baltzer, Dr. Th.A.F. Kuipers and Prof. H.N. Weddepohl.

1. Hamminga (1983), p. 66-71.
2. In my (1983) I have dealt with "conceptual retreats" more precisely in terms of a meta-philosophical language (p. 111 etc.).
3. One of the rare examples is Samuelson's proof of univalence (factor price equalization) with the help of the implicit function theorem (Samuelson (1953), p. 16; Gale and Nikaido (1965)).
4. Ohlin (1933), preface.
5. Hamminga (1983), p. 18-28.

6. Metzler (1949a).
7. Samuelson (1947), p. 3-4.
8. Samuelson (1947), p. 3-4.
9. This phrase, where I substituted "economist" for "scientist", is Popper's (1963), p. 241.
10. Samuelson (1948), p. 164.
11. Papandreou (1958), calls FEA's "basic theories" (p. 137 etc.).
12. By "falsificationism" I mean the philosophical tradition that started with K.R. Popper (1934), in which, besides Popper, Imre Lakatos is perhaps the best known (vid. Lakatos (1970)).
13. Gordon (1955): "... an additional foundation of economic analysis will be the hypothesis that the functions used are not unstable ..." (p. 307). The point is that if you claim that $f_i(x_1, x_2, \alpha_1, \alpha_2)=0$ $(i=1,2)$, and that $\frac{\partial x_1}{\partial \alpha_1} > 0$, then you can always "explain" a prima facie inconsistent displacement like $(\Delta x_1, \Delta x_2, \Delta\alpha_1, \Delta\alpha_2)=(-, 0, +, 0)$ by a "shift", or, what is the same, "instability" of f_i.
14. Since it is a priori conceivable that FEA, EI and S conditions yield partial derivatives of positive, negative, or zero sign, independently of each other, there are $3^4=81$ possibilities for having operationally meaningful theorems of all partial derivatives, in the case m=n=2. Table 2 represents only one of these 81 possibilities (column (13)), where

$$\text{sign}\begin{bmatrix}\frac{\partial x_1}{\partial \alpha_1} & \frac{\partial x_1}{\partial \alpha_2}\\ \frac{\partial x_2}{\partial \alpha_1} & \frac{\partial x_2}{\partial \alpha_2}\end{bmatrix} = \begin{bmatrix} + & + \\ + & + \end{bmatrix}$$

15. There is a priori reason enough to do so: real number variables of economic theories change continuously with only very few exceptions. For all i, $\Delta\alpha_i \neq 0$ seems a very plausible a priori assumption.
16. The existential clause is inserted only to facilitate reading. The clause is already contained in the definition of E.
17. This calculation - available on request - runs as follows. We have 729 rows in R. $E_0=26$. The special conditions (62), (Samuelson (1947), p. 277), including

$C_i \approx 0$ (p. 278) allow the exclusion of 168 rows. This means we have even more than E_0 without any operationally meaningful theorem derived! The empirical force is $168/_{729}$ or about 23%, practical testability zero. Special condition (69) (p. 279) excludes 52 more rows, resulting in an empirical force of $220/_{729}$ or about 30%. Practical testability remains zero. Special condition (71) (p. 279) adds 112 rows to the empirical content. Empirical force is now $332/_{729}$ or about 46%. For the first time, 36 rows with non zero displacements for all exogenous variables are excluded: practical testability is now $36/_{216}$ or about 17%. Finally, the special condition "normally savings out of a given income increase with the interest rate, or, if they do decrease, do so not so much as does investment" (p. 280), adds 122 rows, and thus the total empirical content yielded by the procedures in Case 1 is 454 rows out of 729 logically possible displacements, yielding an empirical force of $454/_{729}$ or about 62%, and a practical testability of $72/_{216}$ or about 34%.

18. Popper (1963), p. 241.
19. Hamminga (1983), p. 18-28.
20. Main results in this programme came from Lancaster (1962), Gorman (1964), Lancaster (1965) and Lunghini (1970).
21. Gordon (1955), p. 307: "If a theory is composed of stable functions, then arbitrary or controlled shifts in the independent variables - which are necessary to test any operational hypothesis of these types - will trace out functional relationships .... From these, not only qualitative, but quantitative results can be obtained .. ..".
22. Generally: a linear regression system of n+m first order differences in n equations yields quantitative information if we dispose of more than m-1 comparative observations. The matter is associated with the "identification problem" (vid. Fisher (1966)).
23. Now it may seem some empirical power derives from the fact that for refuting the qualitative systems we only need the signs of the observed differences, whereas for obtaining the quantitative linear information from (7) we need the values of the observed variables. But why try to do without information we usually have?
24. Hamminga (1983), p. 71-97.
25. Mill (1874), p. 149.
26. Keynes (1973), p. 296 f.

REFERENCES

Fischer, F.M.: 1966, The Identification Problem in Econometrics, New York.
Gale and Nikaido, D. and H.: 1965, "The Jacobian Matrix and Global Univalence of Mappings", Mathematische Annalen, 159, pp. 81-93.
Gordon, D.F.: 1955, "Prof. Samuelson on Operationalism in Economic Theory", Quarterly Journal of Economics, pp. 305-310.
Gorman, W.M.: 1964, "More Scope for Qualitative Economics", Review of Economic Studies, XXXI, pp. 65-68.
Hamminga, L.: 1983, Neoclassical Theory Structure and Theory Development, Berlin, etc.
Keynes, J.M.: 1973, The Collected Works of John Maynard Keynes, Vol. XIV, London.
Lakatos, I.: 1970, "Falsification and the Methodology of Scientific Research Programmes", Criticism and the Growth of Knowledge, Lakatos I. and Musgrave A. (ed.), Cambridge Un. Press, Aberdeen.
Lancaster, K.; 1962, "The Scope of Qualitative Economics", Review of Economic Studies, XXIX, No. 2, pp. 99-123.
Lancaster, K.: 1965, "The Theory of Qualitative Linear Systems", Econometrica, Vol. 33, No. 2, pp. 395-408.
Lunghini, G.: 1970, "Qualitative Analysis, Determinacy and Stability", Quality and Quantity, IV, No. 2, pp. 299-324.
Metzler, L.A.: 1949a, "Tariffs, the Terms of Trade, and the Distribution of National Income", Journal of Political Economy, Vol. LVII, pp. 1-29.
Mill, J.S.: 1844, Essays on Some Unsettled Questions of Political Economy 1874, London.
Nagel, E.: 1961, The Structure of Science, London.
Ohlin, B.: 1933, Interregional and International Trade, Cambridge, Mass.
Papandreou, A.G.: 1958, Economics as a Science, Chicago, etc.
Popper, K.R.: 1934, The Logic of Scientific Discovery, London.
Popper, K.R.: 1963, Conjectures and Refutations, London, 1974, Fifth ed.
Samuelson, P.A.: 1947, Foundations of Economic Analysis, Cambridge, Mass.
Samuelson, P.A.: 1948, "International Trade and The Equalization of Factor Prices", The Economic Journal, 58, pp. 163-184.
Samuelson, P.A.: 1953, "Prices of Factors and Goods in

General Equilibrium", Review of Economic Studies 21, pp. 1-20.

Haim Gaifman

WHY LANGUAGE?

I had planned to deal mainly with the concept of translation, but after listening to some talks I have decided that I should present a different point of view and move in a different direction. The direction I will point out is perhaps orthogonal to the Stegmüller-Sneed structural approach based on the advice to ignore language and to use Bourbaki-style structures. So here I want to make a point for language analysis in the philosophy of science. Technically the Bourbaki-style structures can be very useful for bringing out certain points and sometimes one gets very elegant formulations. But when it comes to basic philosophical questions that are related to the philosophy of science, I do not see that you can get around linguistic considerations. One of the reasons is that one does not want to commit oneself ahead of time to a particular interpretation. When trying to analyse a theory without commiting ourselves to a particular ontology, the most obvious handle that we have is the theory's language. The linguistic-oriented approach is the most promising when we want to do philosophy of science in a way that applies not only to classical physics but to very different disciplines like mathematics, physics, psychology and others, and which brings out and illuminates analogies. I am not saying that I am opposed to Bourbaki-style structures; I am saying only that this might restrict us in ways that we do not want to be restricted. In some cases linguistic analysis cannot be avoided.

Note that when doing analysis via Bourbaki-style structure we are working in intuitive set theory, so we are committed to a very particular kind of interpretation. With linguistic analysis one might even consider a different logic altogether; and this is such a radical step that the only way to do it is to approach language "from the outside" and see what kind of logic we have there.

Now the question I want to discuss is a central philosophical question which is naturally related to the philosophy of science - the question of ontology. What is the ontology of a scientific framework ? In other words: What is the reality presupposed by the framework ? What is the reality this theory

W. Balzer et al. (eds.), Reduction in Science, 319–330.

claims to describe ? The question about ontology can be framed in two ways:

(1) What are the entities ?

(2) What constitutes a fact ?

The first way is the one that Quine has focused on and he offered to explicate (1) by rephrasing it as: What do you quantify on ? But the second is, I think, the truer question. That ontology has to do with facts not objects has been stated explicitly by Wittgenstein and it corresponds to the traditional concept of ontology . Now the way I propose to explicate (2) is by the well-known criterion of objective truth values, that is to say bivalence. (2) becomes: What are the statements that have objective truth values ? That is to say, when considering some scientific theory, pay attention to those sentences in the language which are considered by the one who uses the theory as objectively true-or-false. Not knowing the truth-value of a sentence he still believes the sentence to be either true or false. This belief indicates his ontological standing, <u>what he conceives reality to be over and above what he knows of it</u>. Let me illustrate this by means of a small example. Consider the sentence:

"Hamlet killed his uncle."

Suppose someone hears a discussion about Hamlet in which this sentence is asserted. He believes the sentence to be true and that Hamlet was a real person. Now consider:

"Hamlet married Ophelia."

Everyone agrees that this is false; he did not marry Ophelia and he did kill his uncle. Then somebody poses the question:

"Was Hamlet born around midnight?"

And this is a kind of question for which there is no answer. The sentence "Hamlet was born around midnight" is neither true nor false, unless we adopt some specific convention of how to assign it a truth value. The reason is that there is no indication in the play as to whether he was born around midnight or not. The first two statements, one true, the other false, can be decided by what we know from the play. It tells us this part of the story. But the play tells us nothing about the third statement and since Hamlet did not exist outside the text (so we believe) we say that it has no truth value. You see the difference if you compare it with another similar sentence:

"Shakespeare was born around midnight."

I don't know the answer to this one; I do not think that anyone might know the answer in the future, even if he tried very hard. But we believe the sentence to be true-or-false because we think that Shakespeare was born; and when we think that there was such an event it is part of our conceptual framework to be committed to this sort of belief. I do not want to enter here into the intricate question of borderline cases, i.e., cases where it is not clear whether a certain time is midnight or not. We can ignore them because we can see clearly the difference between the two sentences, namely that the first cannot have a truth value, except by some arbitrary convention, and that the second must have a dictated-by-facts truth value, except for the uncommon borderline occurrence.

Now in this example the belief in objective truth values has to do with the belief in somebody's existence. The first sentence has no truth value because Hamlet did not exist, the second sentence has some (unknown) truth value because Shakespeare existed. But sometimes the problem cannot be reduced to the existence of particular objects. Sometimes the having or not having of objective truth values is a feature of the whole system and cannot be traced to particular facts which the system describes. These are the ontologically significant cases.

Here is an example. Suppose somebody asserts that event e_1 precedes event e_2. In Newtonian physics such a statement is objectively true-or-false. But in relativistic physics precedence may depend on the coordinate system from which the events are viewed. If in some coordinate system the events are separated by distance d and time Δt and $d > c.\Delta t$ (c = light velocity) then their temporal order depends on the coordinate system and the above mentioned statement lacks a truth value; thus to ask: does e_1 precede e_2 ? is to ask a factually meaningless question. And this shows the ontological difference between Newtonian and relativistic physics. The difference is often described by saying that in Newtonian physics we have absolute time ordering and in relativistic physics we do not. But the more precise way of expressing the difference is by pointing to the sentences whose true-or-false status differs from one theory to another.

This sort of divergence between two theories, or frameworks, occurs in different disciplines and one cannot but note how fundamental the analogy is. Newtonian versus relativistic physics is one example. Here are some others. In mathematics we have classical against intuitionistic mathematics, where again there is a difference of ontology, that is to say, of what is considered to be objectively true-or-false. Then we have ontological differences between higher order and lower order mathematical theories. A richer ontology is committed to the bivalence of sentences which make assertions concerning sets of natural numbers, sets of sets, etc. For example one might regard the (special) continuum hypothesis as objectively true-or-false, though perhaps we shall never know for sure its truth value. Another might have this sort of belief only with respect to first-order statements in the language of arithmetic; thus he will regard Fermat's conjecture, or the Riemann hypothesis (which is equivalent to an arithmetical first-order statement) as true-or-false but will argue that in the case of the continuum hypothesis there is no factual question of truth but only a formal question of choosing an axiom. Hilbert's Programme was an attempt to reduce mathematical ontology to a certain minimal kernel; the ontological commitment would, by such a reduction, assign the true-or-false status to a narrow class of so-called finitistic statements; the rest of the apparatus would become a tool for organizing, condensing and abbreviating proofs in the finitistic class which are too long and too complicated to be handled directly.

Note the crucial role played by language: the ontology can be given by indicating those linguistic creatures that have objective truth-values. The thing will have to be further clarified through the notion of translation, of which I hope to say something at the end of my talk. Let me pause on another example, this time from psychology. The question concerns the reality of sensations. In its primitive form it appears as:

"Are sensations real things ?"

This gives the general direction, but it is far from a precise question. Consider sensations of colours (to be distinguished of course from the colours themselves which are defined via wavelength). On an everyday level everyone will agree that colour-sensations are real in as much as we experience them. But being " real things" implies that you can compare them like other objective properties: colours, shapes, degrees of hardness, etc. And here we can see at least two different ontologies concerning sensations. One might find a comparison of sensations of people to be factually meaningful and this is the richer ontology, or one might limit the comparison to sensations experienced by the same person and this is the more restricted one. Hence we get the following test question:

"Is a's sensation of x similar to b's sensation of y ?"

In the richer ontology the question has an objectively meaningful answer (whenever a and b are people and x and y - colours). In the more restricted ontology we can compare only sensations of the same person, hence factual meaning is granted only if it is the same person, i.e. only if a = b. The first system admits a 4-place predicate P(, , ,) such that P(a,b,x,y) asserts that a's sensation of x is similar to b's sensation of y. In the second system we have a corresponding 3-place predicate P'(, ,) defined by:

$$P'(a,x,y) \Leftrightarrow_{Df} P(a,a,x,y).$$

By means of this definition some, but not all, sentences of the first system can be translated into the second. And it is only these sentences which are granted an objective true-or-false status by the second ontology. Now one of the arguments against the first system is the apparent difficulty of assigning P(a,b,x,y) its truth value. Suppose, for example, a to be colour-blind, so that he does not distinguish between green and red. So there is an agreement that P(a,a,g,r) is true. Let b have normal vision, so we agree that $\neg$P(b,b,g,r) is the case. Now in the first system it makes perfect sense to ask if a's sensation of green (and red) is similar to b's sensation of green, or whether it is similar to b's sensation of red; perhaps it is similar to neither.

P(a,b,g,g,) ? P(a,b,g,r) ? P(a,b,g,x) ?

These questions are regarded in the second system as factually meaningless and this view sees confirmation in the apparent impossibility of getting them answered (even if we believed in their meaning). But I have read somewhere that, as a matter of fact, researchers did succeed in getting some answers. They found people who are colour-blind in only one eye and they asked them to compare the sensations of the colour-blind eye with those of the normal eye. (Thus you may find in some books that certain colour-blindness consists in seeing both red and green as a kind of yellowish.) This success may count in favour of the first view for, one can argue, it goes to indicate its fruitfulness as a research programme. But this is another matter that does not concern me here.

Let me now take up my main example to illustrate how linguistic analysis can help us in the philosophy of science. It is the Leibniz-Newton debate on the absoluteness of space. The discussion is found in the Clarke-Leibniz letter exchange.

(Newton being too grand a person, Clarke did his debating for him, but Newton supplied the arguments.)

Newton, as is well known, believed in absolute space. To visualize it imagine an empty space without any physical bodies, where each point has an identity of its own. You may stick imaginary labels on these points, naming each. After putting bodies into the space, each will occupy at any given time a certain region, i.e. - a certain set of space-points. Suppose you move the whole body configuration in the empty space, keeping all distances between bodies and body-parts constant. For Newton it will be a different world because the bodies occupy a different region in absolute space. But for Leibniz it will be the same world, for space is determined by physical body configurations and there is no change in as much as the relative positions of all the bodies, with respect to each other, are the same: "Space is nothing else than the order of existing things", "something merely relative", "the order of bodies among themselves" "...has no absolute reality" etc. Thus Leibniz regards the picture of body-configurations that keeps all relative positions and moves in empty space (as, say, a ship moves in water) to be an imaginary creation that has nothing to do with the real world. This is all well known, but the question still remains concerning the real meaning of the difference. Is it more than a debate about the favourable way of visualizing things? Pictures, metaphors, visualizations are common tools for grappling with reality, but the difference may exist only on the metaphorical, or psychological, level. (What difference does it make if I imagine the natural numbers 0,1,2,3,...,n,... in a left-to-right row, or in a right-to-left one ?) We have to reconstruct the Newton-Leibniz debate on a more precise level, being of course as faithful as we can to the historical picture. Here is where the formal language approach comes in.

Let us first consider some language for describing physical bodies in space and time. There are many ways of doing it, of which I shall use the following one:

(I) First there is a purely mathematical part, L_0, in which we can speak about real numbers, addition, multiplication and whatever mathematics that we wish to do.

(II) Next we add to L_0 an apparatus for describing the space-time structure. It is common to represent space points as 3-tuples of real numbers and time points by single numbers. But the representation is not unique, for the coordinate system can be fixed arbitrarily. I shall later use such representations but for the present purpose it is more suitable to have a language which treats space and time directly as primitives. This reconstruction is also more faithful to the historical context of the Leibniz-Newton debate. So let L_1 be obtained from L_0 by adding to it the following items: Variables $u, u_1, \ldots, v, v_1, \ldots$ ranging over time-points; variables $x, x_1, \ldots, y, y_1, \ldots$ ranging over space-points; equality symbols '$=_t$' and '$=_s$' for time and for space ('$u =_t v$' states that the time-point u is equal to v and '$x =_s y$' - that the space-point x is equal to y); '$<_t$' - for temporal ordering and two 4-place function symbols, 'r_t' and 'r_s' whose interpretation is: $r_t(u_1, u_2, u_3, u_4)$ is the real number equal to the ratio of the time interval $[u_1, u_2]$ to the time interval $[u_3, u_4]$ and $r_s(x_1, x_2, x_3, x_4)$ is the number equal to the ratio of the segment $[x_1, x_2]$ to the segment $[x_3, x_4]$. Thus we have a many-sorted language with $r_t(\; , \; , \; , \;)$ and $r_s(\; , \; , \; , \;)$ taking as arguments time-point terms and space-point terms, upon which they become numerical terms. In L_1 we have statements expressing all the structural properties of Euclidean space with an additional directed line. The question of a deductive system is not relevant here because the debate is about true-or-false sentences not about deductions. Both Newton and Leibniz would regard the Euclidean axioms as general truths. Note that in L_1 we cannot refer to any particular spatio-temporal points. Consequently Newton and Leibniz should have no disagreement about the statements of L_1; they agree as far as the general laws of the space-time structure are concerned. The difference appears when we introduce physical bodies and events:

(III) Extend L_1 to L by adding a list $b_1, b_2, \ldots$ of names of physical bodies, a list $e_1, e_2, \ldots$ of names of events and two function symbols τ and σ. Here $\tau(e)$ is supposed to read "the time-point of the event e" and $\sigma(b,u)$ reads "the space-point of the body b at time u". Thus the events are supposed to occur at precise time-points and the bodies are mass-points. This is a simplified picture; but you can regard $\tau(e)$ as the mid-point of the time interval associated with e and $\sigma(b,u)$ as b's centre of gravity at time u.

L is the language I shall consider.

To simplify the notation I shall use the same symbols for linguistic entities and for their interpretation in the model. A model for L is a structure, M, of the form:

$$(M_0, T, S, e_1, \ldots, b_1, \ldots, =_t, =_s, <_t, r_s, r_t, \tau, \sigma)$$

where M_0 is a purely mathematical structure, say $M_0 = (\mathbb{R}, +, \ldots)$. $\mathbb{R}$ = real line; T = set of time-points, S = set of space-points; $e_1, e_2, \ldots$ is a list of events, $b_1, b_2, \ldots$ - a list of bodies;

$=_t$ and $=_s$ are the equality relations for T and S, $<_t$ is the time-ordering of T; $r_t:T^4 \to \mathbb{R}$, $r_s:S^4 \to \mathbb{R}$ are functions; τ is a function mapping $e_1, \ldots$ into T and σ maps $\{b_1, \ldots\} \times T$ into S. Spelled out in detail it may look complicated but in fact it is simple. The language also has connectives and quantifiers but I shall not go into that. For the sake of simplicity I shall use '=' for both '$=_s$' and '$=_t$'. Imagine M to represent our actual world where the e_i's and b_i's are certain unspecified events and bodies.

Now if b is a physical body and e is an event then for Newton $\tau(e)$ is an absolute time-point and $\sigma(b,\tau(e))$ is an absolute space-point; hence he will consider sentences of the following form as objectively true-or-false:

$$\sigma(b_1,\tau(e_1)) = \sigma(b_2,\tau(e_2))$$

$$r_s(\sigma(b_1,\tau(e_1)),\ \sigma(b_2,\tau(e_2)),\ \sigma(b_3,\tau(e_3)),\ \sigma(b_4,\tau(e_4))) = z$$

(where z is a real number). The first asserts that the space-points of b_1 at time $\tau(e_1)$ and of b_2 at time $\tau(e_2)$ are equal; the second - that the ratio of two segments is z, where the end points of the segments are given in the form: the space-point of b_i at time $\tau(e_i)$. But for Leibniz space-points indicate only relative positions at a given time. Hence the first statement is factually meaningful if and only if $\tau(e_1) = \tau(e_2)$. As for the second, Leibniz's view requires that $\tau(e_1) = \tau(e_2)$ and $\tau(e_3) = \tau(e_4)$. (In the way I explicate him he does not require $\tau(e_2) = \tau(e_3)$, because he allows for a comparison of segments defined at different times; e.g. the segments determined by the end-points of the same rigid rod at different times are of equal length.) Unless this condition is satisfied the statement is factually meaningless. Note the analogy with the preceding example from psychology where the equality of certain coordinates is a condition for bivalence.

So far I have considered very elementary sentences, but one can give a general definition for the whole of L which determines the class S_L of sentences which are true-or-false in Leibniz's ontology. Newton's class S_N consists of all sentences of L and the difference $S_N - S_L$ shows the extent to which his ontology is richer. (Leibniz can use some conventions for assigning truth values to sentences in $S_N - S_L$, but no convention is acceptable to Newton just as no convention can fix for us the truth value of "Shakespeare was born around midnight".)

The language and the model can be simplified if we represent space-points by 3-tuples and time-points by real numbers.

Our world is then represented by the following structure $\hat{M}$:

$$(M_0, \mathbb{R}, \mathbb{R}^3, e_1, e_2, \ldots, b_1, b_2, \ldots, \hat{\tau}, \hat{\sigma}).$$

T and S are replaced here by $\mathbb{R}$ and $\mathbb{R}^3$; $\hat{\tau}(e_i)$, the time of e_i, is now a real number and, for each $t \in \mathbb{R}$, $\hat{\sigma}(b_i, t)$, the place of b_i at time t, is in $\mathbb{R}^3$. But M gives rise to infinitely many $\hat{M}$'s according to the choice of coordinates. Different choices yield different $\hat{\tau}$'s and $\hat{\sigma}$'s. If $\hat{L}$ is the language of $\hat{M}$ then each sentence φ of L has a translation $\hat{\varphi}$ in $\hat{L}$. A sentence, ψ, in $\hat{L}$ is objectively true-or-false in Newton's ontology iff it is logically equivalent to a translation, $\hat{\varphi}$, of some sentence of L. One can show that this is equivalent to saying that the truth-value of ψ is not changed if we change the coordinate system. We can change the coordinates by changing the following items: the 0 of the time line, the unit of time, the (0,0,0) of the space-coordinates, the axes and the unit of distance. Such a change will transform $\hat{M}$ to $\hat{M}_1$ in which $\hat{\tau}$ and $\hat{\sigma}$ are replaced by $\hat{\tau}_1$ and $\hat{\sigma}_1$, as follows: $\hat{\tau}_1(e) = \rho(\hat{\tau}(e))$, where ρ is a linear direction-preserving transformation of $\mathbb{R}$ ($\rho(x) = a.x + b$, $a>0$); and $\hat{\sigma}_1(b,t) = \lambda(\hat{\sigma}(b, \rho(t)))$, where λ is an orthogonal transformation followed by a translation. Let G_N be the group of all such transformations. Let $\widehat{S_N}$ be the Newtonian true-or-false sentences of $\hat{L}$, then:

$\psi \in \widehat{S_N}$ <u>iff</u> ψ <u>has the same truth-value in all</u> $g\hat{M}$, <u>where</u> $g \in G_N$.

We can now go beyond the language which we started with and generalize the last equivalence:

A statement, or property, is factually meaningful in Newtonian space-time ontology iff it is invariant under the transformation group G_N.

We do not specify here the language in which the statement is to be stated. At the limit we can identify "property" with a class of models. Note however that the language will be reflected through the type of the models which we consider (the so-called signature).

Now a sentence of $\hat{L}$ is true-or-false in Leibniz's ontology iff it is equivalent to a translation $\hat{\varphi}$ <u>such that</u> $\varphi \in S_L$. It turns out that Leibniz's ontology can also be characterized by means of a transformation group; call it G_L. A transformation in G_L transforms $\hat{\tau}$ and $\hat{\sigma}$ to $\hat{\tau}_1$ and $\hat{\sigma}_1$, where $\hat{\tau}_1(e) = \rho(\hat{\tau}(e))$, as before, but $\hat{\sigma}_1(b,t) = \lambda_t(\hat{\sigma}(b, \rho(t)))$ where λ_t is as λ, but for different t's we can have different λ_t's; the only restriction is that all λ_t's multiply the Euclidean distance by the same factor (i.e. the distance unit is not changed with time).

If we replace G_N by G_L we get a characterization of Leibniz's ontology.

SOME REMARKS CONCERNING TRANSLATION

We have just seen that Newton's point of view can be expressed either by means of the language L or by means of $\hat{L}$ and that the same is true for Leibniz. In L every sentence is according to Newton true-or-false, in $\hat{L}$ only sentences invariant under the transformation group, G_N, have this property. The equivalence is established by pointing to a translation from L into $\hat{L}$; we can also obtain translations in the other direction but I do not want to enter into these details. Instead, let me make some points about translation in general.

Imagine two languages L and L*. Their statements, at least part of them, are believed by the languages' users to have factual meaning. There is also a deductive apparatus for proving things, that is to say for establishing the _a priori_ truth of certain statements. I shall not enter here into what "_a priori_" can mean and what a proof consists of. The essential point is that this apparatus is accessible to the translator and he can check for himself whether such proofs are correct. I shall say that φ is valid in L, or L-valid, if its truth can be proved in L. Thus, L-validity and L*-validity are concepts that can be applied by the translator. Moreover, each system regards certain sentences as _a priori_ true-or-false and the deductive apparatus can also be used to establish _a priori_ truth-or-falsity. I shall therefore speak of the L-validity of " φ is true-or-false". This as well is fully accessible to the translator.

Now a translation is a function which transforms sentences of L into sentences of L*. (The function need not be defined for all sentences but its domain should include those that are _a priori_ true-or-false in L.) Let φ^t be the translation of φ . Then the first requirement is:

$$\varphi \text{ is L-valid} \Leftrightarrow \varphi^t \text{ is L*-valid.}$$

If we assume both languages to share the same logic, then an obvious requirement is that the negation of φ, $\neg\varphi$, be translated into something which is equivalent in the L* system to $\neg(\varphi^t)$. Equivalence in L* means the L*-validity of the biconditional $(\neg\varphi)^t \leftrightarrow \neg(\varphi^t)$; or, if $\leftrightarrow$ is unavailable, that from each sentence we can derive in L* the other. Similar equivalences are required for the other sentential connectives. We also require that

$$\text{"}\varphi \text{ is true-or-false" is L-valid} \Leftrightarrow$$
$$\text{"}\varphi^t \text{ is true-or-false" is L*-valid.}$$

Now a crucial point, which is often ignored, concerns the language in which the translation itself is defined. Consider for example the following trivial translation from L into a language which contains only two statements "0 = 0" and "0 = 1": Translate φ to "0 = 0" if φ is true; translate φ to "0 = 1" if φ is false. This translation is unacceptable because it is defined using the truth concept with respect to L. It makes sense only in as much as one is committed to the factual meaningfulness of L's statements. Moreover, in order to carry it out one has to find the truth-values of the statements that one wants to translate. Such translations are ruled out if we require that the definition itself involve only a minimal ontological commitment. We should imagine the translator as someone who understands both systems without necessarily commiting himself to the point of view of either. A translation might mention unknown parameters, which in schematic form, appear as free variables. For example the reduction of thermodynamics to the molecular theory of gases defines macroscopic magnitudes in terms of the momentum and energy distributions in very large collections of molecules. When it comes to translating particular sentences these free-hanging variables are replaced by definite descriptions. Thus the translation of "The gas pressure in container A is 2 atmospheres" is a sentence containing expressions such as "the distribution of momentum in the gas molecules in container A".

I interpret the minimal commitment condition as the requirement that the translation be recursive, i.e. effectively performable by some(purely mathematical)algorithm. This indeed is a natural condition, for we do want the translation to be effective and the recursive level seems to involve no undue ontological commitment. The existence of an algorithm is not, in itself, sufficient to ensure the possibility of carrying out the task. (For example, the question of the side which has a winning-or-drawing strategy in chess is solvable by a known algorithm, but we still don't know the answer.) The translation should belong to a reasonable level of computational complexity. At least one should require performance in polynomial time; but linear time (implying also that the length of φ^t is bounded by a linear function of the length of φ) is the more realistic condition.

When complexity is explicitly considered new light is shed on some old problems. In the elementary examples of translations the primitive predicates of L are translated into formulas of L* and the sentence-structure is preserved. But then logic and mathematics supply many cases where much more sophisticated forms are necessary. The bound on complexity imposes, together with the other above mentioned requirements, quite strong restrictions on the class of available translations. Such restrictions depend of course on the source and target languages. Quine's thesis concerning the indeterminacy of translation rests for its appeal on simple examples of isolated sentences. This is certainly not enough. Even for the sake of an example one should consider rich enough languages. It may then turn out that some global constraints are quite restrictive and the indeterminateness is much less severe than appears at first sight.

Haim Gaifman
Department of Mathematics and Computer Science
The Hebrew University
Jerusalem
Israel

W. Balzer[1)]

ON THE COMPARISON OF CLASSICAL AND SPECIAL RELATIVISTIC SPACE-TIME

One way of comparing two theories T,T′ is to reduce T to T′ in a formal sense. Much has been written about different intuitions on reduction and several meta-scientific concepts of reduction have been proposed; but few examples have been used by way of detailed examination in order to throw light on those meta-scientific concepts.[2)] Of course, there are numerous examples of reduction in the ordinary (i.e. "non-meta") scientific literature. But usually meta-scientific concepts of reduction cannot be directly applied to such examples: the "ordinary" treatments may be too vague, too sloppy, too incomplete, or they may use special assumptions so that actually only very small fragments of the theories are involved.

This problem of application is well known from ordinary science, and usually part of its solution consists in an interplay between reality (as given by examples) and scientific concepts. At the beginning there are usually a few examples which an author uses as paradigms in order to introduce his concepts. But once the concepts are presented there are attempts to apply them to other "new" cases as well. If difficulties arise then either the concepts may be kept unchanged and the new examples have to be "twisted", or the new example can be taken as "experimentum crucis", and the concepts have to be adjusted.

This interplay also takes place at the meta-level of the philosophy of science, and I believe that with respect to reduction we are still at a rather early stage of it. Much attention will have to be given to examples, and the present volume is only a first attempt in that direction.

W. Balzer et al. (eds.), Reduction in Science, 331–357.

My aim in this paper is to present a formally elaborated example of reduction in which the theory of classical space and time (CT) is reduced to the theory of special relativistic space-time (RT). Since the reduction to be employed will be strict, i.e.not approximative,the question of adequacy arises very pressingly because "every physicist" will say that the appropriate reduction relation of CT to RT has to be an approximative one.In the presence of this considerable opposition I will try to defend my example as a genuine case of reduction by considering the various objections that might be raised.In this way I hope to shed some light on the general concept of reduction without subscribing to any one of the existing formal notions.Also,the discussion will contribute to a clarification of the concept of classical space-time and its relation to Galilei-invariance.As far as I know,this is the first comparison of space-times on the axiomatic level (as opposed to the "group theoretic" level).The surprisingly easy way of defining a reduction relation ρ in this setting should be regarded as an argument for paying more attention to axiomatic analysis which in investigations of space-time at the moment is completely suppresed in favour of group theoretical methods.

I GENERAL NOTIONS

Today in physics space-time structures are characterized with respect to their corresponding invariances.Roughly and generally, one starts with some structure $x=\langle D,R_1,\ldots,R_m\rangle$ consisting of a set D and relations R_i on D.Automorphisms of x are those bijective functions $\varphi : D \rightarrow D$ which preserve all R_i,i.e. $\forall a_1 \ldots a_n \in D(R_i(a_1,\ldots,a_n) \leftrightarrow R_i(\varphi(a_1),\ldots,\varphi(a_n)))$,provided R_i is n-ary.The set of all automorphisms of x together with the concatenation operation of functions is a group, called the <u>transformation group</u> of x. If $D=\mathbb{R}^n$ and if the R_i are specified (e.g. for n=1, $R_1=<$,$R_2=+$ etc.) then the corresponding transformation groups are well known and can be characterized easily.

These characterizations are then "transferred" to non-mathematical structures by means of group isomorphisms. A structure x is identified by means of its transformation group being isomorphic to some well known transformation group of a given mathematical structure. For instance, some structure is a Galileian space-time iff its transformation group is isomorphic to the group formed by Galilei-transformations on $\mathbb{R}^3$ plus affine transformations of $\mathbb{R}$. In order to demonstrate that some "direct" characterization (as opposed to an indirect via transformation groups) is adequate it is sufficient to show that the transformation group of a model thus characterized is isomorphic to the corresponding mathematical group accepted by physicists.

It turns out that such direct proofs are complicated, and it is easier to show that any structure x under consideration is isomorphic to a given mathematical structure y which has the known transformation group. For if this is so then the two automorphism groups (of x and of y) are isomorphic, too.

I will use a slightly more general set-up which is a version of Bourbaki's "species of structures".[3] What has just been outlined then takes the following form.

A theory T consists of a class of <u>potential models</u> M_p and a class of (proper) <u>models</u> M:

$$T=\langle M_p, M\rangle \quad \text{where } M \subseteq M_p.$$

All potential models have the form

$$\langle D_1,\ldots,D_k;A_1,\ldots,A_l;R_1,\ldots,R_m\rangle$$

where $k,l,m \in \mathbb{N}$ are fixed, $D_1,\ldots,D_k$ are sets, called <u>base sets</u>, $A_1,\ldots,A_l$ are sets of mathematical objects (called <u>auxiliary base sets</u>) and $R_1,\ldots,R_m$ are relations of given set-theoretic types $\tau_1,\ldots,\tau_m$ "over" $D_1,\ldots,D_k,A_1,\ldots,A_l$.[4] For instance, A_1 may be $\mathbb{R}$ and $R_1: D_1 \times D_2 \to \mathbb{R}$ a function. The auxiliary base sets represent some mathematical "part" of the model which always has the same (standard) interpretation. Let T and T′ be given so that the types of the potential models of T and T′ and the mathematical parts involved are the same, and let $x=\langle D_1,\dots,D_k;A_1,\dots,A_l;R_1,\dots,R_m\rangle$ εM_p and $x'=\langle D'_1,\dots,D'_k;A_1,\dots,A_l;R'_1,\dots,R'_m\rangle \varepsilon M'_p$. We say that x and x′ are <u>isomorphic</u> iff there are bijective functions $\varphi_i:D_i\to D'_i$ $(i\leq k)$ such that for all appropriate arguments $a_1,\dots,a_n$ and all $j\leq m$:

$$R_j(a_1,\dots,a_n)\leftrightarrow R'_j(\varphi_{i_1}(a_1),\dots,\varphi_{i_n}(a_n)).$$

$\varphi=\langle\varphi_1,\dots,\varphi_k\rangle$ is called an <u>automorphism</u> of x if φ is an isomorphism from x to x. By Aut(x) we denote the group of automorphisms of x (with group operation defined by $\varphi\circ\psi=\langle\varphi_1\circ\psi_1,\dots,\varphi_k\circ\psi_k\rangle$. The result indicated above holds in this more general setting, too: if x and x′ are isomorphic then so are Aut(x) and Aut(x′).

II CLASSICAL THEORY OF SPACE AND TIME (CT)

<u>D1</u> x is a <u>potential model of</u> CT $(x\,\varepsilon\, M_p(CT))$ iff $x=\langle S,T;\mathbb{R};\lessdot,\tau,\delta\rangle$ and
1) S and T are non-empty sets, and disjoint
2) $\lessdot\subseteq T\times T$
3) $\tau:T\times T\to\mathbb{R}$
4) $\delta:T\times S\times S\to\mathbb{R}$

S is the set of points of space, T the set of instants. The intended meaning of $\lessdot$, τ and δ is this. $t\lessdot t'$ means that t is earlier than t′, $\tau(t,t')=\alpha$ means that the period of time between t and t′ (as measured by some clock) has length α, and $\delta(t,a,b)=\alpha$ means that at time t the distance between a and b is α.

If N is a set and $d:N\times N\to\mathbb{R}$ then $\underline{bet}_d\subseteq N^3$ and $\equiv_d\subseteq N^4$ are defined by

$$\underline{bet}_d(a,b,c) \text{ iff } d(a,b)+d(b,c)=d(a,c)$$

$$ab\equiv_d a'b' \text{ iff } d(a,b)=d(a',b').$$

If $\delta:T\times S\times S\to\mathbb{R}$ and $t\,\varepsilon\, T$ then $\delta(t):S\times S\to\mathbb{R}$ is defined by $\delta(t)(a,b)=\delta(t,a,b)$. The meaning of $\underline{bet}_d(a,b,c)$ is that b is between a and c, and $ab \equiv_d a'b'$ means that the pairs $\langle a,b\rangle$ and $\langle a',b'\rangle$ are congruent.

D2 x is a model of CT ($x \in M(CT)$) iff
1) $x=\langle S,T;\mathbb{R};\lessdot,\tau,\delta\rangle \in M_p$
2) for all $t \in T$: $\langle S,\delta(t)\rangle$ and $\langle T,\tau\rangle$ are metric spaces, and $\lessdot$ is a linear order
3) $\langle T,\underline{bet}_\tau,\equiv_\tau\rangle$ is a 1-dimensional Euclidean geometry
4) for all $t \in T$: $\langle S,\underline{bet}_{\delta(t)},\equiv_{\delta(t)}\rangle$ is a 3-dimensional Euclidean geometry[5)]
5) for all $t,t' \in T$: $\delta(t)=\delta(t')$

We can best imagine a model as a "series" of identical copies of 3-dimensional spaces where T provides the indices. T can be visualized by a straight line on which an ordering $\lessdot$ and a distance τ is given. At each instant t the corresponding space $\langle S,\underline{bet}_{\delta(t)},\equiv_{\delta(t)}\rangle$ satisfies all the axioms of Euclidean geometry. If we omit T from δ then we would just have two metric spaces put together. This couldn't be called a "space-time" because in such a structure we could not formulate expressions of the form "at t the distance of a and b is α". By making δ dependent on t we obtain the possibility of formulating such expressions. On the other hand the time-dependence of δ is immediately withdrawn by means of D2-5) which requires δ, in fact, not to depend on t properly. The effect is a "rigid" space-time consisting essentially of the cartesian product of "space" and "time".

Some further comments may be helpful. First, in a model the (relativistic) set E of events could be explicitly defined by $E=S\times T$. I have chosen not to use E as a primitive in order to do justice to the historical situation before Minkowski. Second, I have not included any notions and requirements concerning the orientation of space. So, reflections are not excluded from the corresponding transformation group. The system could be easily adjusted to obtain the proper transformation groups. I have chosen not to exclude reflections because this would make things more complicated without adding new aspects to the reduction relation. Third, if at $t \in T$ we choose coordinate systems K for $\langle S, \delta(t)\rangle$ and K′ for $\langle T, \tau\rangle$ then the content of D2) is represented equivalently by the structure $\langle S \times T, \psi\rangle$, where $\psi : S \times T \times \mathbb{R}^3 \rightarrow \mathbb{R}$, $\psi(\langle b,t\rangle) = \langle \psi_1(b), \psi_2(t)\rangle$ and $\psi_1(b)$, $\psi_2(t)$ are the coordinates of b and t relative to K and K′.[6)]

I will next describe the corresponding transformation group and only afterwards discuss the question of adequacy. Let $\mathscr{E}$ be the "elementary group"[7)], i.e. $\mathscr{E}$ is defined as the direct product $\mathscr{E} = \mathscr{E}_S \otimes \mathscr{E}_T$ where $\mathscr{E}_S$ is the group of dilatations, translations and rotations of $\mathbb{R}^3$ and $\mathscr{E}_T$ is the affine group of $\mathbb{R}$. Let $\mathscr{E}'$ be obtained from $\mathscr{E}$ by including reflections in $\mathscr{E}_S$ and by omitting dilatations from $\mathscr{E}_T$ and $\mathscr{E}_S$.

<u>T1</u> If $x \in M(CT)$ then Aut(x) is isomorphic to $\mathscr{E}'$.

Proof: Let < be the usual "smaller than" relation on $\mathbb{R}$, and let $|\cdot|, \|\cdot\|$ be the Euclidean distance functions on $\mathbb{R}$ and $\mathbb{R}^3$, respectively. Then the structure $y = \langle \mathbb{R}^3, \mathbb{R}; \mathbb{R}; <, |\cdot|, \delta_R\rangle$ has the same type as our models of CT if we define

$\delta_R: \mathbb{R} \times \mathbb{R}^3 \times \mathbb{R}^3 \rightarrow \mathbb{R}$ by $\delta_R(\alpha, \mathfrak{x}, \mathfrak{y}) = \|\mathfrak{x} - \mathfrak{y}\|$.

(The second occurrence of "$\mathbb{R}$" in y indicates the use of $\mathbb{R}$ as range of $|\cdot|$ and δ_R in the status of an auxiliary base set). It is well known that Aut(y) is isomorphic to $\mathscr{E}'$, so by what was said in Sec. I it is sufficient to show that any model $x \in M(CT)$ is isomorphic to y. Let $x = \langle S, T; \mathbb{R}; \lessdot, \tau, \delta\rangle \in M(CT)$. Then an isomorphism $\varphi = \langle \mu, \eta\rangle$ with $\mu : S \rightarrow \mathbb{R}^3$ and $\eta : T \rightarrow \mathbb{R}$ is obtained by introducing coordinate systems for S and T respectively in the well known way ∎

The physical meaning of the differences between $\mathscr{E}$ and $\mathscr{E}'$ is clear. Spatial reflections cannot be actively performed in reality, and the passive possibility of looking at physical systems through some mirror has played no role up to now. Dilatations correspond to the freedom of choice of a unit. A treatment including dilatations in the transformation group of CT would have to start at the qualitative level of *bet* and $\equiv$. It is achievable, but more complicated than the present formulations. I think in spite of these small deviations one can say that CT is "essentially" represented by $\mathcal{E}$, and thus is "essentially" Newtonian space-time.

The immediate objection now is that "classical space-time" has to be Galilei-invariant so that the full group of Galilei-transformations, and not its sub-group $\mathcal{E}$, is the appropriate transformation group.The objection has three parts.First,as a sociological statement,one simply observes that most physicists today hold that classical space-time is Galilei-invariant. Second, from an historical point of view, one may argue that in the period leading to and including the introduction of classical mechanics space-time was regarded as Galilei-invariant. Third, from a systematic point of view, a comparison of CT with RT (or other theories) may suggest that we look for Galilei-transformations as a counterpart to Lorentz-transformations. I will consider the three items in turn.

As to the first point, I agree that physicists today require classical space-time to be Galilei-invariant. But philosophy of science is not the same as sociology of science and what constitutes an unshakable fact for the latter may be of less importance for the former. I believe that this first part of the objection is the least important one, and is outweighed by the other two. I will argue that with respect to the other parts classical space-time should *not* be Galilei-invariant but only be invariant under the elementary group.

From an historical point of view it seems to me that Galilei was the first to point out that *mechanical* (i.e.dynamical) events will be the same if taking place in or being perceived from two different frames of reference moving relative to each other with constant velocity.During the development of Newtonian mechanics,too, Galilei-invariance in this special sense always turned up with considerations of mechanical systems ("dynamics"). In the course of such considerations space-time was always presupposed,i.e. the properties of space and time were assumed to be already known.

Space was represented by Euclidean geometry and time by a straight line (if at all), and there was no idea of the relevance of the state of motion of an observer for the properties of space and time. The latter statement is compatible with the fact that the discussion of inertial frames of reference as a matter of dynamics preceded the advent of special relativity. I am not in a position to give a detailed historical account of this topic. But unless historical arguments to the contrary are put forward I conclude that, historically, classical space-time is adequately represented by the elementary group.

Third, my formulation of RT in the next section is especially suited to making clear why there is a systematic drive for Galilei-invariance on the classical side. Any model of RT "implicitly" contains some frame of reference W. But W is not uniquely determined by the other parts of the model, and a change of W in general will not leave unaffected the validity of the axioms. So in RT it is natural to consider transformations of coordinates relative to different frames of reference which are possible in one model. This leads to Lorentz-transformations. One is tempted to look for a similar feature at the side of CT. Things look differently from different frames of reference, and one would like to know how the coordinates transform under changes of the frame of reference. In order to perform such investigations it is necessary on the classical side to introduce different frames of reference. This can be done, but only at the price of introducing a new basic concept. In the models of CT only one frame of reference can be defined in analogy to W in RT, namely $\{\{\langle a,t\rangle / t \in T\} / a \in S\}$. If we want to talk about different classical frames we are forced to use further concepts not available in CT. Thus there is a formal distinction between RT and CT. For a potential model of RT there are many different possible frames of reference W which make it into a model. For a potential model of CT there is only one possible frame, namely the one defined above, and this frame is not necessary for stating the axioms. Intuitively, in RT the basic stuff the models are formed of (E and $\prec$) has to be enriched by further entities (W) if we want to express the full complexity of the models. In CT no such additional entities are needed. Again, what was said here is compatible with stating that the status of Galilei-transformations in mechanics is independent of relativistic theories.

One way of obtaining a Galilei-invariant theory from CT is to enrich CT by frames of reference. Models then would have the form $\langle S,T;\mathbb{R};\prec,\tau,\delta,F\rangle$ where F is a partition of $S \times T$ satisfying further requirements to the effect that F is just a "bundle" of parallel straight lines so that the lines are not "orthogonal" to T. It is clear that different F's can make some given model of CT into a model of the new theory, and all these F's can be obtained from each other by Galilei-transformations. It is not difficult to show that this theory in fact is represented by the group of Galilei-transformations in the sense of Sec.I. But it is also clear that the new concept of a frame of reference is not linked in any interesting way to the "old" concepts; it is added ad hoc. There is no intrinsic connection between F and the other concepts, in contrast to the connection between W and $E,\prec$ in RT. The only systematic reason for Galilei-invariance of the classical theory comes from the search for an analogue to relativistic frames of reference. But any theory created by this analogy is an artificial construct which has no standing on its own -in contrast to CT.

As far as I can see all other arguments for Galilei-invariance of classical space-time can be traced back to the three just mentioned. For instance, it may be said that space-time theory and the full theory of classical mechanics form an inseparable unit, so that the invariances of mechanics are also relevant for the underlying space-time. This is the same kind of reasoning by analogy from RT as we just met before. Again, on closer inspection, this view imposes features on the classical theory which seem to be added ad hoc after the invention of RT.

To summarize, then, I would say that historical and systematical considerations favour classical space-time as being represented by the elementary group, that these two aspects are more important than the sociological one, and that, therefore, CT is an adequate reconstruction of classical space-time.

III THE SPECIAL RELATIVISTIC THEORY OF SPACE-TIME (RT)

D3 x is a potential model of RT ($x \in M_p(RT)$) iff $x = \langle E; \prec \rangle$ and
1) E is a non-empty set
2) $\prec \subseteq E \times E$

E is the set of events and $\prec$ is the so called "causal relation". $e \prec e'$ means that a signal can be sent from event e to event e'.[8] Some notation needs to be fixed for the following. Let $x = \langle E; \prec \rangle \in M_p(RT)$. If $X \subseteq E$ and $e_1, e_2 \in E$ we write "$e_1 \preccurlyeq e_2$" for "$e_1 \prec e_2$ or $e_1 = e_2$". We say that e is an upper (lower) bound of X iff for all $e_1 \in X$: $e_1 \preccurlyeq e$ ($e \preccurlyeq e_1$) We write $e = \inf_\prec X$ ($e = \sup_\prec X$) iff e is a lower (upper) bound of X and for all lower (upper) bounds e_2 of X: $e_2 \preccurlyeq e$ ($e \preccurlyeq e_2$). X is called bounded iff X has an upper and a lower bound. We write "$e_1 \rightsquigarrow e_2$" for "$\exists e, e' \in E (\neg e \prec e' \wedge \neg e' \prec e \wedge e \neq e' \wedge e_1 \prec e \prec e_2 \wedge e \prec e' \prec e_2)$" which means that a signal slower than light can be sent from e_1 to e_2.

If $x \in M_p(RT)$ we say that W is a frame for x iff (1) W is a partition of E, (2) for each $w \in W$ there are functions $f_w: E \to w$ and $g_w: E \to w$ so that for all $e \in E$:

$$f_w(e) = \begin{cases} e & \text{if } e \in w \\ \inf_\prec \{e' \in w / e \prec e'\} & \text{if } e \notin w \end{cases}$$

$$g_w(e) = \begin{cases} e & \text{if } e \in w \\ \sup_\prec \{e' \in w / e' \prec e\} & \text{if } e \notin w \end{cases}$$

If W is a frame for x then $\underline{bet}_x \subseteq W^3$ and $\equiv_x \subseteq W^4$ are defined by

$\underline{bet}_x(u,v,w)$ iff $f_u \circ f_v \circ f_w = f_u \circ f_w$ and

$uv \equiv_x u'v'$ iff $f_{u'} o f_u o f_v o f_u = f_{u'} o f_{v'} o f_{u'} o f_u$.

Intuitively, W can be imagined as a bundle of parallel straight lines running "time-like" (wrt. $\prec$) through E. Each line $w \varepsilon W$ represents a possible path of some free particle and is called a world line. $f_w(e)$ is the event of arrival of a flash of light at world line w which is emitted at e. Similarly, $g_w(e)$ is the event on w determined by the condition that a flash of light omitted at $g_w(e)$ would hit e. $\underline{bet}_x$ and $\equiv_x$ have the usual meaning of relations of betweenness and congruence, respectively: only that the objects they are defined for are world lines.

<u>D4</u> a) x is a <u>model of</u> RT <u>relative to</u> W iff
1) $x = \langle E; \prec \rangle \varepsilon M_p(RT)$ and W is a frame for x
2) $\prec$ is transitive and $e \prec e'$ implies $\neg\, e' \prec e$
3) for all $w \varepsilon W$ and all $X \subseteq w$: if X is bounded then there are e_1, e_2 so that $e_1 = \inf_\prec X$, $e_2 = \sup_\prec X$, $e_1 \varepsilon w$ and $e_2 \varepsilon w$
4) for all $w \varepsilon W$ and all $e, e' \varepsilon E$: if $e \varepsilon w$ and $e' \varepsilon w$ then $(e = e' \vee e \rightarrow e' \vee e' \rightarrow e)$
5) for all $w \varepsilon W$ and $e \varepsilon E$ there are e_1, e_2 so that $e_1 \prec e \prec e_2$ and $e_1, e_2 \varepsilon w$
6) for all $v, u \varepsilon W$: if $u \neq v$ then $f_{u/v}$ and $g_{v/u}$ are inverse to each other
7) $\langle W; \underline{bet}_x, \equiv_x \rangle$ is a 3-dimensional Euclidean geometry
8) for all $u, v, u', v' \varepsilon W$:
$(f_w o f_v o f_u) o (f_w o f_{v'} o f_{u'})_{/w} = (f_w o f_{v'} o f_{u'}) o (f_w o f_v o f_u)_{/w}$
and $(f_w o f_v o f_u)_{/w} = (f_w o f_u o f_v)_{/w}$
9) for all $w \varepsilon W$ and $e, e' \varepsilon w$: if $e \prec e'$ then there is $v \varepsilon W$ so that $v \neq w$ and $f_w(f_v(e)) \prec e'$
10) for all $w \varepsilon W$ there is a countable and dense (wrt. $\prec$) subset of w
11) for all $u, v, w \varepsilon W$: if there is $e \varepsilon w$ so that $f_u(f_v(e)) = f_u(e)$ then $\underline{bet}_x(u, v, w)$

b) x is a <u>model of</u> RT ($x \varepsilon M(RT)$) iff there is some W so that x is a model of RT relative to W

The present axiomatization is essentially due to A.Kamlah who made precise Reichenbach's original version in (Kamlah,1979),pp.436. Three deviations from Kamlah's system should be mentioned.First, I require W to exhaust all of E, second I have added D4-a-3) which I cannot prove in Kamlah's system. Third, I have added D4-a-11) which is essential for the proof of T2).

Axiom 3) is needed in order to prove that the defining conditions for f_w and g_w, in fact, guarantee uniqueness,that is, f_w and g_w, in fact, are functions. 4) requires each world line to run through the time-like sections of the light cones, and 5) rules out absolute boundaries wrt. $\prec$. $f_{u/v}$ in 6) denotes the restriction of f_u to v. Requirement 6) is of more technical character. The functions f_w are well defined (on the basis of the previous axioms) and uniquely determined in x. So the betweenness and congruence relations $\underline{bet}_x$ and $\equiv_x$ in 7) are well defined,too. 8) expresses a kind of invariance. It makes no difference whether a signal travels via world lines u',v',w,u,v to w or alternatively via u,v,w,u',v': the event of arrival at w will be the same in both cases. Similarly, it makes no difference to go to w by way of u and v or via v and u. These conditions also guarantee that the world lines of W are "straight". Condition 9) requires that the world lines are dense in E (wrt. $\prec$): in each neighbourhood of each f_w there is another f_v. 10) guarantees that world lines have the right cardinality (used for mapping them on $\mathbb{R}$ bijectively). Requirement 11), finally, enforces that all world lines of W are "parallel" to each other.

In models of RT we can introduce clocks, simultaneity and a "space-like" metric as follows.

<u>D5</u> Let $x=\langle E;\prec\rangle \varepsilon M_p(RT)$, let W be a frame for x and $w \varepsilon W$.

a) Φ_w is a <u>clock for</u> x (relative to w) iff

$\Phi_w : w \to \mathbb{R}$ is bijective and for all $u \in W$ and $e, e' \in w$: 1) $e \prec e'$ iff $\Phi_w(e) < \Phi_w(e')$
2) $\Phi_w(f_w(f_u(e))) - \Phi_w(e) = \Phi_w(f_w(f_u(e'))) - \Phi_w(e')$

b) If Φ_w is a clock for x relative to w then $sim_{\Phi,w} : w \to Pot(E)$ is defined by: $e' \in sim_{\Phi,w}(e)$ iff $e=e'$ or there exist $v \in W$ and $e_1, e_2 \in w$ such that (1) $e' \in v$, (2) $e_1 \prec e \prec e_2$, (3) $e' = f_v(e_1)$ and $e_2 = f_w(e')$, (4) $\Phi_w(e) = 1/2(\Phi_w(e_2) + \Phi_w(e_1))$

c) If $v \neq w, v \in W$ then $d_{x,v,w}$ is <u>a metric for</u> x relative to v,w iff $d_{x,v,w} : W \times W \to \mathbb{R}$ is a metric such that for all $u_1, \ldots, u_4 \in W$:
1) $\underline{bet}_x(u_1, u_2, u_3)$ iff $d_{x,v,w}(u_1,u_2) + d_{x,v,w}(u_2,u_3) = d_{x,v,w}(u_1,u_3)$
2) $u_1u_2 \equiv_x u_3u_4$ iff $d_{x,v,w}(u_1,u_2) = d_{x,v,w}(u_3,u_4)$
3) $d_{x,v,w}(v,w) = 1$

$\Phi_w(e)$ is intended to denote the time (as measured on w) at which event e takes place. $sim_{\Phi,w}(e)$ is the class of all events of E which are simul taneous to e (with respect to Φ_w), and is called the simultaneity class of e (wrt. Φ_w). $d_{x,v,w}(u,u')$ is the spatial distance between world lines u and u'.

With respect to RT the question of adequacy is easier to settle. There is common agreement that the causal Minkowski-structure

$\langle \mathbb{R}^4; <_c \rangle$ with $\langle \alpha_1, \ldots, \alpha_4 \rangle <_c \langle \beta_1, \ldots, \beta_4 \rangle$ iff

$$\alpha_4 < \beta_4 \wedge \left(\sum_{i \leq 3} (\alpha_i - \beta_i)^2 \right)^{1/2} \leq c \cdot (\beta_4 - \alpha_4)$$

is a model (indeed, <u>the</u> model) of RT. The automorphism group of this structure is the group of Lorentz-transformations, as was indicated already by Weyl.[9] Our scheme of Sec. I, however, cannot be directly applied to these structures. For $\mathbb{R}^4$ gets its standard meaning only through additional relations (like $<, +, \cdot, 0, 1$) which are not mentioned in $\langle \mathbb{R}^4; <_c \rangle$, and these relations are kept fixed when Lorentz-transformations are considered. So $\mathbb{R}^4$ here has the status of an auxiliary base set, and therefore no proper base set at all is involved. But without a base set there are no automorphisms in the sense of Sec. I.

Fortunately, it is easy to modify the structure $\langle \mathbb{R}^4; <_c \rangle$ in an equivalent way so that the modified version will fit into the Bourbaki scheme. Consider the structure $x = \langle R; <_R \rangle$ where R is a set and the axiom for $<_R$ is

$$\exists c\, \exists \varphi : \mathbb{R} \to \mathbb{R}^4 (\ \varphi \text{ is bijective} \wedge \forall a,b \in R\ (a <_R b \leftrightarrow \varphi(a) <_c \varphi(b))).$$

Let me call x a <u>Bourbaki model</u> of RT. If L is the group of automorphisms of $\langle \mathbb{R}^4; <_c \rangle$, i.e. the group of bijective mappings $\varphi: \mathbb{R}^4 \to \mathbb{R}^4$ preserving $<_c$, then L is isomorphic to Aut(x). An isomorphism $\Delta_\varphi : L \to \text{Aut}(x)$ is given by $L \ni \lambda \mapsto \Psi = \varphi^{-1} \circ \lambda \circ \varphi \in \text{Aut}(x)$ where φ is as required in the definition of $<_R$. For, by the definition of L, $<_R$ and Ψ, we have $\varphi(a) < \varphi(b)$ iff $(\lambda \circ \varphi)(a) <_c (\lambda \circ \varphi)(b)$ iff $(\varphi^{-1} \circ \lambda \circ \varphi)(a) <_R (\varphi^{-1} \circ \lambda \circ \varphi)(b)$ iff $\Psi(a) <_R \Psi(b)$. So by what was said in Sec. I it is sufficient to show that each model of RT is isomorphic to some Bourbaki model $\langle R; <_R \rangle$.

<u>T2</u> If $x \in M(RT)$ then x is isomorphic to some Bourbaki model $\langle R; <_R \rangle$.

Proof: If $x = \langle E; \prec \rangle$ let $R := E$ and $<_R := \prec$. From the proof of TIV-9-b) of (Balzer, 1982) (p.251) it follows that there is some c and some bijective $\varphi: E \to \mathbb{R}^4$ so that for all $e, e' \in E$: $e \prec e' \leftrightarrow \varphi(e) <_c \varphi(e')$. So by the definition of $<_R$ and R, $y = \langle R; <_R \rangle$ is a Bourbaki model of RT and, trivially, x and y are isomorphic ■

This shows that our axiomatization of RT is adequate.

IV REDUCTION OF CT TO RT

We define a reduction relation ρ as follows.

D6 a) If $x=\langle S,T;\mathbb{R};\lessdot,\tau,\delta\rangle \in M_p(CT)$, $y=\langle E;\prec\rangle \in M_p(RT)$ and W is a frame for y then we set $x\ \rho_W y$ iff there are v,w,Φ_w and $d_{y,v,w}$ so that

1) $v,w \in W$ and $v \neq w$
2) Φ_w is a clock for y relative to w
3) $d_{y,v,w}$ is a metric for y relative to v,w
4) $S=W$
5) $T=\{sim_{\Phi,w}(e)/e \in E\}$
6) for all $t,t' \in T$: $t \lessdot t'$ iff $\exists e,e' \in w(t=sim_{\Phi,w}(e) \wedge t'=sim_{\Phi,w}(e') \wedge e \prec e'$
7) for all $t,t' \in T$ and $\alpha \in \mathbb{R}$: $\tau(t,t')=\alpha$ iff $\exists e,e' \in w(t=sim_{\Phi,w}(e) \wedge t'=sim_{\Phi,w}(e') \wedge |\Phi_w(e)-\Phi_w(e')|'=\alpha$
8) for all $t \in T$ and $a,b \in S$: $\delta(t,a,b)=d_{y,v,w}(a,b)$

b) A relation $\rho \subseteq M_p(CT) \times M_p(RT)$ is defined by $x\rho y$ iff there is a frame W for y such that $x\ \rho_W y$

Note that, once v,w,Φ_w and $d_{y,v,w}$ in D6-a) are given, requirements 4)-8) have the form of explicit definitions of $S,T,\lessdot,\tau$ and δ. Intuitively, S is identified with the set W of world lines, and T with the set of simultaneity classes. That is, classical points of space are identified with world lines (paths of free particles) and classical instants with classes of simultaneous events. The ordering of classical instants in x is given by the ordering induced in simultaneity classes by $\prec$ (D6-a-6). Classical time-distance is defined by the time-distance read off from the relativistic clock on w (D6-a-7), and classical spatial distance is defined by the distance function induced on W in y. Roughly, for given v,w Φ_w and $d_{y,v,w}$, x is defined in terms of the components of y. We obtain the following theorems.

<u>T3</u> If y is a model of RT relative to W and $x \rho_W y$ then x is a model of CT.

<u>T4</u> For all $y \in M(RT)$ there is x so that $x \rho y$ and $x \in M(CT)$.

<u>T5</u> For all $x \in M(CT)$ there is $y \in M(RT)$ so that $x \rho y$.

<u>T6</u> Not: for all x,y, if $y \in M(RT)$ and $x \rho y$ then $x \in M(CT)$.

<u>T7</u> Not $(\forall x,x',y(\langle x,y\rangle \in \rho \wedge \langle x',y\rangle \in \rho \rightarrow x=x'))$ and not $(\forall x,y,y'(\langle x,y\rangle \in \rho \wedge \langle x,y'\rangle \in \rho \rightarrow y=y'))$.

For the proofs let $y, W, v, w, \Phi_w, d_{y,v,w}$ be given as in D6) and let y be a model of RT relative to W.

<u>Lemma 1</u> For each $w \in W$ there is a clock Φ_w for y relative to w, and Φ_w is uniquely determined up to linear transformations.

Proof: See (Kamlah, 1979), pp. 448 in a slightly different notation ■

<u>Lemma 2</u> $\{sim_{\Phi,w}(e)/e \in E\}$ is a partition of E. For all $e \in E$ and $v \in W$: $sim_{\Phi,w}(e) \cap v$ is a singleton, and in particular $sim_{\Phi,w}(e) \cap w=\{e\}$.

Proof: (1) It is easy to show that the relation defined by $e \approx e'$ iff $\exists e_1(e,e' \in sim_{\Phi,w}(e_1))$ is an equivalence relation on E. (2) $e' \in sim_{\Phi,w}(e_1)$ and $e' \in sim_{\Phi,w}(e_2)$ imply $e_1=e_2$, because the events required to exist in D5-b) are uniquely determined and because Φ_w is bijective. (3) We show that there is an e' such that $sim_{\Phi,w}(e) \cap v=\{e'\}$.

Case 1) v=w. From $e' \in sim_{\Phi,w}(e)$ and $e \neq e'$ it follows that there is an e_1 such that $e'=e_1 \prec e$, and in the same way, that there is an e_2 such that $e \prec e_2 = e'$. From these two statements we obtain $e'=e_1 \prec e \prec e_2=e'$ which is impossible. So $e=e'$ and $sim_{\Phi,w}(e) \cap w=\{e\}$

which proves the special case,too.Case 2): $v \neq w$. There are $e_1^+, e_2^+ \varepsilon w$ such that $e_1^+ \prec e \prec e_2^+$ and $\exists e^+ \varepsilon v(f_v(e_1^+)=e^+ \wedge f_w(e^+)=e_2^+)$. Choose $e_1, e_2 \varepsilon w$ so that $e_1 \prec e_2$, $\Phi_w(e_2)-\Phi_w(e_1)=\Phi_w(e_2^+)-\Phi_w(e_1^+)$ and $\Phi_w(e_2)-\Phi_w(e)=\Phi_w(e)-\Phi_w(e_1)=1/2(\Phi_w(e_2)-\Phi_w(e_1))$. The definition of Φ_w implies $e_2=f_w(f_v(e_1))$ and therefore (with $e'=f_v(e_1)$): $e' \varepsilon \mathrm{sim}_{\Phi,w}(e)$, that is, $\mathrm{sim}_{\Phi,w}(e) \cap v \neq \emptyset$. Suppose $e', e'' \varepsilon \mathrm{sim}_{\Phi,w}(e) \cap v$. Then $e'=f_u(e_1)$ and $e_2=f_w(e')$, so $e' \varepsilon u \cap v$, and, because W is a partition, u=v. Thus we obtain $e'=f_v(e_1) \wedge \Phi_w(e)=1/2(\Phi_w(e_2)+\Phi_w(e_1))$, and in the same way: $e''=f_v(e_1'') \wedge \Phi_w(e)=1/2(\Phi_w(e_2'')+\Phi_w(e_1''))$. Suppose $e' \prec e''$. Then $f_v(e_1) \prec f_v(e_1'')$, so $e_2 \prec e_2'' \wedge e_1 \prec e_1''$, from which we obtain $\Phi_w(e_2) < \Phi_w(e_2'') \wedge \Phi_w(e_1) < \Phi_w(e_1'')$. But this together with $\Phi_w(e_2) < \Phi_w(e_2'')$ and $\Phi_w(e_1) < \Phi_w(e_1'')$ is impossible, so not $e' \prec e''$. In the same way we obtain not $e'' \prec e'$ from which it follows,finally, that $e'=e''$ ∎

<u>Lemma 3</u> There is precisely one metric $d_{y,v,w}$ for y relative to v,w.
Proof: This is the well known Representation Theorem for Euclidean geometry ∎

Proof of T3: By lemma 2) $v \cap \mathrm{sim}_{\Phi,w}(e)=\{e'\}$ and by D4-5) $v \neq \mathrm{sim}_{\Phi,w}(e)$. If $S \cap T$ were not empty then for some $b \varepsilon S \cap T$, we would obtain $b=v=\mathrm{sim}_{\Phi,w}(e)$ which yields a contradiction.Again by lemma 2), if $\mathrm{sim}_{\Phi,w}(e_1)=t=\mathrm{sim}_{\Phi,w}(e_2)$ and $e_1, e_2 \varepsilon w$ then $e_1=e_2$. So in D6-a-6), e and e′ are uniquely determined by t and t′. Hence τ,as defined in D6-a-6) is a function.Also, by lemma 3), $d_{y,v,w}$ is uniquely determined, which proves D1-3). D2-1) is trivial. That τ is a metric is checked with the help of lemma 2) and the triangle inequality in $\mathbb{R}$. That $\delta(t)$ is a metric follows from the fact that

$d_{y,v,w}$ is a metric, and that D6-a-7) does not depend on t. $\leqslant$ is a linear order because $\prec$ on w is a linear order and because of lemma 2). We prove D2-3): $<T;\underline{bet}_\tau, \equiv_\tau>$ is a 1-dimensional Euclidean geometry. From lemma 2) and by direct calculation we obtain (1) $\underline{bet}_\tau(t_1,t_2,t_3)$ iff $\exists e_1,e_2,e_3 \varepsilon w$ $(t_i = sim_{\Phi,w}(e_i) \wedge e_1 \preccurlyeq e_2 \preccurlyeq e_3)$ and (2) $t_1t_2 \equiv_\tau t_3t_4$ iff $\exists e_1 \ldots e_4 \varepsilon w(t_i = sim_{\Phi,w}(e_i) \wedge |\Phi_w(e_1) - \Phi_w(e_2)| = |\Phi_w(e_3) - \Phi_w(e_4)|)$. Now let bet and $\equiv$ on w be defined by: bet(e_1,e_2,e_3) iff $e_1 \preccurlyeq e_2 \preccurlyeq e_3$ and $e_1e_2 \equiv e_3e_4$ iff $|\Phi_w(e_1) - \Phi_w(e_2)| = |\Phi_w(e_3) - \Phi_w(e_4)|$. From (1) and (2) it follows that if $<w;bet, \equiv>$ is a 1-dimensional Euclidean geometry then so is $<T;\underline{bet}_\tau, \equiv_\tau>$. Thus it is sufficient to show that $<w;bet, \equiv>$ is a 1-dimensional Euclidean geometry, and this is proved directly by using lemma 1). D2-4) follows immediately from the definitions of $\underline{bet}_{\delta(t)}$, $\underline{bet}_{d_{y,v,w}}$ etc. and from D6-a-7). D2-5) also follows from D6-a-7) directly ∎

Proof of T4: Let $y = <E;\prec> \varepsilon M(RT)$, i.e. there is a frame W for y so that y is a model for RT relative to W. Let $v,w \varepsilon W$. By lemma 1) there is a clock Φ_w for y relative to w and by lemma 3) there is a metric $d_{y,v,w}$ for y relative to v,w. Define $x = <S,T;\mathbb{R};\leqslant,\tau,\delta>$ by conditions 4)-8) of D6-a). By lemma 2) $\{e\} = w \cap sim_{\Phi,w}(e)$. So $sim_{\Phi,w}(e_1) = t = sim_{\Phi,w}(e_2) \wedge e_1,e_2 \varepsilon w$ implies $e_1 = e_2$, i.e. α in D6-a-6) is uniquely determined and therefore τ is a function. By lemma 3) $d_{y,v,w}$ is uniquely determined, so δ is a function, too, i.e. $x \varepsilon M_p(CT)$. By the definition of x: $x \rho_W y$, and so by T3): $x \varepsilon M(CT)$ ∎

Proof of T5: Let $x = <S,T;\mathbb{R};\leqslant,\tau,\delta> \varepsilon M(CT)$. Define $<E;\prec>$ as follows: $E = S \times T$ and $<a,t> \prec <b,t'>$ iff $t \leqslant t' \wedge \tau(a,b) \leq \delta(t,t')$. Define $W \subseteq Pot(E)$ by $w \varepsilon W$ iff $\exists a \varepsilon S(w = \{<a,t>/t \varepsilon T\})$. It is then easily checked that <E;$\prec$ > is a model of RT relative to W (compare (Balzer,1982),pp.222,TIV-5-b) in a slightly different set-up)∎

Proof of T6 and T7: By construction of mathematical counter examples∎

Our claim about ρ is that ρ constitutes a reduction of CT to RT, and therefore ρ is a "reduction relation". Such a claim can be attacked on various lines, and I will consider several objections in turn.

A first objection against ρ as a reduction relation is that it is <u>strict</u> -as opposed to "approximative". In one way this objection may be seen as another version of the objection of Sec. II against CT not being Galilei-invariant. The reasoning seems to be this. <u>If</u> classical space-time is Galilei-invariant <u>then</u> it can only be approximatively reduced to RT.So Galilei invariance on the classical side seems to be sufficient for approximative features of reduction. This is, I think, the intuitive basis of the objection though I do not know how to substantiate it in the absense of generally accepted conditions on all possible forms of reduction. But it is clear why the reasoning has so much credit: because of the approximative relation between the corresponding groups of Galilei- and Lorentz-transformations.

This kind of objection is just a corollary to the one in Sec.II, and <u>if</u> it is conceded that CT is adequate (without being Galilei-invariant as I have argued in Sec.II) then the present objection becomes pointless. To put it differently: if there is a strict reduction relation ρ between CT and RT (as the one just presented) then its strictness need not count as an inadequacy of ρ but can be seen as an inadequacy of CT to represent classical space-time (being not Galilei-invariant). The same point is reinforced by observing that the Galilei-invariant extension of CT mentioned in Sec.II can be reduced to RT in an approximative way. (it is tempting to add "and only in an approximative way" but, again, such a statement seems difficult to substantiate.) Anyway, this possibility again shows that approximation of reduction goes together with Galilei-invariance on the classical side.

Now opponents might concede that CT is an adequate reconstruction and still insist that my ρ is not a reduction relation. This amounts to saying that the ρ employed does not have the properties which an adequate relation should have. Since there is a considerable variety of different concepts of reduction, it will not suffice to show that ρ can be subsumed under one of them: the objection might be sustained by using a different concept of reduction. Given this situation I will go through some of the different requirements proposed by different authors and comment on their bearing on the present example.

First, there is the traditional condition of derivability of the laws of the reduced theory from those of the reducing theory "after translation" which by Adams[10)] was expressed as follows:

$$(1) \qquad \forall x,y(y \in M' \wedge x \rho y \rightarrow x \in M)$$

where ρ reduces T to T', $\rho \subseteq M_p \times M'_p$. T3) above attempts to establish this condition for the present example but it does not succeed completely. Strictly, (1) fails, for the frame W employed in establishing the relation $x \rho_W y$ may be different from the one which makes y a model of RT (see T6) above). If W is chosen perversely enough then "$x \in M(CT)$" does not follow any longer.[11)] This is a puzzling result, and it would be helpful to see whether the difference between (1) and T3) has to do -and if so, in which way precisely- with the difference between "strict" and "approximative" reduction. For someone taking condition (1) as necessary for reduction, my ρ cannot be a reduction relation.

Second, there is Sneed's condition of uniqueness:[12)]

$$(2) \qquad \forall x,x',y\ (x \rho y \wedge x' \rho y \rightarrow x=x').$$

T7) above says that (2) is not satisfied in the present example. But I doubt whether this condition can be imposed generally. Intuitively, Sneed justifies the requirement as expressing that the reducing theory gives a more detailed picture of reality. But this property is not equivalently expressed by (2). Condition (2) may be sufficient for T′ giving a more detailed picture than T but (2) certainly is not necessary for this property to hold. The "finer" picture of T′ may be "coarsened" in different ways so that the outcomes still are of the same structure (as shown by the present example). In general, I see no argument for condition (2) to be satisfied for all reduction relations, and I would hesitate to exclude non-unique relations on apriori grounds.

Third, there is the condition that to each model of the reduced theory there corresponds -via ρ - a model of the reducing theory:

(3) $$\forall x(x \varepsilon M \rightarrow \exists y(y \varepsilon M' \wedge x \rho y)).$$

Requirement (3) is essential for Mayr′s account and can be traced back to Suppes.[13] T5) above shows that this condition is satisfied.

Fourth, there is a kind of "converse" of (3), namely

(4) $$\forall y(y \varepsilon M' \rightarrow \exists x(x \varepsilon M \wedge x \rho y)),$$

i.e. each model of the reducing theory via ρ gives rise to a model of the reduced theory. The spirit of condition (4) is this. Given a model y of the reducing theory T′ we can construct or define in y a structure which is a model of the reduced theory T. This is the basic idea of Bourbaki′s "procedure of deduction of a structure of species Θ from a structure of species Σ ".[14] T4) above shows that condition (4) is satisfied. It should be noted that (4) also can be regarded as expressing a "derivability requirement" as mentioned in connection with (1).

A fifth formal condition is that each model of the reduced theory can be embedded into a model of the reducing theory:

(5) $\forall x\ (x \in M \rightarrow \exists y(y \in M' \wedge x \sqsubset y))$

where "$x \sqsubset y$" means "x is a substructure of y".(5) represents the kernel of Ludwig's notion of "Einbettung".[15] It is obvious that this condition does not apply to the case at hand, but I doubt whether it should. Intuitively, I would interpret (5) as saying that T is a <u>specialization</u> of T', and specialization and reduction are two different intertheoretic relations which can and should be kept apart.[16]

Sixth, there is the condition of ρ "preserving invariances".[17] A weak version of this condition is the following:

(6) $\forall x,y(\ x \rho y \rightarrow [x]\ \bar{\rho}\ [y]'\ )$

where [x] and [y]' denote the equivalence classes of x and y given by the corresponding invariances of T and T', and $X \bar{\rho} Y$ is a shorthand for $\forall x \in X\ \exists y \in Y(x \rho y) \wedge \forall y \in Y\ \exists x \in X(x \rho y)$. In (Pearce & Rantala, 1983a) a condition similar to (6) is considered as an aspect of continuity in scientific change. In my view (6) is a "special law" of the "theory of reduction" which will be satisfied only in special cases. It is possible to construct a pair of models $\langle x,y \rangle \in M(CT) \times M(RT)$ for which [ ] and [ ]' are given by elementary transformations and Lorentz-transformations, respectively, and for which (6) is false. My hypothesis is that typically (6) will be false for pairs of theories which are incommensurable but are nonetheless connected by some reduction relation.

Last but not least there is the condition of translatability of the language of the reduced theory into the language of the reducing theory, inherent in the received view and recently substantiated by Pearce.[18] Without going into technical details this requirement can be nicely illustrated with my example. Consider the atomic expressions "$t \lessdot t'$", "$\tau(t,t')=\alpha$" and "$\delta(t,a,b)=\alpha$" of CT. In D6-a) these are "defined" in terms of the primitives of RT "up to the choice of v,w and the units for the clock". But this "up to" prevents us from finding proper translations of these expressions in the language of RT. Also, the relations among pairs of the form <a,t> in CT "is simultaneous with" and "occupies the same point of space as" can be easily defined in CT but cannot be translated into expressions of RT (this is why RT is called "relativistic"). It is clear therefore, that translatability in the usual sense does not obtain in my example. This might be regarded as an argument against the adequacy of ρ. But it might as well be regarded as an argument against requiring translatability as a condition necessary for <u>all</u> reduction relations. What was said in the last paragraph applies here, too. Translatability characterizes only a special subclass of the class of all reduction-pairs -though a very interesting one. Typically, translatability will not obtain in cases of incommensurable theories, the two theories considered here constituting a commonly accepted example of incommensurability.

To summarize these considerations: It seems to me that condition (4) is the most central one for reduction. It combines the intuition of a derivation of the laws of the reduced theory from those of the reducing theory with the formal achievements of Bourbaki's work.[19] All the other conditions will be satisfied only in special cases but not in general. If this view is not completely misled then it is difficult to see how and why my ρ-relation should be inadequate, i.e. no reduction relation proper.

A last line of attack against ρ is to say that it does not satisfy the informal requirement that all intended applications of the reduced theory correspond via ρ to intended applications of the reducing theory:[20]

$$(7) \qquad \forall x \in I \, \exists y \in I' (x \rho y)$$

where I and I′ are the sets of intended applications of T and T′, respectively. In the present case, A. Kamlah has pointed out that probably among the intended applications of CT there are systems which are considered from <u>accelerated</u>

frames of reference which are ruled out by RT and therefore do not belong to I(RT). Then (7) would fail, provided we accept that the y required there is given by "the same" accelerated system which gave rise to the classical x we start with. The crucial point in this argument is, I think, that reference is made to some systematic concept not available in both theories: "acceleration". And this, in turn, leads to the question of how the intended applications of a theory are determined in general. Is it the case that scientists in the course of achieving agreement on whether some system is an intended application for T use concepts from theories systematically dependent on T? A clear cut "no" would be dogmatic. Scientific practice as far as it is documented by historians will perhaps yield the answer "in most cases not". At least this is the answer one would expect from systematic reflections on the determination of I of I.[21] According to Sneed and Stegmueller[22] I is determined "paradigmatically",i.e. one gives a list of "paradigms" forming a set $I_o \subseteq I$, and systems not in I_o will belong to I if they are sufficiently similar to those of I_o. If "sufficient similarity" cannot be decided on easily then the theory itself will be used as a criterion.Some new system x will be regarded as an intended application iff it can be successfully subsumed under T.

The paradigmatic method, if applied to CT, yields as intended applications systems which are either in direct contact with the earth or consist of stars or planets as seen from the earth. It is hard to come across space-time systems (as distinct from mechanical systems) described from frames of reference which are accelerated relative to the earth. The question certainly deserves a more detailed analysis but for the moment the above remarks will have to suffice. At least they make plausible the claim that (7) need not constitute a definite refutation of CT being reducible to RT.

In total, then, it seems to me that the present example should be regarded as a proper case of reduction. Both theories involved are physically adequate and based on operationally accessible notions. The reduction relation, too, makes physical sense and has formal properties which fit some of the general definitions of reduction already available.

NOTES

1) I am indebted to A.Kamlah,D.Pearce and H.-J. Schmidt for many remarks and suugestions on an earlier draft.
2) Compare the examples enlisted in Moulines' contribution to this volume. Recent notions of reduction can be found in (Ludwig,1978), (Mayr, 1976),(Pearce,1979),(Pearce & Rantala,1983a), (Sneed,1971), and (Balzer & Sneed,1977/78).
3) See (Bourbaki,1968),pp.259.
4) For further explanations compare my set-theoretic (as opposed to Bourbaki's rather idiosyncratic "syntactic") treatment of species of structures in (Balzer,1984).
5) See (Tarski,1959). I assume here that Tarski's A13) is always replaced by the corresponding second-order version, namely the formula on p.18 loc.cit. By an appropriate change of the axioms of dimensionality we easily obtain the axioms for 1-dimensional Euclidean geometry used in D2-3). More precisely,we have to omit A11) and A12) and add (in Tarski's notation): $\forall xyz[\beta(xyz) \vee \beta(yxz) \vee \beta(xzy)]$.
6) Compare (Balzer,1982),Chap.III.
7) Terminology is taken from (Weyl,1923),p.142. According to (Ehlers,1973) this group is characteristic for Newtonian spacetime.
8) For intuitive explanations of the following formalism see (Balzer,1982),Chap.IV.
9) An exact proof is found in (Zeeman,1964).
10) See (Adams,1959).
11) It is possible to modify RT so that each model contains a frame W explicitly (compare (Balzer, 1982),Chap.IV). By using such a modified RT, condition (1) for reduction can be proved for some ρ' modified along the same lines. The resulting version of RT, however, is open to criticism concerning its adequacy for the automorphism groups of its models are not pre-

cisely isomorphic to the Lorentz-group.

12) See (Sneed,1971),p.221,D51-2).

13) See (Mayr,1976),p.289,Definition (2.12-ii) and (Suppes,1957),p.271.

14) See (Bourbaki,1968),p.267.

15) See (Ludwig,1978),p.88.

16) Compare (Balzer & Sneed,1977/78).

17) This condition was first suggested to me by H.-J.Schmidt at an informal meeting.

18) See (Pearce & Rantala,1983a) and (Pearce,1979).

19) The treatment of reduction given in (Balzer & Sneed,1977/78) as essentially covered by (1) -though expressing the first intuition- falls short of exhibiting all the formal advantages of (4).

20) See (Sneed,1971),p.229,D54-A-2).

21) Compare (Balzer,1982),pp.28.

22) (Stegmueller,1973),pp.198.

W.Balzer
Seminar fuer Philosophie,Logik und Wissenschaftstheorie,Universitaet Muenchen
Ludwigstr.31
D-8000 München 22

REFERENCES

Adams,E.W.: 1959, 'The Foundations of Rigid Body Mechanics and the Derivation of its Laws from those of Particle Mechanics',in: The Axiomatic Method, Henkin,Suppes,Tarski (eds.),Amsterdam

Balzer,W.& Sneed,J.D.: 1977/78, 'Generalized Net Structures of Empirical Theories', Studia Logica 36 and 37

Balzer,W.: 1982, Empirische Theorien: Modelle,Strukturen,Beispiele,Braunschweig-Wiesbaden

Balzer,W.: 1984, 'On a New Definition of Theoreticity',to appear in Dialectica

Bourbaki,N.: 1968, Theory of Sets, Paris

Ehlers,J.: 1973, 'The Nature and Structure of Spacetime',in: J.Mehra (ed.), The Physicist's Conception of Nature,D. Reidel

Kamlah,A.: 1979, 'Erläuterungen',in: A.Kamlah and M.Reichenbach (eds.): Hans Reichenbach: Gesammelte Werke,Band 3, Braunschweig-Wiesbaden
Ludwig,G.: 1978, Die Grundstrukturen einer physikalischen Theorie, Berlin-Heidelberg-New York
Mayr,D.: 1976 , 'Investigations of the Concept of Reduction I', Erkenntnis 10
Pearce,D.: 1979, Translation,Reduction and Equivalence: Some Topics in Intertheory Relations, Dissertation,University of Sussex
Pearce,D.& Rantala,V.: 1983a 'Logical Aspects of Scientific Reduction',in: Epistemology and Philosophy of Science,Proceedings of the 7th International Wittgenstein Symposium 1982,Wien
Pearce,D.& Rantala,V.: 1983b, 'The Logical Study of Symmetries in Scientific Change', in: Epistemology and Philosophy of Science, Proceedings of the 7th International Wittgenstein Symposium 1982,Wien
Sneed,J.D.: 1971, The Logical Structure of Mathematical Physics,D. Reidel,second edition 1979
Stegmueller,W.: 1973, Theorie und Erfahrung,Zweiter Halbband,Berlin-Heidelberg-New York
Suppes,P.: 1957, Introduction to Logic,Princeton
Tarski,A.: 1959, 'What is Elementary Geometry?',in: Henkin,Suppes,Tarski (eds.): The Axiomatic Method,Amsterdam
Weyl,H.: 1923, Raum,Zeit,Materie, 5th ed.,Berlin
Zeeman,E.C.: 1964, 'Causality Implies the Lorentz-Group', Journal of Mathematical Physics 5

Othman Q. Malhas

SPACE-TIME GEOMETRIES FOR ONE-DIMENSIONAL SPACE

0. INTRODUCTION

It is now widely accepted that the axiomatization of empirical theories is necessary for giving precise answers to questions concerning their internal structures or their relations to each other and to the external world. It should, however, be noted that there are usually many ways of axiomatizing a theory (depending on the choice of primitive terms and axioms), and that an axiomatization of an empirical theory, while perhaps mathematically elegant (e.g. because the number of primitive terms is small), need not, for example, be adequate for clarifying the relation of the theory to the external world or to a rival theory.

To see this, let us begin by proposing that an axiomatization of an empirical theory should, if only to avoid the paradoxes of incommensurability, recognize the division of physical concepts into observational and theoretical concepts. Observational concepts are those that can be 'defined' only ostensively, i.e. by directly experiencing (or pointing out) examples of their intended interpretations. Examples: The colour 'red', the relation 'is hotter than', tables, chairs, the arrival of a bus at the station, the 'streaks and dashes' in a Wilson cloud chamber. Theoretical concepts are those that are constructed somehow out of observational concepts. Examples: Time, entropy, energy, electric field, proton, quark. It often happens that a theoretical concept will have a denotation if certain consequences of the theory in which it occurs are true, and would lose its meaning (become uninterpreted) if some of these consequences are discovered not to be true, (Giedymin, 1973). Such theoretical concepts deserve to be called theory-dependent concepts, since their denotations depend on the theories in which they occur. By contrast, observational concepts are theory-independent. The history of science provides many examples of physical concepts that lost their meanings upon discovering that some observational consequences of certain classical theories were not true. Examples:

W. Balzer et al. (eds.), Reduction in Science, 359–380.

The phlogiston, the ether, 'up' (a term denoting a unique direction in space that lost its significance when the earth was found not to be flat), and so on.

With this in mind, it seems inappropriate to leave the choice of primitive terms for an axiomatization of an empirical theory to considerations of mathematical elegance and economy alone. This is because the primitive terms of an axiomatic theory determine 'what the theory talks about' in the sense that the denotation of any concept of the theory (and, in view of the above remarks, whether it has any denotation at all) depends on how the primitive terms are interpreted. Indeed, the choice of primitive terms for an axiomatization of an empirical theory determines what the author of the axiomatization would like us to believe to be what the theory ultimately talks about. An empirical theory, however, ultimately talks about a theory-independent fragment of the external world. It would, therefore, seem reasonable to demand that the primitive terms for an axiomatization of an empirical theory should be the set of all its observational concepts (usually a finite set). All other physical concepts of the theory are then logically constructed in some way from these. This requirement can be weakened if two competing empirical theories are being compared. The two theories should, then, be developed from the same set of primitive terms, characterizing the physical structure that they are competing to describe correctly. It is then sufficient to demand that the primitive terms should be independent only of the two given theories, but not necessarily to be pure observational concepts (i.e. concepts independent of all empirical theories).

The difficulties that would otherwise arise can be illustrated by A.A. Robb's axiomatization of special relativity as presented in Winnie (1977). Theorems, in Minkowski's space-time theory, which have the form of iff-statements are expoited to reduce the number of primitive terms to one binary relation (causal connectibility) on the set of all point events. This is achieved essentially by turning these iff-theorems into definitions. Let R, P, Q, S,... denote point events. We shall write $P\overline{'}Q$ to denote that P is causally connected to Q. (In Minkowski's theory $P\overline{'}Q$ means that events P and Q are on a line representing the motion of a free particle.) Now Winnie defines a ternary relation:

(i) R-P-Q iff $R\overline{'}Q$ and for every event S, $S\overline{'}P$ implies $S\overline{'}R$ or $S\overline{'}Q$.

(If we take R-P-Q to mean that the time of occurrence of P,

in any frame, is between the time of occurrence of R and that of Q, then statement (i) is a theorem in Minkowski's theory.) The axioms chosen for $\top$ (and these, naturally, are theorems in Minkowski's theory) will imply that this ternary relation has some of the familiar properties of 'betweenness'. Next Winnie defines 'light connectibility':

(ii) P/Q iff P$\top$Q and, for all events R,S,

P-R-Q and P-S-Q implies R$\top$S.

(Again, if we take P/Q to mean that the events P and Q are on a light-like line, i.e. on a straight line representing the motion of a light pulse, then statement (ii) is a theorem in Minkowski's theory.) Other iff-theorems in Minkowski's theory, similar to (i) and (ii), are exploited in similar fashion in order to exhibit all the basic concepts of relativity as concepts ultimately defined in terms of $\top$.

There can be little doubt as to the great elegance of Robb's theory, but it is not certain that such logical constructions as those given above can help to throw light on the relation of relativity to classical theory or to the external world. Consider statement (ii). If we take this statement seriously as the *definition* of 'light connectibility', then the meaning of this term can be known only through (ii) (or a logically equivalent statement). Now, both definitions (i) and (ii) make sense in classical theory. Let us take P$\top$Q to mean that P and Q are events on the line representing the motion of a free particle. It follows that in classical theory if P and Q are causally connected and R is any event whatsoever, then P-R-Q. Classically, if R and S are two different simultaneous events, then R and S are not causally connected. Furthermore, such simultaneous events exist. Applying these conclusions from classical theory to definition (ii) we conclude that if classical theory is true, then *no two different events are light connectible*. The fact that the reader can *see* this page is, then, evidence against classical theory in favour of relativity! (This is because the reflection of light off the page and its arrival at the eye of the reader are two different light connectible events.) This, naturally, cannot be true, since classical theory recognizes the existence of different light connectible events and makes specific claims about them. For example, classical theory makes its own claims concerning the outcome of a Michelson-Morley experiment. Light connectibility is not a theory-dependent concept (the theory here being relativity), but an observational concept whose denotation can be determined by theory-independent means.

Statement (ii), and many other similar 'definitions' in Robb's theory, should be recognized for what they really are. Statement (ii) is not a definition (i.e. a certain kind of tautology), but a relativistic claim concerning the nature of light connectibility. Experiment can vindicate or refute this claim. The fact that this claim happened to have the form of an iff-statement should not be exploited in order to turn this claim into a definition and to reduce the number of primitive terms in a prospective axiomatization. The necessity of recognizing the existence of observational and theory-dependent concepts in empirical theories implies that, at the risk of appearing inelegant, we should resist the temptation to reduce the number of primitive terms to the absolute mthematically allowable minimum. "Elegance", Einstein once reported, "should be left to the tailor and the cobbler."

The following presentation of both classical and (special) relativistic space-time theories proceeds as follows: In sections 1 and 2 the theory-independent structure, which the two theories are competing to describe, is introduced. We note that 'light connectibility' (here rendered by the binary relation 'is on a photon with') is amongst its constituents. The present axiomatization differs from that of Robb in many respects, notably in the construction of the so-called space-time 'metric'. We show that there is no need to resort to constructions in two or more spatial dimensions (i.e. intersecting light cones) in order to obtain the 'metric'. Thus one can, without much loss of generality, develop all the essential features of special relativity within the confines of one spatial dimension.

1. PARTICLES AND EVENTS

It is reasonable to suppose that we do not have direct experience of either space or time, but that these concepts are, somehow, reducible to our experience of more tangible things. In fact, it is reasonable to suppose that the concepts of space and time derive their physical significance ultimately from relationships amongst the more tangible concepts of material body and event. For after all, one may conjecture, 'space' is something to do with the arrangement of material bodies around us, and 'time' is something to do with the order of events. If, as we claim, material bodies and events are very elementary components of our experience, then they can be 'defined' only by pointing out typical examples of each. Examples of material bodies: trees, TV-sets, typewriters,

Bertrand Russell, tennis balls, flowers, birds,...
Examples of events: The collision of two billiard balls, football matches, the utterance of a word, the utterance of a syllable in that word, a kiss, a lecture, the birth of Bertrand Russell,...

In fact it can be argued that much of our *direct* experience of the physical world consists of very little other than our experience of material bodies and events, and relations among them. Of these relations we shall be particularly interested in the so-called relation of *incidence* between material bodies and events, itself another very elementary component of experience (an observational concept) and one that can only be defined ostensively.
Examples of incidence: If two billiard balls h, k collide, then that collision is an event, call it P, experienced by both balls. If a third ball j does not partake in that same collision, then j does not experience P. We say that h and k are *incident with*, or *are on*, P, but that j is not. If a man h utters a word, then the utterance of the word is an event, call it P. We say that h is *incident with*, or *is on* P.

In general, if a material body k experiences an event P, then we say that k *is on* P or that k *is incident with* P or that P *is on* k or that P *is incident with* k.

Like all ostensively defined concepts, the relation of incidence suffers from partial vagueness: It is not always possible to determine with absolute certainty whether a material body is on a given event or not. For example, we would all agree that a player in a football match experiences the match directly and, hence, is incident with it; but we may have differing views on whther a spectator on the terraces can be said to be incident with the match or not. This vagueness in ostensive definitions can be dealt with in precise ways (e.g. Przelecki, 1969) that need not delay us here. For our purposes it is sufficient to assume that the procedure of ostensive definition is reasonably, though not completely, effective.

Material bodies and events seem to be 'extended' entities. This means that given any material body (or event), it has 'parts' which are themselves material bodies (or events). For instance, a car is a material body. It has many parts, e.g. the engine, the front bumper and a steering wheel. Each of these is itself a material body. Similarly, the building of a house is an event. It has many parts which are themselves events: the laying of the foundations, putting up the roof, installing the plumbing, etc.

Thus, any set of material bodies (or events) is partially ordered by a 'part relation'. We shall say that a material body (or event) is a *particle* (or *point event*) if and only if it has no parts other than itself. (This is, essentially, the same as the definition of an 'atom' in a partially ordered set with a null element, except that here we do not assume a null element.) We do not know, *a priori*, whether particles or point events exist, but sometimes we may not be able to distinguish a material body (or event) from its parts, *because of the limitations on the resolving power of our means of observation*. For example, with the naked eye a speck of dust cannot be distinguished from its parts. It can, then, be considered a particle. Similarly, without resorting to sophisticated instruments of observation, the collision between two billiard balls cannot be resolved into further events. Hence, it can be considered a point event.

Of course, if we use instruments of observation with greater resolving power, then what was previously thought to be a particle or point event can then be resolved further; 'smaller' bodies (or events) would then appear to be particles (or point events). We can then say that the earlier instruments of observation *are coarser than* the later instruments of observation.

Without further ado, let us assume that, due to the limited resolving power of our instruments of observation, it is possible to distinguish two non-empty sets: one of material bodies and one of events that, for all intents and purposes, behave as particles and point events, respectively.

Finally, we need one more, evidently observational, concept, namely that of 'light'. In particular we shall be concerned with *light signals*. Being an observational concept, a light signal cannot be verbally defined. In principle one can point out examples of light signals by actually producing them. Failing that, one can describe procedures for producing them. We may here presume that the intended significance of the words 'light signal' can be sufficiently clarified by saying that a light signal is a 'highly localized' pulse of light, something like a 'light corpuscle'. In fact, a light signal will be considered a special kind of particle, and will also be called a *photon*.

We need to establish some useful logical notation. If A(x) is any expression in which the variable x occurs, then the expression

$$(\exists !x)A(x)$$

is an abbreviation for $(\exists x)(\forall y)(A(y) \leftrightarrow x=y)$, and it says that there exists *exactly one* value of x that makes A(x) true. The expression

$$(\exists^{*} x)\ A(x)$$

is an abbreviation for $(\forall x)(\forall y)(A(x) \wedge A(y) \rightarrow x=y)$, and it says that there is *at most one* value of x for which A(x) is true. The expression

$$(\exists !! x)\ A(x)$$

is an abbreviation for $(\exists x)(\exists y)(x \neq y \wedge (\forall z)(A(z) \leftrightarrow (z=x \vee z=y)))$ and says that there are exactly two different values of x for which A(x) is true.

Now let there be two disjoint non-empty sets $\underline{X}$ and $\underline{Y}$ called *the set of all possible particles* and *the set of all possible events*, respectively. If $P \in \underline{Y}$ and $k \in \underline{X}$ and P is on k, then we shall indicate this by writing

$$P@k \quad \text{or} \quad k@P.$$

If it is not the case that $P@k$, then we shall write $P \not@ k$ or $k \not@ P$. Capital Roman letters P, Q, R,... will always indicate elements of $\underline{Y}$, and small Roman letters h, k, j,... will indicate elements of $\underline{X}$. We shall write $P \perp Q$, and say that P is *detached from* Q, iff $(\forall k \in \underline{X})(k@P \rightarrow k \not@ Q)$. We shall write $P /\!/\!/ Q$, and say that P is *quasi-simultaneous* with Q, iff $(P=Q \vee P \perp Q)$. The negation of $P \perp Q$ is $P \top Q$ (and then we say that P *is tied to* Q). The negation of $P /\!/\!/ Q$ is $P \not/\!/\!/ Q$. The relations $\perp$, $\top$ and $/\!/\!/$ are symmetric. The relation $/\!/\!/$ is also reflexive. We shall write $h /\!/ k$ iff $(h=k \vee (\forall P \in \underline{Y})(h@P \rightarrow k \not@ P))$. Then the particles h and k are either the same particle or they do not have any events in common. $/\!/$ is both reflexive and symmetric. The negation of $h /\!/ k$ is $h \not/\!/ k$.

2. COSMIC PLANES

The aim of this paper is to reconstruct classical and (special) relativistic space-time theories using the concepts introduced in the last section as building blocks. It is helpful to recall that, in both theories, the set of all possible point events is represented by $\mathbb{R}^4$, where $\mathbb{R}$ is the real line; so that if the point event P is represented by (x,y,z,t), then (x,y,z) represents the point, in space, of the occurrence of P aand t represents the time of occurrence of P. Particles, in each theory, are represented by certain continuous curves. In particular, 'free' particles are represented by straight lines, each having an equation of the form

$$(x,y,z,t) = (a,b,c,d) + s.(e,f,g,h),$$

where a, b, c, d, e, f, g, h, are real numbers and s is a real variable. The components of the velocity of the particle in the x, y and z directions are

$$v_x = e/h, \qquad v_y = f/h, \qquad v_z = g/h$$

respectively. In classical space-time theory the only restriction on these components is that they should be finite, whereas in special relativity it is required that they take values from the closed interval $[-c,+c]$, where c is the velocity of light in vacuum. (The interval is closed because we would like to consider light signals to be particles.) The intended interpretation of P@k in this representation is that the point representing P is on the curve representing k.

Given these facts about the two theories, it is easy to verify that our axioms for these theories do in fact reflect, at least some of, their contents.

Definition: Let $\underline{U}$ be a subset of $\underline{X}$. We say that $\underline{U}$ is a complete set of free particles relative to $\underline{Y}$ iff

(i) $(\forall P,Q\in\underline{Y})(P^{\top}Q \rightarrow (\exists k\in\underline{U})(P@k \wedge Q@k))$;

(ii) $(\forall P,Q\in\underline{Y})(P\neq Q \rightarrow (\exists {*}k\in\underline{U})(P@k \wedge Q@k))$.

It is easy to verify that the set of all straight lines in $\mathbb{R}^4$, representing particles with allowable constant velocities, satisfies (i) and (ii). Thus the set of all 'free particles', in the traditional Newtonian sense, is a complete set of free particles relative to $\underline{Y}$ (= $\mathbb{R}^4$). It is not too difficult to verify that (i) and (ii) are also satisfied by the set of all curves in $\mathbb{R}^4$ representing particles moving in a uniform gravitational field. This is not surprising since, from the point of view of an observer falling in the same gravitational field, all these particles would appear to be uniformly moving along straight lines, and would be represented by him as straight lines in $\mathbb{R}^4$. Thus the set of all particles freely moving in a uniform gravitational field is another complete set of free particles in $\underline{Y}$. The above definition may, roughly speaking, be considered to be the result of trying to restate the Newtonian definition of a free particle without resorting to the concepts of force and velocity. The generality achieved in this way is consistent with the general principle of relativity of motion according to which a system of particles falling in a uniform gravitational field cannot 'internally' be distinguished from a system of free particles in the Newtonian sense.

It immediately follows from the definition that if $\underline{U}$ is a complete set of free particles relative to $\underline{Y}$ and $\underline{W} \subset \underline{Y}$, then $\underline{U}$ is a complete set of free particles relative to $\underline{W}$ in the sense that

(i) $(\forall P,Q \in \underline{W})(P \bar{\tau} Q \to (\exists k \in \underline{U})(P@k \wedge Q@k))$:

(ii) $(\forall P,Q \in \underline{W})(P \neq Q \to (\exists {*}k \in \underline{U})(P@k \wedge Q@k)$.

Ideally, we should, in this paper, aim at using a complete set of free particles relative to $\underline{Y}$ in order to coordinatize $\underline{Y}$, to show that $\underline{Y}$ is a four-dimensional vector space over $\mathbb{R}$ in which the elements of the complete set of free particles are represented by straight lines. Without much loss of generality, however, we shall turn our attention to a subset of $\underline{Y}$ that will turn out to be a 2-dimensional vector space over $\mathbb{R}$, in which 'free particles' are represented by certain straight lines. To make these ideas precise we need the following definitions:

Definition: Let $k \in \underline{X}$ and let $\underline{W} \subseteq \underline{Y}$. We say that k lies in $\underline{W}$

iff $(\forall P \in \underline{Y})(P@k \to P \in \underline{W})$.

Definition: Let $\underline{U} \subseteq \underline{X}$ and let $\underline{W} \subseteq \underline{Y}$. We say that

$(\underline{U},\underline{W},@)$ is a cosmic plane iff

A1: $\underline{U}$ is a complete set of free particles relative to $\underline{Y}$ (and, hence, relative to $\underline{W}$);

A2: $k \in \underline{U} \to k$ lies in $\underline{W}$;

A3: $(\forall k)(\forall P)(\exists ! j)(P@j \wedge k//j)$;

A4: $(\exists P,Q,R)(P \neq Q \neq R \neq P \wedge P \bar{\tau} Q \wedge Q \bar{\tau} R \wedge R \bar{\tau} P \wedge Tr(P,Q,R))$,

where Tr(P,Q,R) means that events P,Q,R are not all on the same particle;

A5: $(\forall k)(\forall P)(\exists Q)(Q@k \wedge Q///P)$.

The quantifiers in A2 to A5 range over $\underline{U}$ or $\underline{W}$ according to whether they are followed by a symbol for a particle or an event, respectively. This convention will continue to apply. Our first assumption is that these exists a cosmic plane

$$CP = (\underline{U},\underline{W},@).$$

Thus statements A1 to A5 are true in CP.

If we interpret particles to be 'lines' and point events

to be 'points', then a cosmic plane would become an affine plane with some of the lines, but not the points on them, deleted. In this model, k//j means that lines k and j are parallel. P ///Q means that points P and Q do not have a line on both of them.

These is another geometric realization of a cosmic plane that was probably first given by Karl Menger (1970). Let U be the set of all points in the xy-plane whose x-coordinates lie between -c and +c, where c is any real number on the extended real line. Hence U is the vertical open strip bounded by x=-c and x=+c. Let W be the set of all open chords of U. Thus if P∈W, then P= {(x,y): y = mx+b and -c < x < +c }, where m and b are finite real numbers. In this model k//j means that the points k and j have the same x-coordinate (hence, if they are distinct, they have no open chord on them both). P ///Q means that the open chords P and Q are 'parallel', i.e. they do not intersect in the strip. If c is a finite number, then the geometry of the strip (with open chords interpreted as 'lines') is hyperbolic.

The open strip model is not just a novel illustration of a cosmic plane. It was shown by Menger that the open strip model is the source of much information concerning the structure of space-time from the point of view of special relativity (for one spatial dimension). What we hope to do is to axiomatize this information.

Note that by virtue of A3, the relation //, which is already reflexive and symmetric, becomes an equivalence relation on U. Now we are able to construct the concept of space:

Definition: Let k∈U. The space relative to k, or the k-space, is the equivalence class [k], of the equivalence relation //, to which k belongs. Each element of k-space is called a point of k-space.

A few words of explanation are now necessary. Intuitively speaking, every point in space can be marked by a 'stationary' particle of matter. In order for the term 'stationary' to make sense, however, we must specify a reference particle. Thus every point in the space relative to k can be marked with a particle stationary with respect to k. It follows from A3 and the definition of // (and the intuitive meaning of motion) that in CP two particles are at rest relative to each other iff they bear the relation // to each other (otherwise they would 'cross paths', i.e. they would have a single event common to them). Thus the space relative to k consists of all possible free particles stationary with respect to k.

It is interesting to note that if j is a point in k-space, then k is a point in j-space. As expected, two particles at rest relative to eachother have the same space.

Now we shall introduce light. Let $\underline{F}$ be a 1-place relation on $\underline{U}$ (i.e. $\underline{F} \subseteq \underline{U}$). We give $\underline{F}$ the following interpretation: If $k \in \underline{U}$ and we assert $\underline{F}(k)$, then this says that k is a *photon*. We shall write P/Q, and say that P *is on a photon with* Q iff $(\exists k)(\underline{F}(k) \wedge P@k \wedge Q@k)$. We shall write P/RQ iff $(P/R \wedge P/Q)$. We assume that $\underline{F}$ satisfies the following axioms

L1: $(\forall P)(\exists ! ! k)(\underline{F}(k) \wedge P@k)$;

L2: $(\underline{F}(k) \wedge k//j) \rightarrow \underline{F}(j)$;

L3: $(R/PQ \wedge S/PQ \wedge P \perp Q) \rightarrow R \mathbin{/\!\!\!-\!\!/} S$.

Suppose $(\underline{U}, \underline{W}, @)$ is a cosmic plane, then the structure $(\underline{U}, \underline{W}, @, \underline{F})$, in which L1, L2 and L3 are true, is called *a cosmic plane with photons*, or *an illuminated cosmic plane*. We shall assume that our cosmic plane CP has a prolongation

$$ICP = (\underline{U}, \underline{W}, @, \underline{F})$$

which is an illuminated cosmic plane. Statements L1, L2 and L3 express three of the kinetic properties of light. (Furthermore they express kinetic properties of light that are independent of both classical theory and special relativity. This is because in both these theories, applied to a 2-dimensional space-time, every event can be considered to be the point in space-time at which two light pulses - one travelling to the 'left' and one travelling to the 'right' - cross each other. Moreover, given the two straight lines representing the two pulses on any event, all other pulses of light are given by the set of all lines parallel to one of these two lines. With this interpretation, it is easy to verify L1, L2 and L3. It also follows that the velocity of light in any frame of reference is independent of the velocity of the source of light, since for any observer all the pulses of light travelling to the 'left' will have the same velocity and all those travelling to the 'right' will have the same velocity. Thus, these innocuous looking axioms, L1, L2 and L3, embody some of the consequences of the 'wave nature' of light. A light signal has in common with a particle only that it has a more or less definite position at any time and, if not interfered with, travels uniformally in a straight line.)

Suppose $P \perp Q$. Then it follows from L1 and L2 that there are two unique events R and S such that R/PQ and S/PQ. It follows from L3 that $R \mathbin{/\!\!\!-\!\!/} S$. It also follows that the unique particle on R and S is not a photon (otherwise there would be three different photons on S: the photon on S and P, the photon on S and Q and the photon on S and R, contradicting L1). We shall call this unique particle the centre of P and Q and denote it by c_{PQ}. The centre of two detached events plays vital roles in both classical theory and special relativity, as we shall see.

The illuminated cosmic plane ICP is the physical structure which is the fragment of physical reality about which the two theories, to be developed below, speak and which they compete to describe. (ICP is 'theory-indepedent' at least in the sense that it is independent of these two theories.)

3. CLASSICAL SPACE-TIME

Let us suppose that assumption C, below, is true in ICP:

$$C: \quad (\forall k)(\forall P)(\exists !Q)(Q@k \wedge Q///P).$$

This strengthens A5. It is easy to show that, with C, the relation /// becomes an equivalence relation. (In fact with C the structure CP becomes what Menger (1970) calls a self dual fragment of the affine plane.) The concept of time can now be introduced, but first let us examine two possible models of the theory.

Models: There are two useful models of ICP with C true.
(I) Let W be the set of all points in the xy-plane. Let U be the set of all lines, in the plane, with finite slopes. Interpret P@k to mean that point P is on the line k. Interpret F(k) to mean that line k has slope +1 or -1. In this interpretation k//h means that lines k and h have the same slope, and P///Q means that points P and Q have the same x-coordinate. Statements A1 to A5, L1 to L3 and C are true.
(II) Let W be the set of all straight lines in the xy-plane with finite slopes. Let U be the set of all points in the xy-plane. Interpret F(k) to mean that point k has x-coordinate +1 or -1. In this interpretation P///Q means that line P has the same slope as line Q, and h//k means that points h and k have the same x-coordinate. In this interpretation, statements A1 to A5, L1 to L3 and C are true.

Definition: Two point events P and Q are said to be simultaneous iff P///Q (i.e. iff they are quasi-simultaneous.)

Definition: The equivalence classes of /// are called moments of time. The set T_c of all moments of time is called classical time. If t is a moment of time and $P \in t$, then we say that t is *the moment of occurrence of* P.

Definition: Let $\underline{L} = \underline{U} \cup T_c$. Let $m \in \underline{L}$. We shall write P:m if and only if $((m \in \underline{U} \wedge P@m) \vee (m \in T_c \wedge P \in m))$. If P:m, then we may say P *is on* m or that m *is on* P.

It follows that $P \neq Q$ implies that there is exactly one m in $\underline{L}$ such that m is on both P and Q. If we say that two elements of $\underline{L}$ are *parallel* iff there is no point event on both of them, then it also follows that given any m in $\underline{L}$ and any point event P not on m, then there is exactly one n in $\underline{L}$ such that P:n and n is parallel to m. From this and A4 it follows that *if we think of W as a set of 'points' and L as a set of 'lines', then* $(\underline{W}, \underline{L}, :)$ *is an affine plane*.

Definition: The affine plane $(\underline{W}, \underline{L}, :)$ is called *classical space-time* (for one spatial dimension). The structure $(\underline{W}, \underline{L}, :, \underline{F})$ is called *illuminated classical space-time* (for one spatial dimension).

We need one more assumption, that can be stated entirely in terms of particles, point events and the relation @, but which can be stated much more succinctly and briefly in the following form

PA: Classical space-time has the Pappus property.

(PA stands for 'Pappus axiom'). As an affine plane (in which $\underline{W}$ is thought of as a set of points and $\underline{L}$ as a set of lines), classical space-time can be coordinatized in the normal way (Blumenthal, 1961). With PA true in it, classical space-time becomes a 2-dimensional vector space over a field V. Illuminated classical space-time is merely classical space-time with certain lines, those corresponding to photons, distinguished: On every point there are exactly two different lines corresponding to photons, and the set of all photons is given by two different classes of parallel lines. The set of all moments of time is represented by a third, distinct, parallel class of lines. All other lines represent elements of $\underline{U}$ other than photons. Using the methods of elementary analytic geometry, it quickly follows that *the centres of all pairs of detached point events have the same slopes*. In symbols:

$$(P \perp Q \wedge R \perp S) \rightarrow c_{PQ} // c_{RS} .$$

All these centres represent particles 'at rest relative to each other'. If k is one of these centres, then *the set of all these centres is k-space and may be identified with Newtonian absolute space* ! Thus if assumption C is true, we can use light signals to detect absolute motion: If k is a particle and it is not the centre of two detached events, then k is moving, in an absolute sense.

Not all coordinate systems are of interest, but only those that are usually called 'Galilean frames of reference'.

Definition: A Galilean frame of reference is a set $\{O, t, k, I\}$, where O is a point (i.e. point event), called the origin of the frame, t is the line representing the moment of time on O (the time of occurrence of O), k is the line representing a particle (from U) on O. Line t is called the x-axis and line k is called the t-axis. I is any point that is on neither the x-axis not the t-axis. I is called the unit event.

Given a Galilean frame of reference and a particle in U, then the line representing this particle has the equation

$$x = vt+b, \quad \text{where } v, b \in V.$$

We call v the velocity of the particle in the given frame. The x-axis has the equation t=0 and the t-axis the equation x=0. Given two Galilean frames of reference, sharing the same origin, then they will have the same x-axis. Suppose that a point has coordinates (x,t) in one frame and that the equation of the t-axis of the second frame with respect to the first is $0 = x - vt$. We can provide point coordinates (x', t') in the second frame so that the transformation $(x,t) \rightarrow$ '$x', t')$ is given by

$$x' = a(x - vt)$$
$$t' = bt.$$

If we choose our units of measurements correctly, i.e. if we choose the correct unit events for each frame, then these transformations reduce to the familiar Galilean transformations.

4. RELATIVISTIC SPACE-TIME

We can replace assumption C by other assumptions which, if found true in ICP, would no longer allow us to ascribe absolute motion (or rest) to any particle in U. Let us replace C by R1 to R6, below.

R1: $(P \perp Q \wedge Q \perp R \wedge R \perp P) \rightarrow ((\forall S)(S @ c_{PQ} \wedge S @ c_{QR}) \rightarrow S @ c_{RP})$;

R2: $(Q \perp P \wedge Q \perp R \wedge c_{QP} // c_{QR}) \rightarrow P \perp R$;

R3: $\sim \underline{F}(k) \wedge P \not{@} h \rightarrow (\exists R)(R @ h \wedge R \perp P \wedge c_{RP} // k)$;

R4: $(h \not{//} k \wedge \sim \underline{F}(k) \wedge \sim \underline{F}(h) \wedge R \neq P) \rightarrow (\exists Q)(Q \perp R \wedge Q \perp P \wedge h // c_{QR} \wedge k // c_{QP})$

R5: $(P \perp Q \wedge R @ c_{PQ}) \rightarrow (R \bar{\top} P \leftrightarrow R \bar{\top} Q)$.

We shall state R6 later. All these assumptions, apart from R3 and R4, are true in the classical case. These assumptions may appear indigestible, but they express simple geometrical properties and are necessary for the purpose of restoring order to a world picture from which assumption C is missing.

Models: There are two useful models. (I) Let W be the xy-plane. Let U be the set of all straight lines in that plane, with slopes in the closed interval [-1, +1]. Interpret P@k to mean that the point P is on the line k. In this interpretation, h//k means that h has the same slope as k. $P \perp Q$ means that points P and Q are not on a line in U. The interpretation of F(k) is that k has slope +1 or -1. If $P \perp Q$, then c_{PQ} is obtained by 'drawing' a rectangle whose sides are photons (i.e. lines with slopes +1 or -1), such that P, Q are vertices. Naturally, P and Q cannot be vertices on the same side of the rectangle, but on the ends of the diagonal with slope between -1 and +1. The line joining the other two vertices is c_{PQ}. In this interpretation A1 to A5, L1 to L3 and R1 to R5 (and R6) are all true. (II) Let U be the set of all points in the xy-plane with x-coordinates in the closed interval [-1, +1]. Thus we may call U the vertical closed strip bounded by x=-1 and x=+1. Let W be the set of all straight lines (not chords), in the xy-plane, with finite slopes. Interpret P@k to mean that the point k is on the line P. Interpret F(k) to mean that the point k has x-coordinate -1 or +1. In this interpretation $P \perp Q$ means that lines P and Q do not intersect in the strip. P/Q means that lines P and Q intersect on an edge of the strip. Suppose $P \perp Q$, then c_{PQ} is obtained as follows. Let k_1 and k_2 be the two points where the line P intersects the edges of the strip. Let h_1 and h_2 be the points of intersection of line Q with the edges of the strip. Suppose h_1 and k_2 are on opposite sides of the strip. Then c_{PQ} is the point of intersection of the lines h_1k_2 and h_2k_1. In this interpretation, statements A1 to A5, L1 to L3 and R1 to R5 (and R6) are true. (Verifying that our axioms are true in both of these interpretations is a non-trivial and instructuve excercise in geometry.)

Theorem (4.1): Every point has at least three different particles on it.
Proof: By A4 there exists a 'triangle' of events R,P,Q whose sides are particles. Let S be any point event. By A3, there are three different particles on S each bearing the relation // to one side of the given triangle. □

Corollary: $P \perp Q \to P \neq Q$.
Proof: If P=Q, then there is at least one particle on Q. Hence $P^{\top}Q$. □

Now we can show that according to the new theory absolute space does not exist. A4 guarantees the existence of both particles and events. It also implies that given any particle, there is an event which is not on it (otherwise there would be a particle with all the events on it and the triangle of events declared in A4 would not exist). Thus suppose that P@h, and suppose that there are two different particles k and k' on P such that neither of them is a photon. By R3, there are two distinct events Q and Q' on h such that each is detached from P and $k//c_{PQ}$ and $k'//c_{PQ'}$. Now since $k \neq k'$ and both k and k' are on P, we have $k \not\!/\!\!/ k'$. Thus, since // is an equivalence relation, we have $c_{PQ'} \not\!/\!\!/ c_{PQ}$, i.e. these two centres are moving relative to each other, which contradicts the classical claim that the centres of all pairs of detached events are at rest relative to each other. Thus, in the new theory, absolute space does not exist. (The is another way of proving this based on R4 which will be left as an excercise.)

Definition: We shall write $P\#_kQ$ if and only if

$$(P=Q \vee (P \perp Q \wedge (\exists h)(h//k \wedge h=c_{PQ}))).$$

Lemma (4.2): $(P\perp Q \wedge Q\perp R \wedge R\perp P \wedge c_{PQ}//c_{RP}) \to c_{QR}//c_{PQ}$.

Proof: Suppose, to the contrary, that

1) $P\perp Q \wedge Q\perp R \wedge R\perp P$
2) $c_{PQ}//c_{RP}$
3) $c_{QR} \not\!/\!\!/ c_{PQ}$

} assumptions (1–3)

Then there exists an event S such that

4) $S@c_{QR} \wedge S@c_{PQ}$ by step 3
5) $S@c_{RP}$ by 1 & 4 & R1
6) $S@c_{PQ} \wedge S@c_{RP}$ by 4 & 5

Step 6 contradicts step 2. □

Theorem (4.2): $\#_k$ is an equivalence relation.
Proof: From its definition $\#_k$ is both reflexive and symmetric. We need only prove that it is transitive. If R,P,Q are not pairwise distinct, then $R\#_kP \wedge P\#_kQ \to R\#_kQ$ follows quickly from the definition. Thus suppose

1) R,P,Q are distinct and $R\#_kP$ and $P\#_kQ$

2) $R \perp P \wedge P \perp Q \wedge c_{RP}//k//c_{PQ}$ from 1 by definition of $\#_k$

3) $R \perp Q$ by 2 and R2

4) $c_{RQ}//k$ by 2,3 & lemma (4.2)

5) $R \#_k Q$ by 3,4 and definition of $\#_k$. □

Definition: If $P\#_kQ$ then we shall say that P is k-simultaneous with Q. The equivalence classes of $\#_k$ will be called k-moments of time. The time of occurrence of P relative to k is the k-moment t_{Pk} to which P belongs. Let T_k be the partition of W induced by $\#_k$. Let

$$T = \{t : t \in T_k \text{ for some } k \in \underline{U}^*\},$$

where $\underline{U}^* = \underline{U} - \underline{F}$.

Theorem (4.3): $(h \not\parallel k \wedge \sim\underline{F}(h) \wedge \sim\underline{F}(k)) \rightarrow (\forall R,P)(\exists Q)(Q\#_kR \wedge Q\#_kP)$.

Proof: Immediate from the definition of # and from R4. □

Definition: Let $\underline{L} = \underline{U} \cup \underline{T}$. Let $m \in \underline{L}$. We shall write P:m if and only if $((m \in \underline{U} \wedge P@m) \vee (m \in \underline{T} \wedge P \in m))$. If P:m we shall say that P is on m or that m is on P. Let $m,n \in \underline{L}$. If m=n or there is no P on both m and n, then we shall say that m and n are parallel.

Lemma (4.4): If $Q\#_kP$, then $t_{Pk} = t_{Qk}$ and if $h//k$, then $t_{Pk} = t_{Ph}$.

Proof: Directly from the definition of $\#_k$ and the fact that $\#_k$ is an equivalence relation. □

Theorem (4.4): $P \neq Q \rightarrow (\exists! m \in \underline{L})(P:m \wedge Q:m)$.

Proof: Suppose $P \neq Q$. Then either $P \perp Q$ or $P \top Q$. Suppose $P \top Q$. Then, by A1, there is exactly one particle on both P and Q. Of course, if $P \top Q$, there is no k-moment, for any k, on both P and Q since this would imply $P \perp Q$ (by the definition of $\#_k$). Thus suppose $P \perp Q$. Let $k = c_{PQ}$. Then $P\#_kQ$ and both P and Q are on t_{Pk}. Suppose that, for some particle h, P and Q are on an h-moment. Then $P\#_kQ$. Thus $h//c_{PQ}$ (by the definition of $\#_h$). Hence $t_{Ph} = t_{Pk}$, by lemma (4.4). □

Thus if point events are thought of as 'points' and elements of L as 'lines', then we have exactly one line on any two different points, as in affine geometry. We would like the notion of parallelism to be equally well-behaved.

Lemma (4.5): $\sim\underline{F}(k) \wedge P \not@ h \rightarrow (\exists R)(R@h \wedge R\#_kP)$.

Proof: Directly from R3 and the definition of $\#_k$. □

Theorem (4.5): If h and k are any two particles and k is not a photon, then there is an event on both h and any k-moment of time.

Proof: Directly from lemma (4.5) and the definition of k-moment. □

Theorem (4.6): If $h \not\parallel k$ and neither h nor k are photons, then there is an event Q on both t_{Ph} and t_{Rk}, where R,P are any two events.

Proof: From theorem (4.3) and definitions of t_{Ph} and t_{Rk}. □

Theorem (4.7): For every $m \in \underline{L}$ and for every event P, there is exactly one n which is on P and is parallel to m.

Proof: Suppose m is a particle, then by A3 there is exactly one particle on P parallel to m. All other elements of $\underline{L}$, by theorem (4.5), are not parallel to m. The theorem is proved for particles. Now suppose m to be a k-moment. Since k-moments are equivalence classes of $\#_k$, the only k-moment parallel to m and on P is t_{Pk}. By theorem (4.5), no particle on P is parallel to m, and by theorem (4.6) no h-moment is parallel to m unless $h // k$. But if $h // k$, then $t_{Ph} = t_{Pk}$, by lemma (4.4). □

Theorem (4.8): If we think of $\underline{W}$ as a set of 'points' and $\underline{L}$ as a set of 'lines', then $(\underline{W},\underline{L},\underline{:})$ is an affine plane.

Proof: From theorems (4.4), (4.7) and A4. □

Definition. The structure $(\underline{W},\underline{L},:,\underline{F})$ is called relativistic space-time (for one spatial dimension).

Our last assumption can be stated entirely in terms of particles, events and @, but can be stated much more briefly as

R6: Relativistic space-time has the Pappus property.

Before coordinatizing relativistic space-time, we shall develop a theory of congruence.

Definition: An ordered pair (P,Q) of point events is called a vector. Every vector of the form (P,P) is called a null vector (Blumenthal, 1961).

We would like to give meaning to the idea that two vectors have the 'same length'.

Definition: Suppose $P \neq Q$. If P/Q, then the vector (P,Q) is called a light vector, but if $P^{\intercal}Q$ and the particle on both P and Q is not a photon, then (P,Q) is called a time vector. If $P \perp Q$, then (P,Q) is called a space vector. The 'line' on P and Q, whether it is a particle or a k-moment (for some k), is called the carrier of (P,Q).

To define a binary relation on the set of all vectors is to give the conditions of membership of that relation. The conditions of membership, for the required congruence relation, call it 'eq', for vectors on parallel carriers, are well known (Blumenthal, 1961). A certain Desargues property (in fact a consequence of the Pappus property) is necessary to show that these conditions are consistent. We shall continue the proceedings by giving conditions of membership of eq for certain nonparallel time vectors. The following definition is essentially an adaptation of a definition by Menger (1970,p.231).

Definition: Let O be a fixed point event. Suppose $O^{\mathsf{T}}R$ and $O^{\bar{\mathsf{T}}}P$ and R//P. Then (O,R)eq(O,P) iff R=P or $O@c_{RP}$.

Theorem (4.9): eq is an equivalence relation on the set of all vectors with O as first element.

Proof: It is clear that eq is reflexive and symmetric. We prove that it is transitive. Suppose that (O,R)eq(O,P) and (O,P)eq(O,Q). If R,P,Q are not pairwise distinct, then it quickly follows that (O,R)eq(O,Q). Thus suppose that (O,R)eq eq(O,P) and (O,P)eq(O,Q) and that R,P,Q are distinct. It follows

1)	$(R \perp P) \wedge P \perp Q \wedge R \perp Q$	from the definition.
2)	$O@c_{RP} \wedge O@c_{PQ}$	
3)	$O^{\mathsf{T}}R \wedge O^{\bar{\mathsf{T}}}P$	
4)	$O@c_{RQ}$	1,2 and R1
5)	$R \perp Q \wedge O@c_{RQ}$	1,4
6)	$O^{\mathsf{T}}R \wedge R \perp Q \wedge O@c_{RQ}$	3,5
7)	$O^{\bar{\mathsf{T}}}Q$	6, R5
8)	(O,R)eq(O,Q)	6,7 and definition of eq. □

Relativistic space-time can be coordinatized in the usual way. Not all coordinate systems are of interest, however, but only those given by the following definition.

Definition: A Lorentz frame of reference is a set {O,k,t,I}, where O is an event called the origin, k is a particle on O, called the t-axis. $t = t_{Ok}$ and is called the x-axis. $I \neq O$ and I is an event on one of two photons on O, and it is called the unit event.

Once relativistic space-time is coordinatizaed, it becomes a 2-dimensional vector space over a field V. Given a Lorentz frame of reference, if h is any particle, then it has the equation $x = vt+b$, where $v,b \in V$. We call v the velocity of the particle in the given frame. If h//j, then both h and j have the same velocity. In particular, the photon on O and I has the equation $0 = x - t$ (since we choose (1,1) as the coordinates of I). Every photon parallel to this also has velocity 1. Now we shall state some consequences of coordinatization which follow from the definitions by the mthods of ordinary analytic geometry. The proofs will only be given in broad outlines.

Theorem (4.10): The photon on O which does not pass through I has velocity -1.

Proof: Let k' be the line on I parallel to the t-axis. Construct the parallelogram of photons whose diagonal is k' such that one of its vertices is O. The point I, of course, is another vertex. Let the vertex opposite O be R. Then $k' = c_{OR}$ and the line on O and on R defines t_{Ok}, where k is the t-axis. Thus the line of O and R is the x-axis. The vertices have coordinates (0,0), (1,1), (a,0) and (1,b). (The first component is the x-component and the second is the t-component.) Calculate the slopes. □

Theorem (4.11): If particle h has the equation $x = vt$, then t_{Oh} has the equation $vx = t$.

Proof: Let h' be the particle on I parallel to h. Draw the parallelogram of photons whose diagonal is h'. Let R be the vertex opposit O. Then line OR is t_{Oh}. Calculate the equation of OR using theorem (4.10). □

Theorem (4.12): If (O,(0,1)) eq (O,(x,t)) then $t^2 - x^2 = 1$.

Proof: Draw a parallelogram of photons so that one of the vertices is the point (0,1) and such that one of its diagonals passes through O. Let R be the vertex opposite O. Calculate the coordinates (x,t) of R using theorems (4.10) and (4.11). □

The Lorentz transformations can be obtained as follows. Let $\{O,k,t_{Ok},I\}$ and $\{O,k',t_{Ok'},I'\}$ be two such frames and suppose that the equation of k' in the first frame is $0 = x-vt$, then, by theorem (4.11), the equation of $t_{Ok'}$ is $0 = t-vx$. (We have, for simplicity, chosen the same origin for both frames.) We shall stipulate that I' is on the line OI and that the two frames use 'the same time standard', i.e. so that the event R on k' for which $(x',t') = (0,1)$ should have (x,t) coordinates satisfying $t^2 - x^2 = 1$ (see theorem (4.12)). By the theory of coordinatization (Blumenthal, 1961), we have

$$x' = a(x-vt) \quad \text{and} \quad t' = b(t-vx).$$

From this and the stipulations made above, we obtain the familiar Lorentz transformations.

5. APPROXIMATIVE REDUCTION

In this final section we offer suggestions concerning the relationship between the two theories expounded above. In particular, we would like to discuss the question: What does it mean to say that classical theory is approximately true ? We would like to give an answer without resort to numerical concepts, e.g. the speed of light, but entirely in terms of the concepts of material body, event and the incidence relation (together with the other qualitative concepts defined in terms of these). We need a theorem in relativistic space-time theory.

Theorem (6.1): Suppose that Q is a point event, k is a particle such that $Q \not@ k$, and Q@h and Q@j, then there are at least two different point events R,P on k with Q///R and Q///P.

Proof: Suppose $\sim \underline{F}(h)$ and $\sim \underline{F}(j)$, $h \neq j$, and Q@h and Q@j. Clearly Q is on t_{Qj} and on t_{Qh}. Since $h \not\parallel j$ (because Q is on both particles) we have $t_{Qj} \neq t_{Qh}$. By theorem (4.5), there are point events R,P on k such that R is on t_{Qj} and P is on t_{Qh}. Hence R///Q and P///Q. By theorem (4.4), $R \neq P$ (or else there would be two different moments on Q and P). □

Clearly, this consequence of relativity contradicts assumption C of the classical theory. Now we already argued, in section 1, [illegible] point events are essentially a manifestation of the coarseness [illegible] our means of observation. With better means of observation, what previously appeared to be a point event could be resolved further and be seen to have parts. Conversely, what appears to be a collection of point events would, if coarser means of observation were employed, appear to be a single point event.

Assuming relativity to be true, then theorem (6.1) would also be true. Our means of observation, however, may not be sufficiently keen to resolve the different point events on k that bear the relation /// to Q. We would then conclude that there is a single point event on k bearing the relation /// to Q. If the same conclusion is arrived at after many varied similar observations, we might obtain C as a generalization and classical theory would appear to be true.[1]

1. I should like to dedicate this paper to Misatoshi Ikeda whose advice and encouragement led me to complete the research reported here.

The quotation from Einstein, in the first section, is attributed by him to L. Boltzmann, annd appears in the preface of Einstein's 'Relativity', Crown Publishers, 1961.

Othman Q. Malhas, Department of Mathematics, Yarmouk University, Irbid, Jordan

REFERENCES

Blumenthal, L.M. (1961): A Modern View of Geometry, Freeman

Giedymin, J. (1973): 'Logical Comparability and Conceptual Disparity between Newtonian and Relativistic Mechanics', British Journal for the Philosophy of Science 24, 270-6

Menger, K. (1970): Chapter 5, of L. Blumenthal and K. Menger, Studies in Geometry, Freeman

Przelecki, M. (1969): The Logic of Empirical Theories, Routledge and Kegan Paul

Winnie, J.A. (1977): 'The Causal Theory of Space-Time', in J. Earman et al. (eds), Foundations of Space-Time Theories, Minnesota Studies in the Philosophy of Science, VIII, Univ. of Minnesota Press

ERRATA: 1) The definition of the relation 'eq' in Sec.4 should contain the condition that (O,R) and (O,P) are time vectors.

2) Our axioms are not strong enough to justify the claim $R \perp Q$ in step (1) in the proof of theorem 4.9. For this, we have to replace R2 in the text by:

R2': $(Q \perp R \wedge Q \perp P \wedge (\forall O)(O @ c_{QR} \wedge O @ c_{QP} \rightarrow O*Q)) \rightarrow R \perp P$,

where $O*Q$ means $Q \bar{\,} Q$ but not O/Q. We note that R2' implies R2 and, hence, all consequences obtained with the help of R2 are obtainable from R2'. We also note that both models of axioms R1 to R6, given in Sec.4, are also models of our system when R2' is substituted for R2. Now $R \perp Q$ in step (1) in the proof of theorem 4.9. follows from the definition of 'eq' and R2' and R5. The details are straightforward.

Pekka J. Lahti

QUANTUM THEORY AS A FACTUALIZATION OF CLASSICAL THEORY

1. INTRODUCTION

Quantum theories of physics are based on the physical fact of the existence of the universal quantum of action, symbolized by the Planck constant $\hbar$. Classical theories of physics ignore this fact. According to Bohr (1949) they are idealizations which can be unambiguously applied only in the limit where all actions involved are large compared with the quantum. This state of affairs suggests that there are some natural relationships between the two types of theories. First of all, it suggests that quantum theory is a factualization of classical theory where the counterfactual idealizing assumptions are replaced by their factualizations. Secondly, one is led to ask whether classical theory can be inferred from quantum theory and some additional, obviously counterfactual, assumptions. In other words, is classical theory a reduction of quantum theory, or, to reverse it, is quantum theory a generalization of classical theory? If this would be the case one could then argue that classical theory is in a correspondence relation to quantum theory.

It is a historical fact that the correspondence principle of Bohr played a key role in the early developments of quantum mechanics. The inventions of Heisenberg (matrix mechanics), Schrödinger (wave mechanics), and Dirac (quantum Poisson brackets) are important examples of these developments. In them the correspondence principle was used to suggest, by way of a formal analogy, new quantum concepts as natural generalizations of the relevant classical concepts. It was also a central aim in Bohr's work on the interpretation of quantum theory to show that it is in every respect a generalization of the classical physical theories (Bohr (1929)). The Wigner transformation, which leads to a phase space formulation of quantum mechanics, seems to accomplish, in a formal sense, this program (see e.g. Bruer (1982), O'Connell (1983)). However, according to Bruer (1982), it does not help in solving the main interpretative problems of the quantum theory.

In the course of its development quantum mechanics

W. Balzer et al. (eds.), Reduction in Science, 381–396.

acquired also its autonomy and internal coherence. The idea of correspondence is no longer needed to explain the concepts of quantum mechanics. Moreover, the developments of the so-called axiomatic approaches to quantum mechanics during the last two decades have revealed an entirely new point of view: The two types of theories share a remarkable common ground, and the usual formulations of classical and quantum mechanics appear as special cases of the underlying generalized theories.

The aim of the present paper is to discuss the above mentioned relationships between classical mechanics and quantum mechanics within the so-called quantum logic approach to axiomatic quantum mechanics. With respect to this scheme we shall argue that quantum mechanics is not a generalization but rather a factualization of classical mechanics. The two theories, classical and quantum, are particular reductions of the underlying generalized theory. They might be referred to as a counterfactual reduction and a factual reduction, respectively. Any particular relation of correspondence between classical and quantum mechanics is not explicitely studied here. For that see Lahti and Rantala (1984).

We shall begin with a short sketch of the quantum logic approach to axiomatic quantum mechanics which is here used as the general theoretical framework. The classical phase space mechanics and the quantum Hilbert space mechanics will then be reformulated within the quantum logic setting. Thereafter we proceed to analyse some idealizing measurement theoretical assumptions which underly the usual formulations of classical and quantum mechanics. With respect to these assumptions axiomatic reconstructions of classical and quantum mechanics will then be considered. This then leads to the mentioned results.

2. A QUANTUM LOGIC - THE GENERAL FRAME

In this preliminary section we shall briefly recall some basic notions and results of the so-called quantum logic approach to (non-relativistic) axiomatic quantum mechanics which we shall now accept as a general framework for our considerations. In this approach it is assumed that the minimal mathematical structure of any (nonrelativistic) probabilistic (irreducibly or otherwise) physical theory is properly reflected in a couple (L,M), where L, the set of all <u>propositions</u> on or <u>properties</u> of the physical system concerned, carries as a natural structure that of an orthomodular σ-orthocomplete partially ordered set and M is an order-determining set of probability measures, states, on L. Such a prestructure or skeleton (L,M) is called a quantum logic. $\mathcal{L}$ denotes the family of all such pairs (L,M). Occasionally one may specify L to be a complete lattice and M to be the set of all probability measures on L.

According to the standard terminology, any two elements a and b in L are

(1) disjoint if and only if (iff) $a \wedge b = 0$, i.e. the meet of a and b, $a \wedge b$, exists in L and equals to 0, the least element of L;
(2) orthogonal, to be denoted as $a \perp b$, iff $a \leq b^{\perp}$;
(3) compatible or commutative, to be denoted as aCb, iff there are pairwise orthogonal elements $a_1, b_1, c \in L$ such that $a = a_1 \vee c$ and $b = b_1 \vee c$.

All these relations are symmetric. Moreover, one easily finds that any two orthogonal elements are disjoint, i.e. if $a \perp b$ then $a \wedge b = 0$ for any $a, b \in L$. Furthermore, the elements a_1, b_1 and c in (3) are unique, and $c = a \wedge b$. Hence in particular, any two orthogonal elements are compatible, and any two compatible disjoint elements are orthogonal. These important relations can also be used to characterize Boolean lattices as we have:

Theorem 1: An orthomodular σ-orthocomplete partially ordered set L is Boolean σ-algebra iff any two elements a and b in L are mutually compatible (Gudder (1970)).

Consequently we also have:

Theorem 2: An orthomodular lattice L is Boolean (σ-algebra) iff any two disjoint elements are orthogonal.

Proof: Suppose that L is Boolean, but there exist a and b in L which are disjoint but not orthogonal. But as $a = a \wedge 1 = a \wedge (b \vee b^{\perp}) = (a \wedge b) \vee (a \wedge b^{\perp}) = a \wedge b^{\perp}$, we have $a \leq b^{\perp}$, which contradicts the assumption $a \nleq b^{\perp}$. Thus in a Boolean L any two disjoint elements are also orthogonal. Suppose now that in L it holds true that any two disjoint elements are orthogonal. But for any a and b in L we have $a = (a \wedge b) \vee ((a \wedge b)^{\perp} \wedge a)$ and $b = (a \wedge b) \vee ((a \wedge b)^{\perp} \wedge b)$ which now shows that a and b are compatible. Thus L is Boolean.

In the present approach the observables of the considered physical system are described as L-valued measures on $(R,B(R))$, where $B(R)$ denotes the set of all Borel subsets of the real line R. The probability measure

$$\alpha \circ A: B(R) \to [0,1], \quad X \to \alpha(A(X)) \tag{4}$$

is interpreted as the probability distribution of the observable A in the state α. Moreover, the set of all observables $\underline{O}$ of the physical system concerned is surjective: for each a in L there is an A in $\underline{O}$ and X in $B(R)$ such that $a = A(X)$.

For a general exposition of this approach, as well as for the standard definitions of the concepts not defined but used in the text, the reader may consult e.g. Gudder (1970), Piron (1976), Mittelstaedt (1978), and Beltrametti and Cassinelli (1981).

3. TWO MODELS

3.1 Phase space classical mechanics

The classical phase space description of a physical system S is carried out in a phase space Ω, which is here considered as a compact Hausdorff space (cf. the end of section 4.1). The dynamic variables of the system are represented as real valued Borel functions on Ω, i.e. as mappings $f: \Omega \to R$, $\omega \to f(\omega)$ with the property $f^{-1}(X) \in B(\Omega)$ whenever $X \in B(R)$. The canonical position and momentum variables are, in particular, the co-ordinate functions $f_q: \Omega \to R$, (q,p) $f_q((q,p)) = q$ and $f_p: \Omega \to R$, $(q,p) \to f_p((q,p)) = p$ (which are explicitely written for the case $n = 1$). The σ-homomorphisms $B(R) \to B(\Omega)$, $X \to f^{-1}(X)$ induced by the dynamic variables $f: \Omega \to R$ are called the observables of the system S. In particular, we have the canonically conjugate position and momentum observables $Q: B(R) \to B(\Omega)$, $X \to Q(X) := f_q^{-1}(X)$ and $P: B(R) \to B(\Omega)$, $X \to P(X) := f_p^{-1}(X)$, respectively. The set $B(\Omega)$ of all Borel subsets of Ω represents thus, in a natural way, the set of all properties of or propositions on the system S. The points ω of Ω represent the (pure) states of S. They can be identified with the point measures on $B(\Omega)$. To allow also nontrivial probability measures on $B(\Omega)$ as states of the system we take a state to be a probability measure on $B(\Omega)$. We denote the set of all probability measures on the phase space Ω as $M_R(\Omega)_1^+$. The pure states of S are then exactly the extreme elements of $M_R(\Omega)_1^+$ (with respect to its natural convex structure), the nonextreme elements are the mixed states of S. The family $M_R(\Omega)_1^+$ determines the natural (set-theoretical) order on $B(\Omega)$, and the probability measure $\alpha \circ A: B(R) \to [0,1]$, $X \to \alpha(A(X))$ is the probability distribution of the observable A in the state α.

The classical phase space description of a physical system S (with n degrees of freedom) can thus be based on a quantum logic (L,M) where L is the set of all Borel subsets of a phase space Ω and M is the set of all probability measures on $L = B(\Omega)$. We define $\mathcal{L}_{cl} := \{(L,M) \in \mathcal{L}: L = B(\Omega),$ Ω is a phase space, M is the set of all probability measures on $B(\Omega)\}$.

In the classical case, i.e. when $(L,M) \in \mathcal{L}_{cl}$ we thus have:

$$L = B(\Omega) \text{ is Boolean} \tag{5}$$

so that aCb for any two a and b in L, and the notions of disjointness and orthogonality are equivalent. Moreover, in this case

$$M = M_R(\Omega)_1^+ \text{ is simplical,} \tag{6}$$

so that each state can uniquely be decomposed into its pure components, i.e. each $\alpha \in M$ can uniquely be expressed as a (countable or generalized) convex combination of some extreme elements of M (, for details see Alfsen (1971)). This means, in particular, that the probabilities $\alpha \circ A$ admit an epistemic ignorance interpretation.

3.2 Hilbert space quantum mechanics

The quantum Hilbert space description of a physical system S is carried out in a complex separable (generally infinite dimensional) Hilbert space H. The physical quantities of the system are represented as linear, self-adjoint, not necessarily bounded, operators on H. The spectral measures $P_A: B(R) \to P(H)$, $X \to P_A(X)$ induced by the physical quantities $A: dom(A) \to H$, $\varphi \to A\varphi$ are the observables of the physical systems S. The set P(H) of all orthogonal projections on H represents thus, in a natural way, the set of all properties of or propositions on the system S. The unit vectors φ of H represent the (pure) states of S. They can be identified (modulo a complex multiplier of absolute value one) with the (extreme) probability measures on P(H). To include also mixed states one takes a state to be a probability measure on P(H). We denote the set of all probability measures on H as $M_R(H)_1^+$. The pure states of S are exactly the extreme elements of $M_R(H)_1^+$ (with respect to its natural convex structure), the nonextreme elements are the mixed states of S. We recall also the Gleason theorem according to which the family $M_R(H)_1^+$ can be identified with the family $T_s(H)_1^+$ of all self-adjoint positive normalized trace-class operators on H, whenever the vector space dimension of H is at least 3. The family $M_R(H)_1^+ \cong T_s(H)_1^+$ ($\dim(H) \geq 3$) determines the natural order on P(H), and the probability measure $\alpha \circ P_A: B(R) \to [0,1]$, $X \to \alpha(P_A(X))$ is the probability distribution of the observable P_A in the state α.

The quantum Hilbert space description of a physical system S can thus be based on a quantum logic (L,M) where L is the set of all orthogonal projections on a complex separable Hilbert space H and M is the set of all probability measures on $L \cong P(H)$. We define $\mathcal{L}_{qu} := \{(L,M) \in \mathcal{L}$: $L \cong P(H)$, H is a Hilbert space, M is the set of all probability measures on P(H)$\}$.

In the quantum case, i.e. when $(L,M) \in \mathcal{L}_{qu}$ we thus have:

$$L \cong P(H) \text{ is non-Boolean (whenever } \dim(H) \geq 2) \quad (7)$$

so that "aCb for any two a and b in L" does not hold, and the notions of disjointness and orthogonality are not equivalent; moreover, in the quantum case

$$M \cong M_R(H)_1^+ \cong T_s(H)_1^+ \quad \text{is nonsimplicial} \quad (8)$$

so that the decomposition of a state into its pure components is not unique. Actually, any $\alpha \in M_R(H)_1^+$ admits infinitely many representations as a (countable) convex combination of some pure states in $M_R(H)_1^+$. This means, in particular, that the probabilities $\alpha \circ P_A$ do not admit the epistemic ignorance interpretation.

4. SOME IDEALIZATIONS

4.1 Classical ideal

With the slogan classical ideal we mean a set of idealizing (measurement theoretical) assumptions which underly classical theories of physics. The two best known of such assumptions are the compatability assumption C and the assumption on the unique decomposability of mixed states, UDM.

Intuitively, the compatability assumption says that the values of any two physical quantities can simultaneously be measured (within any accuracy). With respect to the chosen theoretical frame this assumption is most directly formalized as:

C. aCb for any two a and b in L.

Let us note that this condition can also be gained through formalizing the notion of measurement as a transformation of states of the system and reducing the notion of simultaneous measurability of physical quantities to the order-independence of the relevant sequential measurements (see e.g. Beltrametti and Cassinelli (1981)). Obviously, the assumption C imposes some further structure on those quantum logics $(L,M) \in \mathcal{L}$ that fulfil this condition. More explicitely we have:

<u>Corollary 1</u>. Let (L,M) be a quantum logic. If L satisfies C then L is Boolean.
We denote $\mathcal{L}_c = \{(L,M) \in \mathcal{L}: \text{L satisfies C}\}$.
Clearly $\mathcal{L}_{cl} \subset \mathcal{L}_C$, and $\mathcal{L}_{qu} \cap \mathcal{L}_C = \emptyset$.

Intuitively, UDM says that in our description of physical systems we have to refer to mixed states only when we ignore some relevant physical conditions in preparing the state of the system - an ignorance which, in principle, is thought to be eliminable. The assumption of the unique decomposability of mixed states, or on the validity of the ignorance interpretation, has a twofold consequence of the set M of all states of the system. Firstly, the set Ex(M) of pure state in M should be sufficiently rich so that each state α in M can be expressed as a convex combination (countable or generalized) of the elements in Ex(M). Secondly, this decomposition should be unique. This then allows one to say that if the system S is in a mixed state $\alpha \in M$ then it is actually in one of the pure states $\alpha_i \in Ex(M)$, the weights w_i with which the pure states α_i participates in the decomposition $\alpha = \Sigma w_i \alpha_i$ describing our knowledge on the actual state of the system. The assumption UDM reads now:

UDM. Each state α in M can uniquely be decomposed into its pure components.

For a further discussion of the formalization of the assumption UDM see Lahti and Bugajski (1983). The assumption UDM imposes a rather strong structure on the set M of states of S. A convex set M possesses the property UDM iff it is a Bauer simplex (Alfsen (1971)). Hence we have:

Corollary 2. Let (L,M) be a quantum logic. If M satisfies UDM then M is a Bauer simplex.

We denote $\mathcal{L}_{UDM} = \{(L,M) \in \mathcal{L}$: M satisfies UDM$\}$. As any Bauer simplex M can be identified with the set $M_R(\Omega)_1^+$ of all probability measures on a compact Hausdorff space (phase space) Ω (Alfsen (1971)), we see that $\mathcal{L}_{cl} \subset \mathcal{L}_{UDM}$, and $\mathcal{L}_{qu} \cap \mathcal{L}_{UDM} = \emptyset$.

The assumptions C and UDM above are given their natural L- and M-theoretical formulations and corollaries 1 and 2 express the immediate implications these assumptions have on the structures of L and M, respectively. However, in a pair $(L,M) \in \mathcal{L}$ the structures of L and M are interrelated: M is an order-determining set of probability measures on L. Thus one is faced with a question about the mutual dependence of the assumptions C and UDM.

Consider a quantum logic $(L,M) \in \mathcal{L}_{cl}$ associated with the two-dimensional phase space R^2. For a given positive constant $h > 0$ we define a subset M_h of M as $M_h := \{\alpha \in M: \Delta(Q,\alpha) \cdot \Delta(P,\alpha) \geq h\}$, i.e. we pick up those states α in M for which the product of the standard deviations $\Delta(Q,\alpha)$ and $\Delta(P,\alpha)$ of the canonical position and momentum observables Q and P is greater than or equal to the given constant $h > 0$. The set M_h is convex, and it determines the order on $L = B(R^2)$ (Lahti (1980)). Thus $(L,M) \in \mathcal{L}_C$, but $(L,M_h) \notin \mathcal{L}_{UDM}$ as the set M_h of states of the description fails to be a Bauer simplex. This example demonstrates that C does not imply UDM.

Consider next a quantum logic (L,M) in $\mathcal{L}_{UDM}$. As M can now be identified with the set $M_R(\Omega)_1^+$ of all probability measures on a compact Hausdorff space Ω one may introduce a mapping $L \to B(\Omega)$, $a \to \hat{a}$ through the relation $\hat{\alpha}(\hat{a}) = \alpha(a)$ for any $\alpha \in M$, $a \in L$. Here $\hat{\alpha}$ denotes the unique counterpart of $\alpha \in M$ in $M_R(\Omega)_1^+$. The mapping $a \to \hat{a}$ is well defined and it preserves the order and the orthocomplementation. Moreover, as for any two states α and β in M $\alpha \neq \beta$ iff $\hat{\alpha} \neq \hat{\beta}$, we have that any two probability measures $\hat{\alpha}$ and $\hat{\beta}$ on $B(\Omega)$ differ (i.e. $\hat{\alpha} \neq \hat{\beta}$) iff their restrictions $\hat{\alpha}_{\hat{L}}$ and $\hat{\beta}_{\hat{L}}$ on $\hat{L} := \{X \in B(\Omega): X = \hat{a}$ for some $a \in L\}$ differ (i.e. $\hat{\alpha}_{\hat{L}} \neq \hat{\alpha}_{\hat{L}}$).

But this then means that L can be identified with $B(\Omega)$, which is Boolean. Thus $L_{UDM} \subset L_C$. Actually this also shows that $L_{UDM} = L_{cl}$ within the present scheme. In that we accept the convention to replace the usual (locally compact Hausdorff) phase space with some of its relevant compactifications. (For a full discussion of this topological question see Lahti and Bugajski (1983)).

We may now conclude that within the quantum logic approach the assumption of the unique decomposability of mixed states implies the compatability assumption, but not the other way round.

4.2. Projection postulate

The projection postulate contains a set of measurement theoretical assumptions which might verbally be summarized as follows:

> PP Each property of a physical system admits a pure, ideal, first-kind measurement.

To formalize the postulate within our theoretical frame we have to define therein the notion of measurement and its possible characteristics purity, ideality, and first-kindness. The postulate should then express the requirement that a given property $a \in L$ of the system admits such a measurement.

In the quantum logic approach the notion of measurement is most properly formalized as an affine transformation of states of the system: $\alpha_i \to \alpha_f$, where the post-measurement (final) state α_f depends, in general, on the pre-measurement (initial) state α_i, on the measuring result, and on the measuring instrument employed. Such state transformations are usually called operations.

Leaving aside the technical details for the present, we shall briefly explain what do the adjectives pure, ideal, and first-kindness aim to characterize. (For a further discussion see Bugajski and Lahti (1980), Lahti (1983)).

Purity. A pure operation (measurement) leaves the system in a "maximal information" state whenever it was in such a state. In short, it takes a pure state onto a pure state, with a possible loss in strength. To formalize this idea within the present scheme it is convenient to extend the notion of state $\alpha \in M$ to an ordered pair $(\lambda,\alpha) \in [0,1] \times M$ whereby the original states $\alpha \in M$ can be identified with the pairs $(1,\alpha)$.

Purity of an operation ϕ then means that $\phi\alpha \in [0,1] \times Ex(M)$ whenever $\alpha \in Ex(M)$.

Ideality. An ideal operation (measurement) changes the state of the system only to an extent necessary for the measuring result, so that the post-measurement state contains all the information of the premeasurement state that is compatible with the measuring result. Such a requirement, known as the principle of least disturbance or minimal interference, can be formalized e.g. by maximizing the 'overlap' or the 'transition probability' between the pre- and post-measurement states.

First-kindness. An operation (measurement) is of the first-kind, if the state of the system remains unchanged whenever the measurement result is probabilistically certain (i.e., if $e(\phi\alpha) = 1$ then $\phi\alpha = \alpha$), and if the repeated application of the operation (measurement) does not lead to a new result ($e(\phi^2\alpha) = e(\phi\alpha)$). (Here $0 \leq e(\phi\alpha) \leq 1$ measures the strength of the state $\phi\alpha \in [0,1] \times M$.)

Let O_f denote the set of all pure, ideal, first-kind operations of a quantum logic (L,M). It may be trivial. Projection postulate expresses now the requirement for the existence of i) a sufficiently rich family O_f of pure ideal first kind operations, and ii) a natural one-to-one correspondence $\Phi: L \to O_f$, with property $a(\alpha) = e(\Phi(a)\alpha)$ for all a in L and α in Ex(M). We write $\mathcal{L}_{pp} = \{(L,M) \in \mathcal{L}: (L,M)$ satisfies the projection postulate$\}$.

We shall now give two examples of physical theories which satisfy the projection postulate. The first of them concerns the quantum Hilbert space description of a physical system whereas the second one concerns its classical phase space description.

Example 1. The quantum Hilbert space description $(P(H), T_s(H)_1^+) \in \mathcal{L}_{qu}$ satisfies the projection postulate with respect to the couple (O_f, L), where
$O_f = \{\phi_P: T_s(H) \to T_s(H),\ \alpha \to \phi_P\alpha := P\alpha P,\ P \in P(H)\}$ and
$L \simeq P(H)$. Note that the operations ϕ_P are the usual von Neumann-Lüders operations.

Example 2. The classical phase space description $(B(\Omega), M_R(\Omega)_1^+) \in \mathcal{L}_{cl}$ satisfies the projection postulate with respect to the couple (O_f, L) where
$O_f = \{\phi_X: M_R(\Omega) \to M_R(\Omega),\ \alpha \to \phi_X(\alpha)$, with

$\phi_X(\alpha)(Y) = \alpha(X \cap Y) \forall X,Y \in B(\Omega)\}$ and $L \simeq B(\Omega)$. Note that the operations ϕ_X are the usual conditionals of the standard probability theory.

These examples demonstrate that the projection postulate is satisfied not only in the Hilbert space quantum mechanics but also in the phase space classical mechanics, i.e. $\mathcal{L}_{cl} \cup \mathcal{L}_{qu} \subset \mathcal{L}_{pp}$.

Though PP is a general measurement theoretical assumption which does not distinguish between classical and quantum theories it anyway imposes very strong structural properties on a quantum logic $(L,M) \in \mathcal{L}$ satisfying it. More explicitely we have (Bugajski and Lahti (1980)):

Corollary 3. Let (L,M) be a quantum logic. If it satisfies the projection postulate then L is atomic and has the covering property, provided that L is a complete lattice.

Note that the reservation of L being a complete lattice in this corollary is a harmless assumption of a technical character as we can always embed L in a natural way in a complete lattice which preserves all the essential structural features of L without introducing any new atoms (Bugajska and Bugajski (1973)). Moreover, for any $(L,M) \in \mathcal{L}_{cl} \cup \mathcal{L}_{qu}$ L is a complete lattice. Thus without any loss in generality we may assume from the outset that L is a complete lattice. Finally, let us note that within the present scheme UDM implies PP, as $\mathcal{L}_{UDM} \subset \mathcal{L}_{PP}$. Note also that the "semiclassical" example (L,M_h) which was used to demonstrate that C does not imply UDM does not satisfy PP. It is our conjecture that $\mathcal{L}_C \cap \mathcal{L}_{PP} = \mathcal{L}_{UDM}$ but we leave it open here as the result is not needed below.

5. ON AXIOMATIC RECONSTRUCTIONS OF $\mathcal{L}_{cl}$ and $\mathcal{L}_{qu}$

Classical phase space mechanics and quantum Hilbert space mechanics can be reformulated within the quantum logic (Sec. 3). The two types of theories appear as special cases of the more general theory:

$$\mathcal{L}_{cl} \cup \mathcal{L}_{qu} \subset \mathcal{L}. \tag{9}$$

This then leads one to consider the possible axiomatic reconstructions of classical and quantum theories based on the general theory of quantum logic.

In the previous section we analysed three measurement

theoretical assumptions, the compatibility assumption C, the assumption UDM on the unique decomposability of mixed states, and the projection postulate PP. These assumptions led to single out the subfamilies L_C, L_{UDM}, and L_{PP} of L with the following properties:

$$L_{cl} \subset L_C \cap L_{PP} \subset L_C \subset L'_{qu} \ (:= L - L_{qu}) \tag{10}$$

$$L_{cl} = L_{UDM} = L_{UDM} \cap L_{PP} \subset L'_{qu} \tag{11}$$

$$L_{cl} \cup L_{qu} \subset L_{PP} \subset L \tag{12}$$

$$L_{qu} \subset L_{PP} \cap L'_C \subset L_{PP} \cap L'_{UDM} \subset L'_{cl} \tag{13}$$

The set theoretical relations (10) show that some essential features, namely the Boolean character, of the classical theory can be reduced to the compatibility assumption C. However, this assumption does not fully determine L_{cl}. Further specification can be gained through the projection postulate PP, which would then give a one-one correspondence between the sets At(L) and Ex(M) of atoms of L and pure states of M.

The second set of the above relations (10)-(13) reveal that within the quantum logic approach an axiomatic reconstruction of the classical theory can be based on the requirement of the unique decomposability of mixed states.

The set theoretical relations (12) emphasize the general character of the projection postulate. It is an idealizing measurement theoretical assumption which does not distinguish between classical and quantum theories. In spite of that it imposes very strong structural properties on a quantum logic satisfying it. Some of its consequences might be read already in (10). Its relevance to axiomatic reconstructions of classical theories is, however, minor, as explained by (11). On the other hand, this postulate may be given a fundamental role in an axiomatic reconstruction of the quantum theory. This becomes clear from below.

The relation $L_{PP} \cap L'_C \subset L_{qu}$ contains, in particular, the celebrated Piron-McLaren representation theorem according to which a complete, irreducible, orthomodular, atomic lattice L which has the covering property and which is of the length ≥ 4 can (almost) be identified with the set P(H) of all projections on Hilbert space H of vector space dimension ≥ 3 (Piron (1976), Beltrametti and Cassinelli (1981)). What is open here is the question of which scalar field the vector space is defined over. If it is a finite extension of the real numbers then the Hilbert space realization of L is obtained. (See Beltrametti and Cassinelli (1981) for an extensive discussion of the problem of choosing the scalars.) Let us recall that, assuming the complete lattice property (cf. 4.2), for any pair (L,M) in L_{PP} L is a complete, orthocomplemented, orthomodular, atomic lattice with the covering property. We note also that the requirement of L being of the length ≥ 4 simply means that there are in L also other nontrivial elements (properties) than atoms. As for an (L,M) in $L_{PP} \cap L'_C$ cent(L):= $\{b \in L\colon bCA \text{ for all } a \in L\} \subsetneq L$ we may then conclude that an essential part of the Hilbert space quantum theory can be reduced to the projection postulate and to the invalidity of the compatibility assumption.

6. QUANTUM THEORY AS A FACTUALIZATION OF CLASSICAL THEORY

Either $h > 0$ or $h = 0$, exclusively. Physical actions cannot have negative values. The equality

$$h = 0 \tag{14}$$

symbolizes the ideal counterfactual assumption that physical actions can be (made) arbitrarily small, whereas the strict inequality

$$h > 0, \tag{15}$$

(which is not a generalization of (14)) symbolizes the physical fact of the existence of the universal quantum of action.

The compatibility assumption C and the assumption UDM on the unique decomposability of mixed states are two formal aspects, or expressions, of the classical ideal "h = 0". It was shown above that within the quantum logic approach classical theories can be based on these assumption: $L_{cl} = L_{UDM} = L_{UDM} \cap L_C$. Moreover, these theories satisfy the projection postulate: $L_{cl} \subset L_{PP}$. However, the equality (14) symbolizes an idealizing contrafactual case, the factual case being symbolized by the strict inequality (15). In the factual case the compatability assumption C and the assumption UDM on the unique decomposability of mixed states do not hold. They are invalid. But in section 5 it was concluded that, within the quantum logic approach, quantum theories can essentially be based on the projection postulate PP and on the invalidity of either the assumption C or the assumption UDM: $L_{qu} \subset L_{PP} \cap (L'_C \cup L'_{UDM}) \subset L_{PP} \cap L'_C \subset L_{PP} \cap L'_{UDM}$. Referring to Krajewski (1977) we thus come to the following conclusion:

> Quantum theory is a factualization of classical theory where the idealizing assumption "h = 0" is replaced by the factual one "h > 0".

We shall close this section by discussing briefly the problem of incorporating the existence of the universal quantum of action into the quantum logic scheme and the possibilities of recovering then the Hilbert space quantum mechanics. We shall distinguish between three cases.

The first method goes back to Mackey (1963) and it consists of two steps: Exclude the classical case with claiming the invalidity of C and/or UDM, and recover the quantum case with postulating $L \simeq P(H)$ and $M \simeq M_R(H)_1^+$ for some Hilbert space H.

The second method exceeds the first one in reducing its Hilbert space axiom to some simpler and more natural assumptions (Piron (1976), Beltrametti and Cassinelli (1981)). Projection postulate amounts a natural set of such assumptions.

The third method exceeds the second method in reducing the claim for the invalidity of C and/or UDM to some more fundamental assumptions, like the complementarity principle, the uncertainty principle, and the superposition principle. With respect to the natural formulations of these fundamental quantum principles one may show (Bugajski and Lahti (1980)) that a quantum logic (L,M) in L which satisfies any of these three principles belongs to the complement of the family $L_C \cap L_{UDM}$. Thus, accompanied with the projection postulate, any of the the three fundamental quantum principles is essentially enough to lead to the usual Hilbert space formulation of the quantum theory. There are also good reasons to assume that the complementarity principle, the uncertainty principle, and the superposition principle presuppose the projection postulate (Lahti (1983)).

Acknowledgements I wish to thank Professor Dr. Heinz-Jürgen Schmidt for his proposals to improve the manuscript. I am also grateful to the organizers of the Symposium as well as to the Alexander von Humboldt-Foundation for providing me the possibility to attend the interesting meeting in Bielefeld.

Pekka J. Lahti
Department of Physical Sciences
University of Turku

REFERENCES

Alfsen, E.: 1971, 'Compact Convex Sets and Boundary Integrals', Springer-Verlag, Berlin.

Beltrametti, E. and Cassinelli, C.: 1981, 'The Logic of Quantum Mechanics', Addison-Wesley Publishing Company, Reading, Massachusetts.

Bohr, N.: 1929, 'Introductory survey', in: Atomic Theory and the Description of Nature, Cambridge University Press, Cambridge, 1934.

Bohr, N.: 1949, 'Discussion with Einstein on Epistemological Problems in Atomic Physics', in: Albert Einstein, Philosopher-Scientist, the Library of Living Philosophers Inc., Evanston.

Bruer, J.: 1982, 'The classical limit of quantum theory', Synthese 50, 167-212.

Bugajska, K. and Bugajski, S.: 1973, 'The lattice structure of quantum logics', Ann. Inst. Henri Poincare 19, 333-340.

Bugajski, S. and Lahti, P.: 1980, 'Fundamental principles of quantum theory', International Journal of Theoretical Physics 19, 499-514.

Gudder, S.: 1970, 'Axiomatic quantum mechanics and generalized probability theory', in: Probabilistic Methods in Applied Mathematics, Academic Press, New York.

Krajewski, W.: 1977, Correspondence Principle and Growth of Science, D. Reidel Publishing Company, Dordrecht, Holland.

Lahti, P.: 1980, 'Uncertainty and complementarity in axiomatic quantum mechanics', International Journal of Theoretical Physics 19, 789-842.

Lahti, P.: 1983, 'On the role of projection postulate in quantum theory', Reports on Mathematical Physics, in print.

Lahti, P. and Bugajski, S.: 1983, 'Fundamental principles of quantum theory II', Turku, FTL-R51.

Lahti, P. and Rantala, V.: 1984, 'A Correspondence Relation between Classical and Quantum Theories, in preparation.

Mackey, G.: 1963, Mathematical Foundations of Quantum Mechanics, Benjamin, New York.
Mittelstaedt, P.: 1978, Quantum Logic, D. Reidel Publishing Company, Dordrecht, Holland.
O'Connell, R.: 1983, 'The Wigner distribution function - 50th Birthday', Foundations of Physics 13, 83-92.
Piron, C.: 1976, Foundations of Quantum Mechanics, W.A. Benjamin, Inc. Reading, Massachusetts.

Ernst-Walter Stachow

CLASSICAL AND NON CLASSICAL LIMITING CASES OF QUANTUM LOGIC

1. INTRODUCTION

"Quantum logic" is considered here, following the main usage in the literature (Beltrametti and van Fraassen (1981)), to be the formal structure of a language which is appropriate for describing quantum physical systems and their properties. Besides its syntax this quantum language comprises a semantics which takes into account the particular conditions of quantum physical propositions and which establishes an adequate concept of truth for these propositions.

Different approaches to such a quantum semantics have been studied in the literature. One well-known approach is based on the Hilbert space formalism of quantum theory: Certain elements within quantum theory are identified with semantical objects. Having established a Hilbert space semantics in this way, the syntax and logical structure of the language are derived, i.e. the syntax and logic are extracted from quantum theory.[1] This approach depends on the quantum theory under consideration and, hence, on the particular empirical domain of the language.

Another approach to a quantum language, the one which is considered in this paper, departs from general conditions of quantum physical propositions and their proof processes which do not depend on the Hilbert space quantum theory or the particular empirical domain already. It leads to an abstract quantum language which partly exhibits the same logical structures as are obtained within the first approach[2]. In distinction to the first approach, the Hilbert space quantum theory does not play the role of the semantics but now that of a realization of the language by means of (concrete) mathematical structures.

The aim of this paper is to discuss the problem of reduction between quantum physics and classical physics within this language frame. The concept of reduction which usually refers to theories is taken over here to language considerations. In this respect we investigate the question of how the reduction can be rendered on the semantical level and how

W. Balzer et al. (eds.), Reduction in Science, 397–418.

it is conditioned by the foundation of the semantics. Furthermore, the logical relations between the languages under consideration are investigated. Concerning the first approach mentioned above, our investigation would have to take into account the difficult problem of reduction between Hilbert space quantum theory and classical theories. In the following, however, we restrict our consideration to the abstract approach which has the advantage that the problem of reduction is confined to a more ascetic object language for physics in which the conditions of reduction can be established systematically. On the other hand, also the reverse problem of how the classical language can be weakened to the quantum language is considered within our approach. It turns out that the conditions of reduction correspond to certain presuppositions of the language the systematic weakening of which leads to the quantum language.

Our semantics for this abstract object language is a constructive semantics in the sense that the concept of truth for elementary and compound propositions is established by means of the proof conditions for such propositions. It is clear that in the physical context the possibilities for proving propositions depend on which proof procedures are physically realizable. - This constructive point of view can be compared with the constructive view of mathematics: Whereas a constructive semantics for mathematics is justified by the very conditions of (formal) mathematical proof procedures, our constructive semantics for physics must take into account the pragmatic conditions of (material) physical proof processes also. These pragmatic preconditions of the language should finally be re-established by means of a theoretical formulation of the proof processes within an (exhaustive) physical theory about the domain of reality which is comprehended by the language. This means that the pragmatics constituting the abstract object language is, on the other hand, to be legitimated by the physical theory, which makes precise what is physically realizable, in a selfconsistent way.[3] For our purpose to achieve an utmost universal object language, only the most general features of physical reality, as far as they are conceivable by the present forms of physics, are relevant as "ontological" premises. The connection between these ontological premises and the abstract object language is summarized in Fig. 1.

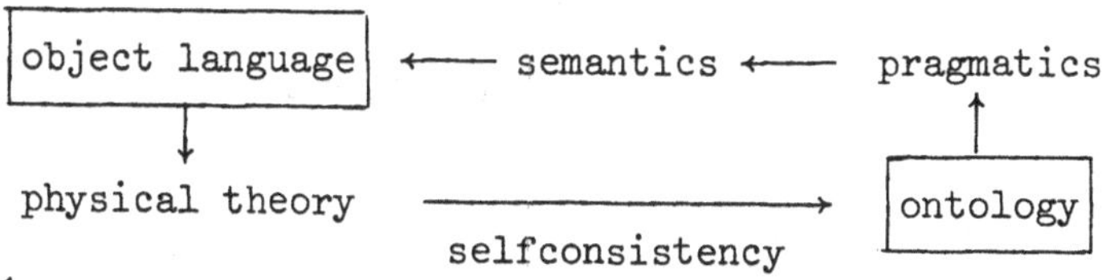

Fig. 1

In part 2 of the paper a brief outline of the constructive approach to the abstract quantum language is given.[4] In this way, the general ontological conditions of quantum physical objects are exhibited, and their consequences for the logical structures of the language are made transparent. It turns out that the structures of quantum logic consist of certain generalizations of well-known classical and non-classical logical systems.

These classical and non-classical logical systems can also be associated with object languages for particular domains of classical physics and mathematics. Starting from ontological suppositions with respect to these domains, the object languages under consideration can be constituted analogously to the foundation of the quantum language as is schematically illustrated in Fig. 1. Therefore, in part 3 of the paper, the reduction between the quantum language and the other languages under consideration will be discussed in a twofold way. On one hand, the formal structural conditions of the reduction, i.e. the logical relations between the languages, will be investigated. On the other hand, the corresponding ontological conditions which establish the reduction within the object languages are considered. The particular cases of classical and non-classical logics which are obtained in this way will be called <u>limiting cases</u> of quantum logic.

Our investigation of the abstract quantum language and the conditions of the reduction between this language and other object languages should also make clear in what sense quantum logic is a <u>universal</u> theory of formally true propositions. It is based on the very weak ontological premises which are consistent with the general conditions of quantum physical proof processes. In comparison with quantum logic, other logical systems like intuitionistic and classical logic involve stronger ontological suppositions, which will be explicitly formulated in part 3 of the paper. It is shown that a systematic weakening of these suppositions, which corresponds to a revision of our knowledge in the domains of physical reality and mathematics, leads to quantum language and logic. Therefore, according to our present state of knowledge in these domains, quantum logic can be considered to be the most universal logic.

2. QUANTUM LANGUAGE AND LOGIC

As it was already indicated in the introduction, the construction of the abstract quantum language is based on the pragmatic preconditions of physically realizable proof procedures. These conditions are considered here as the possibilities of the *speaker* of the language under consideration who, at the same time, is also the physical *observer*. This *speaker-observer* asserts propositions and proves them, partly by material processes like preparations and measurements, partly by formal combinatorial acts.

Those propositions which can be proven or disproven by material proof procedures are called *elementary propositions*. It is assumed here that such a material process, which might be infinite, has two potential results. One of the results defines the proof, the other one the refutation of that elementary proposition. In the following, elementary propositions are denoted by a,b,... and the set of elementary propositions is denoted by $\mathcal{S}_e$. If an elementary proposition a is proven by means of a material process, we also say that it is *materially true*, denoted by a!; in case of a refutation we also say that it is *materially false*, denoted by $\sim$ a!. The material disproof of a defines the *strong negation* $\sim$ a of a.

By means of such a material process performed by the speaker-observer, and its possible results, *properties* of a physical system S can be constituted.[5] Elementary propositions a are then interpreted as propositions a(S) about certain properties of a physical system S. If a(S) is materially true we say that the property E_a pertains to S, if a(S) is materially false we say that the property $E_{\sim a}$ pertains to S. In both of the cases we say that E_a and $E_{\sim a}$ are *objective*, or a and $\sim$ a have *objective* truth-values.

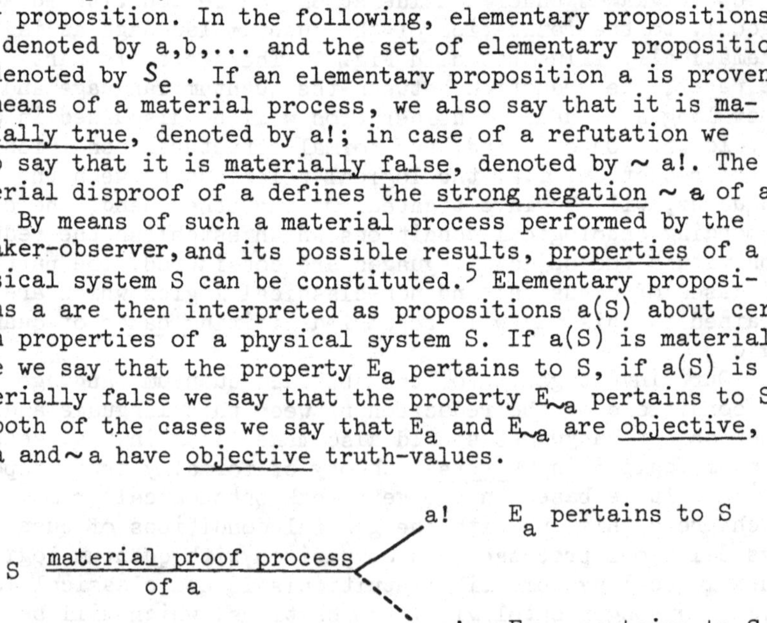

As already mentioned we do not presuppose *finite deci-*

dability for elementary propositions. This has mathematical (the existence of undecidable propositions) as well as physical reasons: There are quantum physical propositions the truth-values of which can only be decided in asymptotic space-time regions which are inaccessible for a finite speaker-observer(examples will be found in my (1985)). If an elementary proposition is finitely decidable we also say that it is value-definite. Moreover, we assume that the material proof processes which define elementary propositions are ideal in the following sense:

(α) If a(S) has been proven at time t, then an immediate repetition of the proof of a(S) at t + δt leads to the same result, provided that δt is sufficiently small.

According to the pragmatic foundation of the language, elementary propositions are primarily associated with objective truth-values only in the situation of proof processes by the speaker-observer. Hence, the objectivity of a property E_a with respect to S is intimately connected with the process of verification of a. We do not assume that, without further conditions, a property E_a is objective prior to the test of a.

The combinatiorial possibilities of the speaker-observer to prove elementary propositions constitute the logical connectives of the language. Starting with a pragmatic time order of all actions of the speaker-observer, the sequential connectives play a primary role. E.g. the sequential conjunction a ⊓ b is defined by the subsequent proofs of first a and then b, represented by the tree:

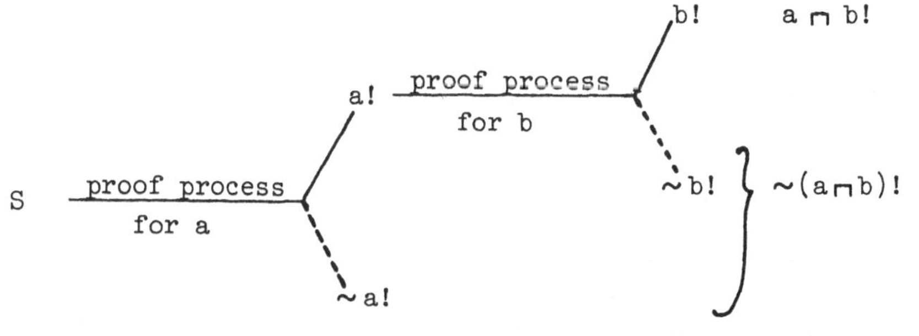

According to the previous remark, the property E_b which is established by means of the proof process for b cannot be assumed to be objective prior to the proof of b. This corresponds to a general restriction of objectivity in the case of quantum physics where this assumption would lead to a contradiction with empirical and theoretical results. The restriction consists in the general <u>incommensurability</u> of quantum physical propositions and can be formulated by:

(β) If a(S) has been proven at the time t, then, after a proof of b(S) at t' > t, the repeated proof of a(S) at t" > t' does not necessarily lead to the same result, even if the time-interval $\delta t = t'' - t$ is arbitrarily small.

Besides combinations of material proofs we also take into account formal derivations of propositions under given hypotheses for the constitution of sequential connectives. In this way the <u>sequential implication</u> $a \dashv b$ might be proven by deriving b from the earlier hypothesis a. In order to establish a constructive semantics for implication, it is useful to introduce a second speaker, whose propositions are considered to be hypotheses. This leads to a <u>dialog-semantics</u> in which the first speaker acts as a <u>proponent</u> P and the second speaker as an <u>opponent</u> O.[6] In a dialog-game P might possess a formal winning strategy for a proposition; if not, he may choose material tests for the elementary subpropositions. The dialog-semantics for the sequential implication is represented by the dialog-scheme:

O		P
		$a \dashv b$
	a	[b]
~a!	a!	a?
		b
	b?	b!

The symbol [b] in the second line stands for the possibility of deriving b from a, either by demonstrating that the hypothesis a leads to a contradiction or by showing that b is implied by the material content of a. The following lines in the above dialog-scheme indicate the further possibility of a material proof of $a \dashv b$, namely that, if at first a is materially true, thereafter b is materially true. a? and b? symbolize demands for proving a and b respectively.

As a further connective we use the <u>negation</u> $\neg a$ which is defined by:

O	P
	$\neg a$
a	[ ]
$\sim a!$	a?

The symbol [ ] means that P can prove $\neg a$ formally only by showing that the hypothesis a is contradictory. In case he does not succeed, he may win the dialog by a proof of the strong negation $\sim a$ which he may demand from O.

The proof processes for the sequential conjunction and disjunction can also be formulated within the dialog-semantics:

O	P
	$a \sqcap b$
a?	a!
b?	b!

O	P	
	$a \sqcup b$	
a?	a!	$\sim a!$
b?		b!

Starting from the pragmatic time order of sequential proof processes, the simultaneity of objective truth-values has to be constituted by means of further conditions. If the pragmatic time is provided with a metric, the condition $\delta t = t' - t \to +0$, where t and t' refer to the proofs of a and b respectively, is not sufficient to guarantee the simultaneity of a and b because of the restriction (β). Only in case the subsequent test of b does not alter the proof result of a and vice versa, the simultaneous objectivity of the truth-values of a and b can be stipulated consistently. In order to incorporate this possibility into the language we introduce two test propositions k(a,b) and $\sim$k(a,b), the commensurability and incommensurability propositions, the proof procedures of which decide on the simultaneous objectivity of the truth-values of a and b:

Def.: a) k(a,b) is materially true iff a and b can be tested in an arbitrary sequence such that the proof results for both of the propositions do not alter.

b) $\sim$k(a,b) is materially true iff the proof result of a might be alterd by a test of b and vice versa.

By means of additional commensurability tests the simultaneous connectives can be defined, e.g. the conjunction $a \cap b$:

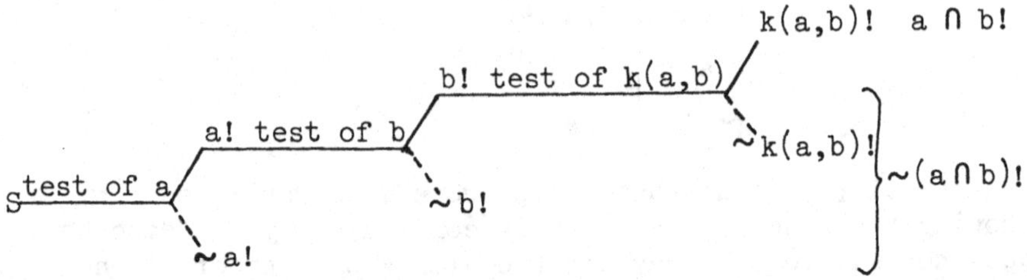

In order to refer the truth-values of a and b to the same time, the time differences between the single tests must be made sufficiently small.

The dialog-semantics is extended to the simultaneous connectives in the following way:

a) conjunction $a \cap b$

O	P
	$a \cap b$
a?	a!
b?	b!
k(a,b)?	k(a,b)!

b) disjunction $a \cup b$

O		P		
		$a \cup b$		
a?		a!	~a	
	b?		b!	~b!
	k(a,b)?			~k(a,b)!

c) implication $a \to b$

	O	P
		$a \to b$
	a	[b]
~a!	a!	a?
		b
	b?	b!
	k(a,b)?	k(a,b)!

The symbol [b] in the last scheme indicates the possibility for P to derive the truth of b from the truth of a such that a and b are simultaneously true, i.e. satisfy the commensurability condition.

Since, in the case of the proof of k(a,b), the truth-value of a is still available after a proof of b in a dialog, k(a,b) is also called an availability proposition.

For the following considerations only the simultaneously connected propositions will be important. The dialog-semantics for the above compound propositions which consist of one or two elementary subpropositions can be extended to finitely connected propositions. The resulting material dialog-game will be given in my (1985). Our language then has the following syntax:

(a) S_e, the set of elementary propositions including V and Λ, the "true" and "false" propositions respectively,
(b) C: = $\{\cap, \cup, \rightarrow, \neg, \sim\}$, the set of simultaneous connectives,
(c) S, the set of logically connected propositions which is inductively defined by
  (i) if $a \in S_e$, then $a \in S$,
  (ii) if $A, B \in S$, then $A \cap B$, $A \cup B$, $A \rightarrow B$, $\neg A$, $\sim A \in S$.

A proposition A is called formally true iff the truth of A does not depend on the material content and, thus, on the truth-values of its elementary subpropositions. All formally true propositions can be established by means of a logical calculus in which they are derivable as theorems, as will be shown in my (1985). If A is a theorem, then $\mathrm{V} \rightarrow A$ is a theorem also. By means of all pairs (A,B) for which $A \rightarrow B$ is a theorem, the implication relation $<$ is defined. The logical calculus which establishes all formally true propositions can equally be used for establishing the implication relation, i.e. for deriving all $A < B$.

This quantum logical calculus may be characterized by its Lindenbaum algebra. According to the standard procedure we form equivalence classes of identical propositions. Thereby we define the identity $A \equiv B$ of two propositions A and B by the indistinguishability of any sequential contexts which differ by substitution of one of the propositions by the other one only. Within the framework of our dialog-semantics this means that for any sequential proposition in wich A or B occur as subpropositions the winning possibility for P in a dialog-game does not alter if one of the propositions is substituted by the other one. In this case we also say that A and B are proof-equivalent. For instance consider a situation in a dialog-game in which the proposition W has been proven. Assume, furthermore, that there exists a conditional probability $\mathbb{P}_{<W>}(B)$ as a quantification of the winning possibility for any subsequent proposition B in the dialog.[7] Then, in general, the equation

$$p_{<W>}(B) = p_{<W \sqcap (A \cup \neg A)>}(B)$$

does not hold since the additional proof of A, formulated in the condition on the right side, might have some influence on B which differs from the sole influence of the proof of W. Indeed, it can be shown that the above equality holds if A and B are commensurable (see Mittelstaedt (1983a)). Therefore, $V \not\equiv A \cup \neg A$ in case A is not commensurable with all propositions of S. An example of a proof-equivalence is $A \cap B \equiv B \cap A$. The substitution of one of these propositions by the other one does not lead to any distinctions within dialog-games.

The Lindenbaum algebra is defined by the set of all $\equiv$-equivalence classes of propositions of S, and by operations and a relation which inherit the structure of the logical connectives and the implication relation. For simplicity we shall not make any distinction between the propositional and the algebraic structures.

It turns out that the Lindenbaum algebra of quantum logic is a generalization of a quasi-pseudo-Boolean algebra (qpB) which is a well-known algebraic characterization of a non-classical logic called constructive logic with strong negation (Rasiowa (1974))[8]. The generalization can easily be classified. It concerns the implication operation only.

In a qpB the condition for the implication $A \to B$ is, according to (Rasiowa (1974)), given by the rule.

$$(qpB_3) \qquad A \cap C < B \;\leftrightarrow\; C < A \to B$$

where the relation $<$ is a quasi-ordering defined by

$$A < B \;\leftrightarrow\; V = A \to B\ .$$

(qpB_3) determines the implication $A \to B$ to be the maximal C which satisfies the premise.

The quantum logical generalization qpQ of a qpB consists of replacing the rule (qpB_3) by the axioms and rule

$$(qpQ_3) \qquad A \cap (A \to B) < B$$

$$A \to B < A \to (A \to B)$$

$$A \cap C < B \;\;\&\;\; C < A \to C \;\sim\; C < A \to B$$

Here the implication $A \to B$ is determined to be the maximal C which satisfies the two premises of the rule, the previous one and, in addition, $C < A \rightarrow C$ which formalizes the <u>commensurability</u> of A and C.

Further axioms and rules for establishing formal commensurabilities which may be used in the above rule have to be added to (qpQ_3) but will not be listed here.

Besides this quantum logical structure of compound propositions another structure which establishes the implication relations between <u>elementary propositions</u> is of great importance for the quantum language. In order to exhibit this structure we first introduce <u>value-equivalence</u> as a second equivalence relation between propositions: Two propositions A and B are called value-equivalent, denoted by $A = B$, iff for each dialog-game in which one of the two propositions is asserted it may be substituted by the other one such that the winning possibilities for A and B are equal. If we assume that the winning possibilities can be quantified by probability measures p according to the different positions of a dialog-game, we have:

$$A = B \quad \text{iff} \quad p(A) = p(B) \quad \text{for all } p.$$

It is obvious that the relation = is weaker than $\equiv$. For instance $V = A \cup \neg A$ holds, although the propositions V and $A \cup \neg A$ are not identical.

We can now formulate as a further important hypothesis:
(γ) To each logically connected proposition A there corresponds a unique (mod $\equiv$) elementary proposition $A^{(e)}$ such that $A = A^{(e)}$.
This hypothesis expresses that in the domain of quantum physics each logically connected proposition A (with complex content) can be considered to be a <u>partition</u> of an elementary proposition (with no complex content) into the elementary subpropositions of A.[9] Since this partition consists of material acts of the speaker-observer, it cannot be expected in general that, if the elementary proposition has an objective truth-value, the value-equivalent compound propositions have objective truth-values also, i.e. that $A \equiv A^{(e)}$.

This hypothesis imposes a remarkable structure on S_e. In order to characterize this structure we define the new operations $\wedge, \vee, \rightarrow, \neg$ inductively by: $a \wedge b := (a \cap b)^{(e)}$, $a \vee b := (a \cup b)^{(e)}$, $a \rightarrow b := (a \rightarrow b)^{(e)}$ and $\neg a := (\neg a)^{(e)}$ for $a, b \in S_e$.

It turns out that the quantum logical structure of the

implication relation between elementary propositions can be algebraically characterized by means of a quasi-implicative lattice $L_{qi} = \langle S_e, \wedge, \vee, \rightarrow, \neg, V, \Lambda \rangle$ with a zero element Λ and $\neg a = a \rightarrow \Lambda$ (Mittelstaedt (1978)). S_e denotes the set of elementary propositions (mod $\equiv$) now. L_{qi} is a generalization of an implicative lattice L_i which is an algebraic characterization of intuitionistic logic (Curry (1977)). Here, again, the generalization concerns the implication and, since $a \equiv a \rightarrow \Lambda$, the negation operation only.

In L_i, the axiom and rule for implication $a \rightarrow b$ are given by:

$L_i(4.1)$ $\quad a \wedge (a \rightarrow b) \leq b$

$L_i(4.2)$ $\quad a \wedge c \leq b \rightsquigarrow c \leq a \rightarrow b$

which distinguishes $a \rightarrow b$ as the maximal element c which satisfies the premise of $L_i(4.2)$.

In L_{qi} this rule is replaced by

$L_{qi}(4.2)$ $\quad a \rightarrow b \leq a \rightarrow (a \rightarrow b)$

$L_{qi}(4.3)$ $\quad a \wedge c \leq b \;\&\; c \leq a \rightarrow c \rightsquigarrow c \leq a \rightarrow b$

which distinguishes $a \rightarrow b$ as the maximal element c which satisfies the previous premise and which, in addition, is commensurable with a. Further axioms and rules for formal commensurabilities between elementary propositions, which may be used for the second premise of $L_{qi}(4.3)$, have to be added:

$L_{qi}(4.4)$ $\quad a \leq b \rightarrow a \rightsquigarrow b \leq a \rightarrow b$ $\quad$ (symmetry)

$L_{qi}(4.5)$ $\quad b \leq a \rightarrow b \;\&\; c \leq a \rightarrow c \rightsquigarrow b * c \leq a \rightarrow (b * c)$

with $* \in \{\wedge, \vee, \rightarrow\}$

3. FORMAL AND ONTOLOGICAL CONDITIONS OF REDUCTION

1. Formal conditions for reduction

The constructive approach to the abstract quantum language led to the following quantum logical structures.

(a) $qpQ = \langle S, \cap, \cup, \ni, \neg, \sim, V, \Lambda \rangle$ is a generalization with respect to the implication $\ni$ of a quasi-pseudo-Boolean algebra qpB. It characterizes algebraically the totality of formally true logically connected propositions and, by means of $A < B \curvearrowright V = A \ni B$, the implication relation on the set S of logically connected propositions.

The formal structure of the set S_e of elementary propositions was obtained by means of the hypothesis (γ) of the value-equivalence between logically connected propositions and elementary propositions. This led us to the introduction of new operations $\wedge, \vee, \rightarrow, \urcorner$ on the set S_e which are value-equivalently related to, but not identical with, the logical connectives $\cap, \cup, \rightarrowtail, \neg$. The strong negation $\sim$, however, did not give rise to a new operation on S_e, since $\sim a \in S_e$ if $a \in S_e$. Therefore we dispensed with the strong negation when constructing $\langle S_e, \wedge, \vee, \rightarrow, \urcorner \rangle$ from the set S' of logically connected propositions without $\sim$.

The structure of the set S' is interesting by itself since, within the framework of constructive logic with strong negation, the restriction of S to S' leads to the intuitionistic logic (Rasiowa (1974)). Starting from our constructive quantum logic with strong negation the restriction of S to S' establishes an intuitionistic quantum logic $pQ = \langle S', \cap, \cup, \rightarrowtail, \neg, V, \Lambda \rangle$. It turns out that this intuitionistic quantum logic is algebraically characterized by a generalization of a pseudo-Boolean algebra pB which characterizes intuitionistic logic (Rasiowa (1974)). The generalization concerns the intuitionistic implication, here denoted by $\rightarrowtail$ also, and is the same as for qpQ. Hence, we have as a further quantum logical structure:

(b) $pQ = \langle S', \cap, \cup, \rightarrowtail, \neg, V, \Lambda \rangle$ is a generalization with respect to the implication $\rightarrowtail$ of a pseudo-Boolean algebra pB. It characterizes algebraically the totality of formally true logically connected propositions without strong negation $\sim$, and by means of $A \leq B \overset{\text{def.}}{\Longleftrightarrow} V = A \rightarrowtail B$, the implication relation on the set S'.

Finally we have for elementary propositions:

(c) $L_{qi} = \langle S_e, \wedge, \vee, \rightarrow, \urcorner, V, \Lambda \rangle$ is a quasi-implicative lattice with zero-element Λ and quasi-pseudo-complement $\urcorner a := a \rightarrow \Lambda$. The lattice relation characterizes algebraically the implication relation on the set S_e of element propositions.

If the condition $a \vee \urcorner a = V$, which is a formalization of the value-definiteness of elementary propositions, is added as an axiom to L_{qi} we have:

(d) $L_q = \langle S_e, \wedge, \vee, \rightarrow, \urcorner, V, \Lambda \rangle$ is an orthocomplemented quasi-modular (or orthomodular) lattice (Mittelstaedt (1978) p. 29).

The above classification of quantum logical algebras already involves a comparison with well-known algebras which characterize non-classical logics, namely qpB, pB and L_i

which, if the operations are treated like the operations in pB, is a lattice theoretic formulation of pB. These algebras can be obtained by adjoining certain additional axioms to the quantum logical algebras.[10]

Starting from qpQ, the further axiom $A < B \rightarrow A$ leads to qpB. It can easily be shown that under this condition the axioms and rule given by (qpQ_3) and the additional commensurability rules reduce to (qpB_3) from which $A < B \rightarrow A$ can be derived. On the other hand, it can be proved that, if $V = A \cup \sim A$ is added as an axiom to qpQ and qpB, these algebras become a Boolean algebra B. This condition, therefore, establishes the connection between quantum logic and classical logic.

If we dispense with the strong negation $\sim$ we have to consider pQ. The additional axiom $A \leq B \rightarrow A$ leads to pB. On the other hand, if the axiom $V = A \cup \neg A$ is added to pQ or pB, we obtain a Boolean algebra B.

The lattice L_{qi} of elementary propositions is related to L_i and L_q by the following conditions. If the axiom $a \leq b \rightarrow a$ is added to L_{qi} we obtain L_i. If, on the other hand, $V = a \vee \neg a$ is added to L_{qi} we obtain L_q. If both of the conditions are added to L_{qi} we obtain a Boolean lattice L_B, the lattice theoretic formulation of a Boolean algebra B.

The formal conditions of the various reductions are summarized in the following diagrams.

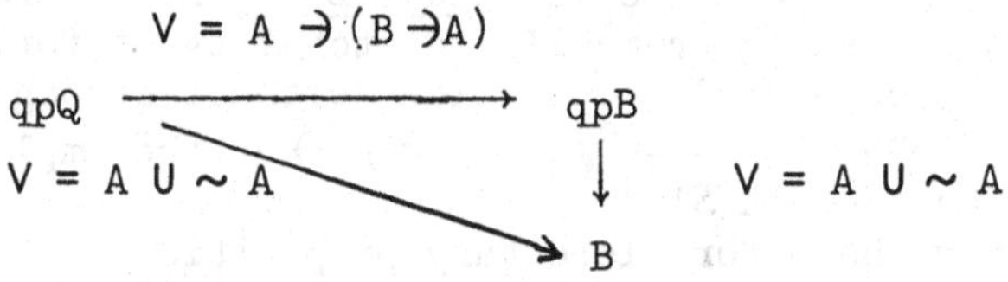
V = A → (B → A)
qpQ
qpB
V = A ∪ ∼ A
B
V = A ∪ ∼ A

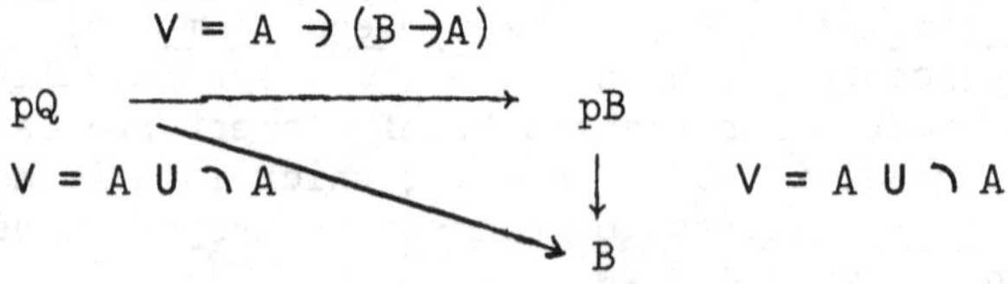
V = A → (B → A)
pQ
pB
V = A ∪ ¬ A
B
V = A ∪ ¬ A

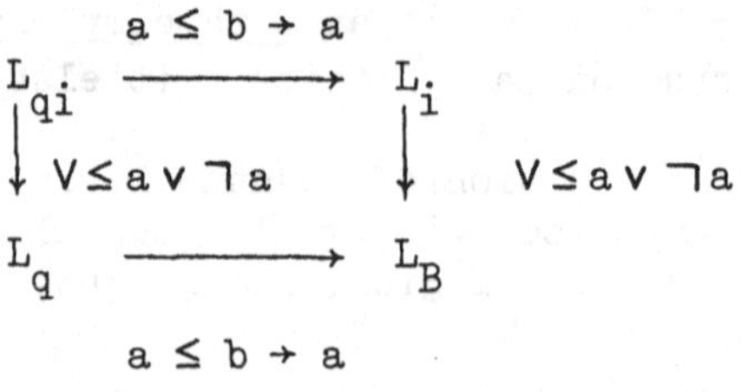
a ≤ b → a
L_qi
L_i
V ≤ a ∨ ¬a
V ≤ a ∨ ¬a
L_q
L_B
a ≤ b → a

The conditions within the diagrams which concern different parts of the language are not independent of each other. Their relations can be made clear by means of their interpretations within the language. In particular, we consider the following formal conditions for elementary propositions: (a) $V = a \rightarrowtail (b \rightarrowtail a)$, (b) $V \leq a \vee \neg a$, (c) $V = a \cup \sim a$. Because of the value equivalence of $a \rightarrowtail (b \rightarrowtail a)$ and $a \rightarrow (b \rightarrow a)$ and of $a \cup \neg a$ and $a \vee \neg a$, and by means of $a \cup \sim a < a \cup \neg a$, we obtain from (a) and (c) that $a \rightarrow (b \rightarrow a)$ and $a \vee \neg a$ are value-equivalent to V. This leads to the conditions $a \leq b \rightarrow a$ and (b) $V \leq a \vee \neg a$ in L_{qi}. Moreover, (a) can be derived from (c). From the conditions (a) and (c) for elementary propositions one can derive the conditions $V = A \rightarrowtail (B \rightarrowtail A)$ and $V = A \cup \sim A$ respectively in qpQ. On the other hand, the condition (b) $V \leq a \vee \neg a$ in L_{qi} does not have any consequences for qpQ. The two conditions (a) and (b), however, reduce pQ to B. Because of the commensurability condition (a), there is no longer any distinction possible between value-equivalent propositions without strong negation $\sim$. Hence, all value-equivalent propositions without $\sim$ are identical. This implies that with $V \leq a \vee \neg a$ we have $V = a \cup \neg a$ also. By qpB' we denote the algebra qpB together with the additional axiom $V = A \cup \neg A$. Therefore, starting from (qpQ, pQ, L_{qi}) we obtain different systems of algebras according to the conditions (a), (b) and (c) as it is shown in Fig. 2.

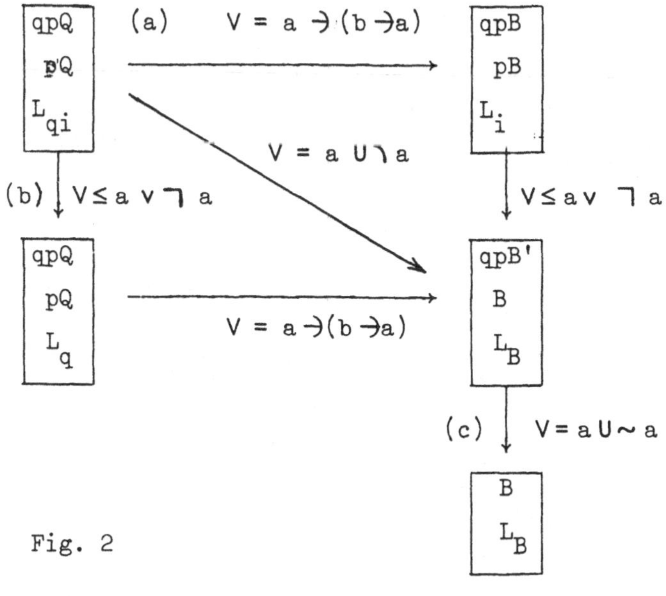

Fig. 2

2. Ontological conditions of reduction

The classical and non-classical logics that are characterized by the algebras B, qpB, pB and the lattices L_B, L_i can also be associated with object languages for certain domains of classical physics and mathematics. B and L_B characterize the abstract languages of classical physics and classical mathematics, qpB characterizes a constructive language for mathematics with strong negation, and pB and L_i characterize the language of intuitionistic mathematics. For simplicity we call a language with the system (qpB, pB, L_i) an *intuitionistic language* and a language with the system (B,L_q) a *classical language*. Furthermore, these languages can also be founded in a constructive way on the pragmatic preconditions for proof procedures in the respective domains, which, on their part, constiture the ontological premises in a selfconsistent way (see Fig. 1 in part 1) It turns out that the formal conditions (a), (b) and (c), which determine the above reduction scheme, correspond to particular ontological premise of these languages. The particular premises legitimate the speaker-observer to stipulate particular proof conditions for propositions which, however, can only be justified by means of the theory of proof processes in the respective domain. Therefore the restriction of these premises to the weak ontological premises of the quantum language, which correspond to the very restrictive proof conditions in the domain of quantum physics, describes the weakening of the classical and intuitionistic languages to the quantum language.

As it was already indicated in part 2 of the paper, the conditions(a) $V = a \rightarrow (b \rightarrow a)$ and (b) $V \leq a \vee \neg a$ formally describe the *commensurability* or *unrestricted availability* and the *value-definiteness* of elementary propositions respectively. It can be shown that these conditions are inherited by all finitely composed propositions. According to the discussion in part 2 of paper, the *commensurability* of a and b guarantees that the property E_a is not "disturbed" by the material proof process which establishes the property E_b with respect to the physical system S. Or in the context of mathematics: The truth-value of a is still available after the proof of b. This means that the objective properties or truth-values, once established, *cumulate* in the course of proof processes. The *value-definiteness* guarantees the *finite decidability* between the properties E_a and $E_{\neg a}$ or the truth-values within proof processes for a and $\neg$ a .

If we postulate value-definiteness for elementary propositions we arrive at a quantum language the logical structure of which is given by the system (qpQ, pQ, L_q). It is well-known that the orthomodular lattice L_q is realized by the lattice of subspaces of a Hilbert space. Since this language underlies the Hilbert space formalism of the usual quantum theory, we also call it *full quantum language*. We can even define a *strong value-definiteness* which guarantees the decidability between E_a and $E_{\sim a}$ or the corresponding truth-values of a. Since we dispensed with strong negation in $\langle S_e, \wedge, \vee, \rightarrow, \neg, \bigvee, \bigwedge \rangle$, we have no formal description of strong value-definiteness. It should be remarked, however, that this condition together with (a) establishes (c) and, thus, a classical language. Condition (c) $V = a \cup \sim a$, which implies each of (a) and (b), expresses the *unrestricted objectivity* of all properties E_a or propositions a. This means that all properties or all propositions are *permanently* decided throughout the course of any proof processes. It can be shown that this condition for elementary propositions is inherited by all finitely composed propositions. The correspondence between the formal and ontological conditions of reduction is summarized in Fig. 3.

(a) $V = a \rightarrow (b \rightarrow a)$ $\leftrightarrow$ (a_o) *cumulative objectivity*

(b'_o) *strong finite decidability*

$\downarrow$

(b) $V \leq a \vee \neg a$ $\leftrightarrow$ (b_o) *finite decidability*

(c) $V = a \cup \sim a$ $\leftrightarrow$ (c_o) *unrestricted objectivity*

$\updownarrow$

(a_o) & (b'_o)

Fig. 3

In the ontological reduction scheme of Fig. 4 we make use of the conditions (a_o), (b'_o) and (c_o) only.

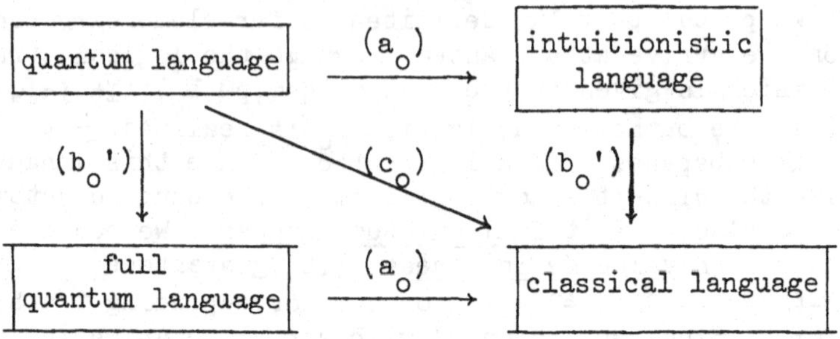

Fig. 4

The classical language involves the strong ontological premises of the strong finite decidability and cumulative objectivity which are equivalent to the unrestricted objectivity. These premises legitimate the speaker-observer to stipulate particular proof conditions with respect to elementary propositions such that the propositions $a \vee \neg a$ and $a \rightarrow (b \rightarrow a)$ are formally true and the logical system of the language is given by (B, L_B). The language still involves the hypothesis (γ) of part 2 . The logically composed propositions A and their corresponding elementary propositions $A^{(e)}$ are, however, identical since under the condition of commensurability any possible contextual difference between value-equivalent propositions vanishes. Therefore, we can associate any propositions $A \in S$ with several proof procedures which differ in the subjective ignorance of the speaker-observer with respect to the truth-values of the elementary subpropositions. In particular, the proof according to $A^{(e)}$ involves maximal ignorance.

If according to his finite possibilities, the speaker-observer dispenses with the postulate of strong value-definiteness, the proposition $a \vee \neg a$ is no longer formally true and the classical logic is weakened to the intuitionistic logic. On the other hand, if, according to the material conditions of his proof processes, the speaker-observer must dispense with the cumulative objectivity of the truth-values of propositions, the proposition $a \rightarrow (b \rightarrow a)$ is no longer formally true and the logic is weakened to the full quantum logic. If both of the postulates can no longer be substantiated, neither $a \rightarrow (b \rightarrow a)$ nor $a \vee \neg a$ are formally true and the logic is weakened to the (intuitionistic) quantum logic. The modifications only concern the implication and negations but not the conjunction and disjunction. In the lattice structures for elementary propositions, however, all operations

$\wedge, \vee, \rightarrow, \neg$ are affected. The possible incommensurability of propositions enables the speaker-observer to distinguish value-equivalent propositions by means of their contexts (e.g. in a dialog). Hence $a \vee \neg a$ and $a \cup \neg a$ are no longer identical if there exists a proposition b which is incommensurable with a. In this case
$b \cap (a \vee \neg a) \neq b \cap (a \cup \neg a) \equiv (b \cap a) \cup (b \cap \neg a) = \Lambda$ and therefore $b \wedge (a \vee \neg a) \neq (b \wedge a) \vee (b \wedge \neg a) = \Lambda$, i.e. the distributivity fails in L_q and L_{qi}. Moreover, because of the possible contextual distiction between value-equivalent propositions, also their simultaneous objectivity is no longer given in general. E.g. if A has an objective truth-value the value-equivalent elementary proposition $A^{(e)}$ does not necessarily have an objective truth-value and vice versa. Hence the proof of $A^{(e)}$ does not in general involve subjective ignorance of the speaker-observer with respect to the elementary subpropositions of A, as is the case in the classical and intuitionistic languages.

The restriction of the premises of the language, although it leads to weaker structures, does not diminish the possibilities of using the language. Since the quantum language incorporates testing procedures for the conditions that can be asked, value-definiteness and commensurability are taken into account whenever they can be substantiated for particular elementary propositions. Therefore, according to our present state of knowledge in the domains of physical reality and mathematics, quantum logic can be considered to be the most universal logic, i.e. the weakest logic from which the other ones can be obtained by means of additional axioms.

Notes

1) This procedure is well-known from the original work of G. Birkhoff and J. v. Neumann (1936) and further investigations by G. Mackey (1963), J.M. Jauch (1968), V. Varadarajan (1970) and others.
2) This approach is used in some axiomatic treatments of quantum logic e.g. by J.M. Jauch (1968) and C. Piron (1976), in the empirical logic approach by D.J. Foulis and C.H. Randall (1981),(1983), and the propositional language approach by P. Mittelstaedt (1978) and the author (1980).
3) For a more detailed discussion of the selfconsistency argument see Mittelstaedt (1983 b).
4) A detailed exposition of this language approach will be given in my (1985).

5) Here we assume that the concept of an individual physical system has already been established such that measuring processes with respect to the system objectively determine certain properties. The procedure of the constitution of individual objects within the quantum language is investigated in a forthcoming paper by the author.
6) A dialog-semantics was first used by P. Lorenzen and K. Lorenz (1978) for a foundation of intuitionistic and classical logics. This method has been applied with certain modifications to quantum physical propositions by P. Mittelstaedt (1976) and the author (1976).
7) For the extension of the quantum language to probabilities see my (1981).
8) The implicative operation which occurs as a further operation in a qpB (see Rasiowa (1974)) is neglected here for simplicity since this operation has no direct interpretation within our dialogic approach.
9) If we consider as particular models of our language the language of classical mechanics and the language of Hilbert space quantum mechanics, the hypothesis (γ)corresponds to well-known postulates. In the first case, it is usually assumed that the measurable sets of the phase space represent measurable physical quantities. The logical connectives correspond to the set theoretical operations. Then it is clear that any connected proposition corresponds to an elementary proposition which is associated with a physical quantity. In the second case, it is usually assumed that the subspaces of the Hilbert space of a quantum mechanical system S represent quantum physical quantities. The logical connectives correspond in a value-definite way to operations in a linear vector space which realize the operations $\wedge$, $\vee$, $\rightarrow$, $\neg$ to be introduced in the following. Hence the above hypothesis (γ) is satisfied.
10) In the following discussion we dispense with all proofs which will be given in my (1985).

Ernst-Walter Stachow
Institut für Theoretische Physik
Universität zu Köln
D - 5000 Köln, West Germany

REFERENCES

Beltrametti, E. and van Fraassen, B.C.: 1981, 'Current Issues in Quantum Logic', Plenum Press, New York

Birkhoff, G. and von Neumann, J.: 1936, 'The Logic of Quantum Mechanics', Ann. Math. 37, 823-843.

Curry, H.B.: 1977, 'Foundations of Mathematical Logic', Dover Publications, New York.

Foulis, D.J. and Randall, C.H.: 1981, Empirical Logic and Tensor Products, in Neumann, H. (ed.): 'Interpretations and Foundations of Quantum Theory', Bibliographisches Institut, Mannheim, 9 - 20.

Jauch, J.M.: 1968, 'Foundations of Quantum Mechanics', Addison-Wesley, Reading, Mass.

Lorenzen, P. and Lorenz K.: 1978, 'Dialogische Logik', Wissenschaftliche Buchgesellschaft, Darmstadt.

Mackey, G.: 1963, 'Mathematical Foundations of Quantum Mechanics', Benjamin, New York.

Mittelstaedt, P.: 1976, 'Philosophische Probleme der modernen Physik', Bibliographisches Institut, Mannheim, English translation: Philosophical Problems in Modern Physics, D. Reidel Publ. Co., Dordrecht, Holland, 1976, Ch. VI.

Mittelstaedt, P.: 1978, 'Quantum Logic', D. Reidel Publ. Co., Dordrecht, Holland.

Mittelstaedt, P.: 1983a, 'Relativistic Quantum Logic', Int. J. of Th. Phys. 22, 293 - 314.

Mittelstaedt, P.: 1983b, 'Wahrheit, Wirklichkeit und Logik in der Sprache der Physik', Zeitschrift für Allg. Wissenschaftstheorie XIV, 24 - 45.

Piron, C.: 1976, 'Foundations of Quantum Physics', Benjamin, Reading, Mass.

Randall, C.H. and Foulis, D.J.: 1983, 'Properties and Operational Propositions in Quantum Mechanics', Found. of Phys. 13, 843 - 858.

Rasiowa, H.: 1974, 'An Algebraic Approach to Non-Classical Logics', North-Holland Publ. Co.

Stachow, E.-W.: 1976, 'Completeness of Quantum Logic', J. of Phil. Logic 5, 237 - 280. Reprinted in C.A. Hooker (ed.): Physical Theory as Logico Operational Structure, D. Reidel Publ. Co., 1979, 203 - 243.

Stachow, E.-W.: 1980, 'Logical Foundation of Quantum Mechanics', Int. J. of Th. Phys. 19, 251 - 304.

Stachow, E.-W.: 1981, 'Die quantenlogische Wahrscheinlichkeitskalkül', in J. Nitsch, J. Pfarr and E.-W. Stachow (eds): Grundlagenprobleme der modernen Physik,

Bibliographisches Institut, Mannheim, 271 - 305.
Stachow, E.-W.: 1985, '<u>Logische Grundlagen der Quantenphysik</u>', Bibliographisches Institut, Mannheim, in preparation.
Varadarajan, V.: 1968, 1970, '<u>Geometry of Quantum Theory</u>' I and II, Van Nostrand, Princeton, N.J.

Reinhard Werner

BELL'S INEQUALITIES AND THE REDUCTION OF STATISTICAL THEORIES

1. INTRODUCTION

A central problem for the interpretation of quantum mechanics is the question in what way this theory transcends the framework of classical probability. In one way or other, every approach to the foundations of quantum mechanics has to answer the challenge of hidden variable theories, i.e. the claim that quantum mechanics can be reduced to a classical theory. With the advent of Bell's inequalities(1) it appears that this claim can be put to an experimental test, at least when the reducing classical theory is subject to certain "locality" assumptions. I shall show how these assumptions are related to a principle which is universally (though usually tacitly) employed in statistical theories (cf. sections 3,4).

A deeper understanding of such experiments and of reduction relations between statistical theories requires the analysis of the relationship between an observational and a theoretical level of theories (exemplified here by "frequency functions" and "statistical dualities" respectively). In this analysis I shall follow Ludwig(2), who takes "theoretical" terms to be defined in terms of (the totality of) observational terms. The apparent contradiction of this approach with the widespread acceptance of the non-definability of theoretical terms is resolved by the observation that in Ludwig's scheme these two levels acquire a certain independence of each other: on the one hand, the observational level cannot be reconstructed from the theoretical level; on the other hand, certain "theoretical" terms cannot be determined on the basis of finitely many observational data. Whether these observations may be extended to a formal reconstruction of the Sneedian "problem of theoretical terms" within the Ludwig approach, remains to be seen in a broader study.

W. Balzer et al. (eds.), Reduction in Science, 419–442.

2. FREQUENCY FUNCTIONS AND STATISTICAL DUALITIES

Bell's inequalities can be seen as a quantitative version of the Einstein-Podolsky-Rosen paradox. Both refer to the general situation of correlation experiments, which may be described schematically as follows: a source s emits pairs of particles, on each of which a measurement is then performed, say, by means of measuring apparatuses b and d. If we denote the outcome sets of these apparatuses by $\bar{b}$ and $\bar{d}$, the result of an individual experiment is a pair (β,δ) with $\beta \in \bar{b}$ and $\delta \in \bar{d}$. (Fig. 1). If this type of experiment is repeated many times, it is assumed that each of the results occurs with a reproducible relative frequency $p(\beta,\delta)$. $(\Sigma_{\beta \in \bar{b}}\Sigma_{\delta \in \bar{d}}\; p(\beta,\delta) = 1)$.

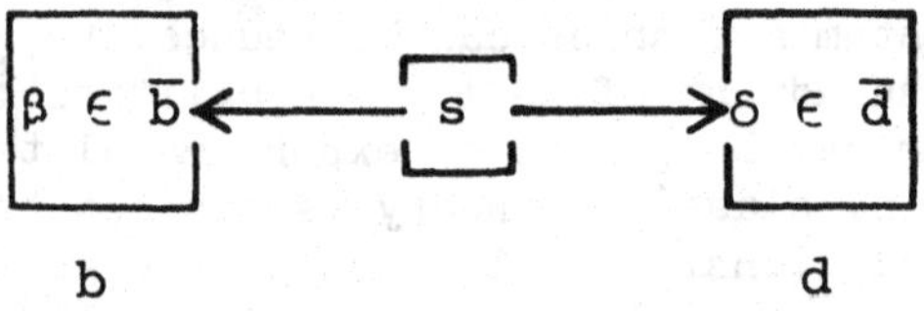

Fig. 1

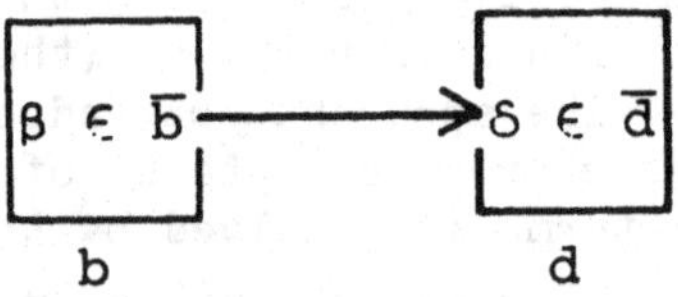

Fig. 2

For our purposes it suffices to consider a class of such experiments with a fixed source s and measuring devices b and d chosen freely from some sets B and D. With the notations $\bar{B} = \bigcup_{b \in B} \bar{b}$ and $\bar{D} = \bigcup_{d \in D} \bar{d}$, we may then consider p as a function $p: \bar{B} \times \bar{D} \to [0,1]$. This whole structure will be abbreviated by the triple (p,B,D) and called a frequency function.

By formally considering the source as part of the apparatus b, we arrive at a more general situation, in which b will be called a preparing apparatus and d a measuring apparatus (Fig. 2). I shall use the same notations in both cases. It is important to note that the apparatuses s, b and d are identified by their macroscopic descriptions alone. The language employed at this level is purely observational, without reference to theoretical terms like "particles" or even "spin".

The statistical theories encountered in physics are usually not given in terms of collections of measuring and preparing apparatuses but in terms of states and observables. A crucial idea in Ludwig's approach to quantum mechanics and other statistical theories is that states and observables can be considered as equivalence classes of apparatuses or procedures under the relation of statistical indistinguishability. The usual formalism of quantum mechanics (using operators on a Hilbert space) is thus understood as a derived structure, in which preparing procedures inducing the same relative frequencies on all available measuring devices are identified. This construction can be carried out for arbitrary frequency functions (p,B,D) and it is useful to encorporate into it also the possibility of taking mixtures of procedures: a mixture of a given set of, say, preparing procedures is a new procedure which is applied in each individual experiment by letting a random generator decide with preassigned weights which of the given procedures to use. It is clear that the frequencies produced by such a mixed procedure will be convex combinations of the frequencies produced by the original procedures. Therefore, a convenient shorthand notation for a mixed procedure is a formal convex combination of procedures(with the weights determined by the random generator).

In order to make mathematical sense out of this notation, let us introduce the free (real) vector spaces lin $\bar{B}$ and lin $\bar{D}$ over $\bar{B}$ and $\bar{D}$, i.e. the spaces of all finite formal linear combinations of elements of $\bar{B}$ and $\bar{D}$. (Thus convex combinations $\Sigma\lambda_i\beta_i$ with $\beta_i \in \bar{B}$, $\lambda_i \geq 0$, $\Sigma\lambda_i = 1$ are well defined elements in lin $\bar{B}$). The frequency function p then induces a bilinear form $\langle\cdot,\cdot\rangle$: lin $\bar{B} \times$ lin $\bar{D} \to \mathbb{R}$ by virtue of

$$\langle\sum_i \lambda_i\beta_i, \sum_j \mu_j\delta_j\rangle = \sum_{i,j} \lambda_i\mu_j p(\beta_i,\delta_j). \quad (\lambda_i,\mu_j \in \mathbb{R})$$

The sets of linear combinations with positive coefficients form natural positive cones $(\text{lin }\bar{B})_+ \subset \text{lin }\bar{B}$ and $(\text{lin }\bar{D})_+ \subset \text{lin }\bar{D}$.
The second, more essential, step of our construction is to identify statistically equivalent elements, i.e. elements $x_1, x_2 \in \text{lin }\bar{B}$ such that for all $y \in \text{lin }\bar{D}$: $\langle x_1,y\rangle = \langle x_2,y\rangle$ (and symmetrically for lin $\bar{D}$). Let us denote the quotient spaces by $\mathcal{B}$ and $\mathcal{D}$ and the canonical projections by $s: \text{lin }\bar{B} \to \mathcal{B}$ and $t: \text{lin }\bar{D} \to \mathcal{D}$. By construction there is a bilinear form $\langle\cdot,\cdot\rangle: \mathcal{B} \times \mathcal{D} \to \mathbb{R}$ such that $\langle sx,ty\rangle = \langle x,y\rangle$. The cones of positive elements in $\mathcal{B}$ and $\mathcal{D}$ are defined as $\mathcal{B}_+ := s((\text{lin }\bar{B})_+) = \{\Sigma\lambda_i s(\beta_i)|\lambda_i \geq 0\}$ and $\mathcal{D}_+ = t((\text{lin }\bar{D})_+)$.
The resulting structure is ubiquitous enough to deserve a name:

Definition 1: A statistical duality $\langle\mathcal{B},\mathcal{D}\rangle$ is given by a pair of real vector spaces $\mathcal{B}$ and $\mathcal{D}$, ordered by (proper and generating) positive cones $\mathcal{B}_+$ and $\mathcal{D}_+$, together with a bilinear form $\langle\cdot,\cdot\rangle: \mathcal{B} \times \mathcal{D} \to \mathbb{R}$, which is positive in the sense that $x \in \mathcal{B}_+$ and $y \in \mathcal{D}_+$ imply $\langle x,y\rangle \geq 0$.

Two particular cases will be of special interest to us: if (Ω,Σ,μ) is a measure space, then $\mathcal{B} := \mathcal{L}^1(\Omega,\Sigma,\mu)$, $\mathcal{D} := \mathcal{L}^\infty(\Omega,\Sigma,\mu)$ and $\langle\rho,f\rangle := \int\mu(d\omega)\rho(\omega)f(\omega)$ with (almost everywhere) pointwise ordering of functions define a statistical duality, called a classical duality. Similarly, a quantum duality over a Hilbert space $\mathcal{H}$ is given by $\mathcal{B} := \mathcal{T}(\mathcal{H})$ = space of hermitian trace class operators, $\mathcal{D} = \mathcal{B}(\mathcal{H})$ = space of bounded hermitian operators (with the usual orderings) and $\langle W,A\rangle = \text{tr}(WA)$.

With respect to these structures we may now ask whether a set of experimental data, given in the form of a frequency function (p,B,D), is consistent with a statistical theory, given in the form of a duality $\langle\mathcal{B}_1,\mathcal{D}_1\rangle$. Certainly we cannot expect that the classes B and D of procedures that we have used in the experiment already contain all procedures possible in the theory, i.e. that (p,B,D) coincides with the frequency function (p_1,B_1,D_1), from which $\langle\mathcal{B}_1,\mathcal{D}_1\rangle$ is presumably constructed. We may ask, however, whether we have chosen a subclass, so that $B \subset B_1$, $D \subset D_1$ and $p = p_1|\bar{B} \times \bar{D}$. In that case the maps s,t from the construction of $\langle\mathcal{B}_1,\mathcal{D}_1\rangle$ would still be defined on $\bar{B} \times \bar{D}$ and satisfy the following definition.

Definition 2: An embedding of a frequency function (p,B,D) into a statistical duality $\langle\mathcal{B},\mathcal{D}\rangle$ is a pair of maps s: $\bar{B} \to \mathcal{B}_+$, t: $\bar{D} \to \mathcal{D}_+$ such that $p(\beta,\delta) = \langle s(\beta), t(\delta)\rangle$

Can a given set of data (p,B,D) be embedded into a classical duality, and thus be given a "hidden variable" description? It is easy to see that the answer is always "yes", quite independently of the nature of those data. (For example, if $\bar{B}$ and $\bar{D}$ are finite, take $\Omega = \bar{B} \times \bar{D}$, μ the measure with $\mu\{(\beta,\delta)\} = p(\beta,\delta)$ and $s(\beta)(\beta',\delta')$ (resp. $t(\delta)(\beta',\delta')$) to be one for $\beta = \beta'$ (resp. $\delta = \delta'$) and zero otherwise. For the infinite case similar constructions are possible, see proposition 1). A quick inspection of the construction just given shows that this is no deep result: the "hidden" variables in this classical theory are nothing but the possible outcomes of individual experiments and hence perfectly observable. Thus the assertion that every set of statistical data may be embedded into a classical theory amounts to the truism that gathering statistical data is an activity carried out within the limits of classical probability. This "reduction" to classical probability is, in fact, a fundamental feature of Ludwig's interpretation of quantum mechanics in terms of frequencies measured on macroscopic apparatuses. So what is really new about quantum mechanics in comparison with its classical predecessors? What have we missed?

One indication comes from the observation that our constructions so far were symmetric with respect to interchange of B and D, whereas the physical situation between preparing and measuring is manifestly asymmetric: the aim of such experiments is after all to learn something about particles and more general physical systems by producing them and modifying them in various ways and then detecting the effect of those operations by means of various measurements. But this interpretation of experiments and the correlations embodied in a frequency function is possible only if the physical systems can be seen as carriers of a directed interaction.[3] The next section is devoted to studying the consequences of this idea for the structure of physical theories.

3. THE PRINCIPLE OF DIRECTED INTERACTION

In order to see consequences of the directed interaction between preparing and measuring apparatus, suppose that the experiments are conducted jointly by two physicists, Berta and Dora. Berta's task is to set up a preparing apparatus $b \in B$ and to record the results $\beta \in \bar{B}$ that occur in each individual experiment and similarly Dora controls the measuring apparatus, but neither is allowed to see the other's choice of apparatus, nor the results obtained by her. Clearly Berta can send a message to Dora by preparing systems with various combinations of (statistical) properties. What we want to exclude, however, is that Dora can communicate messages to Berta. How could she do this? The mere fact that for fixed b and d the frequencies $p(\beta,\delta)$ are correlated won't help, since Berta cannot see the result δ. The only thing Berta can detect is a change in the frequencies $\tilde{p}(\beta,d) := \Sigma_{\delta\in\bar{d}}\, p(\beta,\delta)$. Thus the only chance Dora has, is to use different sorts d of physical equipment, for example polarizers with different orientation. However, if her measuring apparatuses are passive, i.e. cannot be used as transmitters of information, the following condition must be satisfied:

Definition 3: A frequency function (p,B,D) is said to be directed, if for all $\beta\in B$ the expression $\tilde{p}(\beta) := \Sigma_{\delta\in\bar{d}}\, p(\beta,\delta)$ is independent of $d\in D$.

The directedness of (p,B,D) is equivalent to the assertion that, in the duality $\langle\mathcal{B},\mathcal{D}\rangle$ constructed from it, the element $\Sigma_{\delta\in\bar{d}}\, t(\delta)\in\mathcal{D}$ is independent of d. This element will be called the unit $1\in\mathcal{D}$ and corresponds to the "trivial" measurement of setting up any(!) measuring device but ignoring the results obtained by it. Definition 3 now has the following counterpart on the level of dualities:

Definition 4: A duality $\langle\mathcal{B},\mathcal{D}\rangle$ is said to be directed, if there is a distinguished element $1\in\mathcal{D}_+$. A state is an element $W\in\mathcal{B}_+$ such that $\langle W,1\rangle = 1$. An effect is an element $F\in\mathcal{D}$ such that $F\in\mathcal{D}_+$ and $1 - F\in\mathcal{D}_+$.

The classical and quantum dualities are directed, the unit $1\in\mathcal{D}_+$ being given by the constant function 1 and the unit operator respectively. A state can be considered as an equivalence class (i.e. the image under $\underline{s}$) of "unmonitored" preparing procedures with outcome sets $\bar{b}$ consisting of a single element. Similarly an effect[4] $\underline{F}$ corresponds to a yes-no measurement d with outcome set $\bar{d} = \{+,-\}$, such that $t(+) = F$ and $t(-) = 1 - F$. The following definition describes the counterparts of procedures with larger outcome sets on the level of dualities:

Definition 5: Let $\langle\mathcal{B},\mathcal{D}\rangle$ be a directed duality and X a set. Then a preparator over X is a function $W: X\rightarrow\mathcal{B}_+$ such that $\langle\sum_{x\in X} W(x),1\rangle = 1$ and an observable over X is a function

$F: X \to \mathcal{D}_+$ such that $\sum_{x \in X} F(x) = 1$.

This definition[5] was chosen so that for any procedure $b \in B$ of a frequency function (p,B,D) there is a preparator $s: \bar{b} \to \mathcal{B}_+$ of the associated duality $\langle \mathcal{B},\mathcal{D} \rangle$. However, in general not every preparator is of this form, although we may formally construct for every duality $\langle \mathcal{B},\mathcal{D} \rangle$ a frequency function $(\tilde{p},\tilde{B},\tilde{D})$ for which this is true. Therefore, when we assert the existence of some preparator in $\mathcal{B}$, we have not asserted the existence of an actual apparatus realizing it, but only that the existence of such an apparatus is consistent with the theory. It is important to keep this in mind when reading the next definition. This definition describes the situation that several preparators (or observables) can be realized together by one apparatus: the outcome set of such an apparatus is of the form $\bar{b} = \bigtimes_k \bar{b}^k$ and applying this apparatus but ignoring all but the kth component of the outcome is considered as an apparatus with outcome set $\bar{b}^k$.

Definition 6: A finite family $(W_i : X_i \to \mathcal{B}_+)_{i=1\ldots N}$ of preparators is called coexistent, if there is a preparator $W: \left(\bigtimes_{i=1}^{N} X_i \right) \to \mathcal{B}_+$ with i^{th} marginal W_i, i.e.

$$W_i(x_i) = \sum_{x_1 \in X_1} \cdots \sum_{x_{i-1} \in X_{i-1}} \sum_{x_{i+1} \in X_{i+1}} \cdots \sum_{x_N \in X_n} W(x_1 \ldots x_N).$$

The coexistence of a family of observables is defined analogously.

The distinguishing feature of classical theories, which will be important for our discussion of Bell's inequalities, is the property that in a classical duality every family of preparators and every family of observables is coexistent[6]. The latter fact is sometimes expressed by saying that all classical observables can be measured simultaneously; but this terminology is rather unfortunate, since neither "time instants of measurement" nor the necessarily finite time intervals during which the measurements are performed are relevant to the issue.

Let us now return to the question whether a given set of statistical data can be embedded into a classical duality and see how it is modified by the principle of directed interaction. Again we shall consider two frequency functions:

the measured data (p,B,D) and the structure (p_1,B_1,D_1) from which the duality $\langle \mathcal{B}_1,\mathcal{D}_1\rangle$ is constructed. We assume that $B \subset B_1$, $D \subset D_1$ and $p = p_1|\bar{B} \times \bar{D}$ and, in addition now, that (p_1,B_1,D_1) is directed in the sense of definition 3. Then it follows that (p,B,D) is also directed and, moreover, the following definition is datisfied:

Definition 7: An embedding of a frequency function (p,B,D) into a directed statistical duality $\langle \mathcal{B},\mathcal{D}\rangle$ is a pair of maps $s: \bar{B} \to \mathcal{B}_+$, $t: \bar{D} \to \mathcal{D}_+$ such that $p(\beta,\delta) = \langle s(\beta), t(\delta)\rangle$ and for all $d \in \bar{D}$: $\Sigma_{\delta\in\bar{d}}\, t(\delta) = 1$.

It was already noted that a frequency function (p,B,D) can be embedded in this sense only if it is directed. Whether this is the case can be verified directly by looking at the experimental data collection (p,B,D). However, the definition requires much more than this: it is equivalent to the assertion that even if Berta has at her disposal the very large class B_1 of preparing apparatuses, i.e. all preparing apparatuses admitted by the theory $\langle \mathcal{B}_1,\mathcal{D}_1\rangle$ she will not be able to receive a message from Dora. Does this more restrictive definition change the verdict that any set of data can be accounted for by a "hidden variable" theory? This is answered by the following proposition:

Proposition 1: Every directed frequency function may be embedded (in the sense of definition 7) into a classical duality.

Proof: By assumption $\bar{p}(\beta) := \Sigma_{\delta\in\bar{d}}\, p(\beta,\delta)$ is independent of d. As the underlying measure space of the classical theory we take $\Omega = \{\beta\in\bar{B} \,|\, \bar{p}(\beta)\neq 0\}$ with measure μ assigning measure 1 to each point. We define the functions $t(\delta)\in\mathcal{L}^\infty(\Omega,\mu)$ by $t(\delta)(\beta) = \bar{p}(\beta)^{-1} \cdot p(\beta,\delta)$ and $s(\beta)\in\mathcal{L}^1(\Omega,\mu)$ by $s(\beta)(\beta') = \bar{p}(\beta)$ if $\beta = \beta'$ and $s(\beta)(\beta') = 0$ otherwise. The verification of the required properties is straightforward. qed.

A construction resembling the proof of proposition 2 can be used to show that the functions $t(\delta)\in\mathcal{L}^\infty(\Omega,\mu)$ can be chosen to be characteristic functions, taking only the values 0 and 1. Hidden variable theories with this property are sometimes called "deterministic", because the value of the "hidden variable $\omega\in\Omega$" determines the outcome δ with certainty.

As this proposition shows, the principle of directed interaction alone is not powerful enough to make an experimental discrimination between classical theories and quantum mechanics possible. However, there is more structure to a correlation experiment (Fig. 1) than just a preparing and a measuring process (Fig. 2). In the following section we shall study the consequences of applying the principle of directed interaction to both wings of a correlation experiment.

4. BELL'S INEQUALITIES

Consider three physicists, say Berta, Dora and Sam, jointly conducting a correlation experiment (Fig. 1). Sam, who controls the sources s is certainly able to transmit signals to Berta or Dora, by choosing an appropriate apparatus s. (But since we shall consider only experiments with a fixed source we must regretfully deny him this possibility.) Berta and Dora both observe systems from a common source, but we assume that Berta's choice of an apparatus $b \in B$ has no influence on the statistics recorded by any of Dora's devices $d \in D$, and vice versa. This state of affairs is summarized in the following definition:

Definition 8: A frequency function (p,B,D) is called a correlation table, if $\bar{p}(\beta) := \Sigma_{\delta\in\bar{d}}\, p(\beta,\delta)$ is independent of $d \in D$ and $\bar{p}(\delta) := \Sigma_{\beta\in\bar{b}}\, p(\beta,\delta)$ is independent of $b \in B$. A correlation table will be called finite, if the sets B and D are finite.

Rather than continuing with the obvious analogs of definitions 4 and 7, let us formulate directly what "embedding into a classical theory" means in this new context. Again we shall consider the data as part of a classical theory in the sense that $B \subset B_1$, $D \subset D_1$, $p = p_1|\bar{B} \times \bar{D}$ and that the duality $\langle \mathcal{B}_1, \mathcal{D}_1 \rangle$ constructed from (p_1, B_1, D_1) is classical. The crucial assumption is now that (p_1, B_1, D_1) is also a correlation table in the sense of definition 8. Thus the data (p,B,D) must satisfy the following condition:

Definition 9: A correlation table (p,B,D) is called quasi-classical, if there are a classical duality $\langle L^1(\Omega,\Sigma,\mu), L^\infty(\Omega,\Sigma,\mu) \rangle$ with a state $\bar{\rho} \in L^1(\Omega,\Sigma,\mu)$ and two maps $s: \bar{B} \to L^1$, $t: \bar{D} \to L^\infty$ such that

(1) $p(\beta,\delta) = \langle s(\beta), t(\delta) \rangle$

(2) $\Sigma_{\beta\in\bar{b}}\, s(\beta) = \bar{\rho}$ for all $b \in B$

(3) $\Sigma_{\delta\in\bar{d}}\, t(\delta) = 1$ for all $d \in D$

The following proposition shows that we could equivalently

have imposed the stronger condition that the classical theory be "deterministic" (i.e. for given $\omega \in \Omega, b \in B$ and $d \in D$, one and only one of the outcomes $(\beta,\delta) \in \bar{b} \times \bar{d}$ can occur.) The proof provides an explicit description of the convex set of quasi-classical correlation functions with fixed B and D in terms of its extreme points. This makes the verification of Bell's inequalities and their generalizations almost trivial.

Proposition 2: Let (p,B,D) be a finite quasi-classical correlation table. Then an embedding as required in definition 9 can be chosen such that in addition μ is a probability measure, $\bar{\rho} = 1$, and each of the functions $s(\beta)$ and $t(\delta)$ takes only the values 0 and 1. Moreover, p satisfies Bell's inequalities: For $\beta_1,\beta_2 \in \bar{B}, \delta_1,\delta_2 \in \bar{D}$:

$$0 \leq \bar{p}(\beta_1)+\bar{p}(\delta_1)+p(\beta_2,\delta_2)-p(\beta_1,\delta_1)-p(\beta_2,\delta_1)-p(\beta_1,\delta_2) \leq 1$$

Proof: We show that all quasi-classical correlation tables over given B and D can be realized on the same space Ω, with the same embedding (s,t): Let $\Omega = \underset{b\in B}{\mathsf{X}}\, \bar{b} \times \underset{d\in d}{\mathsf{X}}\, \bar{d}$. We may consider $\omega \in \Omega$ as a function ω: $B \cup D \to \bar{B} \cup \bar{D}$, assigning to each apparatus $b \in B$ (resp. $d \in D$) an outcome $\omega(b) \in \bar{b}$ (resp. $\omega(d) \in \bar{d}$). Define $s(\beta)$: $\Omega \to \{0,1\}$ by $s(\beta)(\omega) = 1$ if $\beta \in \bar{b}$ and $\omega(b) = \beta$ and $s(\beta)(\omega) = 0$ otherwise. ($t(\delta)$ is defined analogously). Hence $\Sigma_{\beta\in\bar{b}}\, s(\beta)(\omega) = 1$ and $\Sigma_{\beta\in\bar{d}}\, t(\delta)(\omega) = 1$. Hence every probability measure μ on Ω defines a quasi-classical correlation table p_μ by $p_\mu(\beta,\delta) = \sum_\omega \mu(\{\omega\})s(\beta)(\omega)t(\delta)(\omega)$. In particular, a normalized point measure μ defines a p_μ with $p_\mu(\beta,\delta) \in \{0,1\}$, which is clearly extremal in the convex set of correlation tables and a fortiori in the subset of quasi-classical tables.

We have to show now that, conversely, every quasi-classical correlation table is of the form p_μ. Suppose that $p, \tilde{\Omega}, \tilde{\Sigma}, \tilde{\mu}, \tilde{\tilde{\rho}}, \tilde{s}$ and $\tilde{t}$ satisfy definition 9. Replacing $\tilde{\mu}$ by $\tilde{\tilde{\rho}}\cdot\mu$ and s by $(\bar{\rho})^{-1}s$ (with the convention $^0/_0 = 0$), we may assume that $\tilde{\mu}$ is a probability measure and $\tilde{\tilde{\rho}} = 1$. Define the function ϕ: $\Omega\times\tilde{\Omega} \to \mathbb{R}$ by $\phi(\omega,\tilde{\omega}) := (\prod_{b\in B} \tilde{s}(\omega(b))(\tilde{\omega}))\cdot(\prod_{d\in D} \tilde{t}(\omega(d))(\tilde{\omega}))$.

Then $\sum_\omega s(\beta)(\omega)\cdot t(\delta)(\omega)\cdot\phi(\omega,\tilde{\omega}) = \tilde{s}(\beta)(\tilde{\omega})\cdot\tilde{t}(\delta)(\tilde{\omega})$ and consequently $p(\beta,\delta) = \int\tilde{\mu}(d\tilde{\omega})\tilde{s}(\beta)(\tilde{\omega})\, \tilde{t}(\delta)(\tilde{\omega}) = $
$= \sum_\omega \mu(\{\omega\})\cdot s(\beta)(\omega)\, t(\delta)(\omega) = p_\mu(\beta,\delta)$ with $\mu(\{\omega\}) = \int\tilde{\mu}(d\tilde{\omega})\phi(\omega,\tilde{\omega})$.
(The essential point here is to bring in the joint distribution of the variables $\tilde{s}(\beta),\tilde{t}(\delta)$ on $(\tilde{\Omega},\tilde{\mu})$.[7])

Since Bell's inequalities are linear in p, it suffices to check them on the extremal quasi-classical correlation tables, i.e. under the additional assumption that $\bar{p}(\beta),\bar{p}(\delta) \in \{0,1\}$ and $p(\beta,\delta) = \bar{p}(\beta)\cdot\bar{p}(\delta)$. The sixteen-fold case distinction is reduced considerably by symmetry. qed.

These inequalities have been found to be violated in experiments.(8) Consequently such experiments cannot be accounted for by any hidden variable theory conforming to the independence assumptions of definition 8. The fact that these experiments also provide a beautiful verification of quantum mechanical predictions is quite irrelevant for this argument, since our derivation of the inequalities makes no reference to quantum mechanics. This is a decisive advantage over other arguments against hidden variables of, say Gleason, Kochen-Specker, or von Neumann type.

What reasons can be given for our independence assumptions, beyond their plausibility? A limited experimental conformation comes from the fact that the data (p,B,D) form a correlation table. This is so obvious that it is not even mentioned in the published accounts of the experiments. However, our main assumption that Berta and Dora cannot communicate with each other, even if they have at their disposal every measuring device admitted by the full "hidden variable" theory, can never be verified experimentally. Hence we have to resort to theoretical arguments. We may argue that the principle of directed interaction, and the assumption that correlation experiments lead to results (p,B,D) satisfying definition 8, are necessary conditions for interpreting experimental results as results "about physical microsystems" and not merely results about some other ways in which macroscopical apparatuses may interact. In the above presentation I have implicitly adopted this point of view, according to which the independence assumptions have a normative aspect: if they were found to be violated in an experiment, the experiment would be discarded, or else the experimenter would try to eliminate the unintended interaction between his measuring devices. Note that these principles are implicitly assumed in the usual way of describing experiments in (relativistic or nonrelativistic) quantum mechanics (by assuming the effects $t(\delta)$ to sum up to the unit operator). In this sense the experimental tests of Bell's inequalities, by confirming quantum mechanical predictions, support our independence assumptions rather than casting doubt on them.

There is another powerful argument for our independence assumptions, which is completely independent of the previous ones: Einstein's principle of relativistic causality implies exactly this type of independence in the case that Berta and Dora manage to choose their apparatuses in spacelike separated regions of spacetime. (Even this can be done experimentally[9]!). Usually this argument goes under the heading of locality[10], from which the term "local hidden variable theory", for classical theories admitting a description of correlation experiment in accordance with definition 8, is derived.

We have seen in section III that any set of statistical data can be embedded in a classical theory. This means that the data can also be simulated by a perfectly macroscopic system[11] like e.g. a computer with two terminals B and D. We can see from the above results what kind of "non-locality" must be built into its program if the data are to violate Bell's inequalities. Suppose that Fig. 1 describes the directed flow of information between a central computer s and the two terminals. Then proposition 2 shows that there can be no violation of the inequalities. So let us assume that the channel between s and B allows some two-way communication. We are then in the situation described by Fig. 2, and proposition 1 shows that a simulation of the data is now possible. The question is whether a suitable measurement at D is capable of detecting which setting $b \in B$ has been chosen at the other terminal. Since the data by assumption form a correlation table, this is forbidden by the simulation program. According to that program, however, the terminal D makes only incomplete use of the information in the channel: proposition 2 implies that condition (2) of definition 9 must be violated. Hence a suitable average over the data from the channel $B \to D$ must depend on the choice of $b \in B$. (In the model used in the proof of proposition 1 such averages are given by the characteristic functions of the sets $\bar{b} \subset \bar{B} = \Omega$.)

I shall now give two alternative characterizations of quasi-classical correlation tables, which are based on slightly weaker assumptions and serve to illuminate further aspects of the problem. In the proof of proposition 2 we have not used the full structure of the classical duality. For example, the details of the "measure and integration theory of hidden variables" were irrelevant. Intuitively it is clear that only assumptions about the classical nature of that "part" of the embedding duality should be used, which is in the range of the embedding maps. Using the concept of coexistence introduced in definition 6, this will now be made precise.

Proposition 3: Let (p,B,D) be a finite correlation table, $\langle \mathcal{B}_1 \mathcal{D}_1 \rangle$ a directed statistical duality, and s: $B \to (\mathcal{B}_1)_+$, t: $D \to (\mathcal{D}_1)_+$ an embedding such that $\Sigma_{\beta \in \bar{b}}\, s(\beta) = \bar{W}$ and $\Sigma_{\delta \in \bar{d}}\, t(\delta) = 1$ are independent of b and d. Suppose that either $\{s: \bar{b} \to \mathcal{B}_1\}_{b \in B}$ is a coexistent set of preparators or that $\{t: \bar{d} \to \mathcal{D}_1\}_{d \in D}$ is a coexistent set of observables. Then (p,B,D) is quasi-classical.

Proof: By assumption there is a preparator W: $\times_{b \in B} \bar{b} \to \mathcal{B}_{1+}$ with marginals s: $\bar{b} \to \mathcal{B}_{1+}$. Identify $\times_{b \in B} \bar{b}$ with the set of functions $\omega: B \to \bar{B}$ with $\omega(b) \in \bar{b}$ and set $\Omega = \{\omega \in \times \bar{b} \mid \langle W(\omega), 1 \rangle \neq 0\}$. Let μ be the measure with $\mu(\{\omega\}) = \langle W(\omega), 1 \rangle$ and define the functions $\tilde{t}(\delta): \Omega \to \mathbb{R}$ and $\tilde{s}(\beta): \Omega \to \{0,1\}$ by $\tilde{t}(\delta)(\omega) = \langle W(\omega), 1 \rangle^{-1} \langle W(\omega), t(\delta) \rangle$ and $\tilde{s}(\beta)(\omega) = 1$ iff $\beta \in \bar{b}$ and $\omega(b) = \beta$. Then $\int_\Omega \mu(d\omega) \tilde{s}(\beta)(\omega) \tilde{t}(\delta)(\omega) = \Sigma_{\omega \in \Omega} \tilde{s}(\beta)(\omega) \langle W(\omega), t(\delta) \rangle = p(\beta, \delta)$.
Thus p is quasi-classical. The argument for the case that the observables are coexistent is analogous. qed.

This result contains some advice for the experimenter who wishes to detect a violation of Bell's inequalities: the observables corresponding to his measuring devices b and d must not be coexistent. (In the quantum mechanical case this implies that the associated operators must not commute.) Photon polarizers with different orientations satisfy this condition. He also has to choose a source appropriate for detecting this non-coexistence. This is shown by the next proposition, deducing the quasi-classical nature of a correlation table from a hypothesis about the internal structure of the source (Fig. 3).

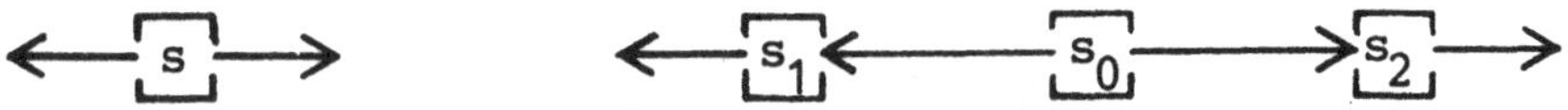

Fig. 3

Suppose that the source s contains a random generator s_0, whose signals determine the choice of preparing apparatuses s_1 and s_2[12]. These apparatuses produce physical systems, on which measurements are performed by means of the procedures from B and D respectively. Suppose that these systems, traveling from s_1 to b and s_2 to d, are described by dualities $\langle \mathcal{B}_1, \mathcal{D}_1 \rangle$ and $\langle \mathcal{B}_2, \mathcal{D}_2 \rangle$. Then the resulting correlation table p satisfies the hypothesis of the following proposition.

Proposition 4: Let (p,B,D) be a finite correlation table and $\langle \mathcal{B}_1,\mathcal{D}_1\rangle$, $\langle \mathcal{B}_2,\mathcal{D}_2\rangle$ directed dualities. Let t_1: $\bar{B}\to\mathcal{D}_{1+}$ and t_2: $\bar{D}\to\mathcal{D}_{2+}$ be maps such that for all b and d: $\Sigma_{\beta\in\bar{b}}\, t_1(\beta) = 1\in\mathcal{D}_1$ and $\Sigma_{\delta\in\bar{d}}\, t_2(\delta) = 1\in\mathcal{D}_2$. Suppose that for i = 1...N there are states $W_i^1 \in \mathcal{B}_1$, $W_i^2 \in \mathcal{B}_2$ and weights $\lambda_i \geq 0$ such that

$$p(\beta,\delta) = \sum_i \lambda_i \langle W_i^1, t_1(\beta)\rangle \cdot \langle W_i^2, t_2(\delta)\rangle$$

Then p is quasi-classical.

Proof: Since the set of quasi-classical correlation tables is convex, we may assume $p(\beta,\delta) = \langle W^1, t_1(\beta)\rangle \cdot \langle W^2, t_2(\delta)\rangle$. Evidently this may be embedded in the sense of definition 9 into a measure space consisting of a single point. qed.

In the case that $\langle \mathcal{B}_i,\mathcal{D}_i\rangle = \langle \mathcal{T}(\mathcal{H}_i),\ \mathcal{B}(\mathcal{H}_i)\rangle$ (i = 1,2) are quantum dualities, the hypothesis is equivalent to the statement that the state $W\in\mathcal{T}(\mathcal{H}_1\otimes\mathcal{H}_2)$ of the composite system which describes the source's is of the form $W = \Sigma\lambda_i W_i^1 \otimes W_i^2$. Thus the violation of Bell's inequalities shows that there are quantum states for a composite system, which are not convex combinations of uncorrelated states $W^1 \otimes W^2$. (This is in sharp contrast to the situation where at least one of the dualities $\langle \mathcal{B}_1,\mathcal{D}_1\rangle$, $\langle \mathcal{B}_2,\mathcal{D}_2\rangle$ is classical.) The existence of such states is an essential feature of quantum mechanics. Often the states of a composite system which are most easily prepared, like the ground state of an atom, are properly "quantum correlated". The time evolution of a composite system always transforms classically correlated states $\Sigma\lambda_i W_i^1 \otimes W_i^2$ into quantum correlated states, unless of course the evolution operator $\mathcal{U}_t$ decomposes into $\mathcal{U}_t^1 \otimes \mathcal{U}_t^2$ and there is no interaction at all(13).

5. REDUCTION OF STATISTICAL DUALITIES

Before defining reduction relations between statistical dualities, it is useful to consider the relationship between the "observational" level of frequency functions and the "theoretical" level of statistical dualities in more detail. Suppose we are given two measuring devices and want to decide whether the associated observables are coexistent or not. We may try to do this by testing the two devices (and the candidates for devices jointly measuring the two) against some class of preparing apparatuses. But no matter what finite or infinite set of preparing procedures we choose, proposition 1 shows that there will be a classical theory to account for the measured data, in which the two observables are necessarily coexistent, and it is equally simple to show that there will be another duality in which they are not. Consequently the coexistence relation cannot be decided operationally, even though it is explicitly defined in terms of measurable quantities.

The reason is that the definition of coexistence contains an equation between effects, and equality of effects was defined as statistical equivalence with respect to "all" preparing procedures. Thus the theoretical concepts "states", "effects", "observables", "preparators" and "coexistence" are all defined relative to an assumed totality of preparing and measuring procedures. Such concepts, whose definition involves universal quantification over basic interpreted sets ("Bildmengen") occur very frequently in the reconstruction of physical theories in Ludwig's scheme. Proposition 3 is perhaps a typical of example how a theoretical relation like the coexistence of observables can nevertheless be tested experimentally: the data (violating Bell's inequalities) must be supplemented by "theoretical assumptions", which by themselves can also not be tested experimentally (in this case that the states $\Sigma_{\beta\in\bar{b}}\, s(\beta)$ are all equal). In that sense every experimental test of the coexistence relation is theory-dependent.

The successive introduction of such theoretical assumptions made it necessary to distinguish between different concepts of embedding frequency functions into statistical dualities. The same will be true for the definition of embeddings from one duality $<\mathcal{B}_1,\mathcal{D}_1>$ into another $<\mathcal{B}_2,\mathcal{D}_2>$. Again we shall start on the level of the frequency functions (p_i,B_i,D_i), from which these dualities are constructed, with the straightforward "homogeneous reduction relations"(14) $B_1 \subset B_2$, $D_1 \subset D_2$ and $p_1 = p_2|\bar{B}_1 \times \bar{D}_1$. As before these inclusions induce an embedding (s,t) of the frequency function (p_1,B_1,D_1) into the duality $<\mathcal{B}_2,\mathcal{D}_2>$, but now (p_1,B_1,D_1) is only indirectly given through $<\mathcal{B}_1,\mathcal{D}_1>$. However, since the maps $s_1: \mathrm{lin}\bar{B}_1 \to \mathcal{B}_1$, $t_1: \mathrm{lin}\bar{D}_1 \to \mathcal{D}_1$ described in section 2 are surjective, we may choose a representative $\tilde{s}_1(W) \in \mathrm{lin}\bar{B}_1$ for each equivalence class $W \in \mathcal{B}_1$ and hence obtain a map $s: W \to s_2\tilde{s}_1(W) \in \mathcal{B}_2$ and a similar map $T: D_1 \to \mathcal{D}_2$ satisfying the following definition:

Definition 10: An embedding of a statistical duality $\langle \mathcal{B}_1, \mathcal{D}_1 \rangle$ into a statistical duality $\langle \mathcal{B}_2, \mathcal{D}_2 \rangle$ is a pair of maps $S: \mathcal{B}_{1+} \to \mathcal{B}_{2+}$, $T: \mathcal{D}_{1+} \to \mathcal{D}_{2+}$ such that for $W \in \mathcal{B}_{1+}$, $F \in \mathcal{D}_{2+}$: $\langle S(W), T(F) \rangle_2 = \langle W, F \rangle_1$.

Again any duality $\langle \mathcal{B}_1, \mathcal{D}_1 \rangle$ can be embedded into a classical duality. It suffices to take for the hidden parameter space Ω the set of positive normalized linear functionals on $\mathcal{D}_1$, for T the evaluation map and for S(W) the multiple $\langle W, 1 \rangle$ of the point measure at the normalized functional $\langle W, 1 \rangle_1^{-1} \cdot W$ (We could also use for Ω the set of extreme points of the above Ω, and for S(W) a suitable boundary measure provided by Choquet-theory.)

The status of a classical duality $\langle \mathcal{B}_2, \mathcal{D}_2 \rangle$ as a hidden variable theory depends crucially on the assumptions about the measurability of effects $F_2 \in \mathcal{D}_2$ not of the form $T(F_1)$. (In the above case these are all non-linear functions on Ω.) If such effects are held to be unobservable in principle, the choice between $\langle \mathcal{B}_1, \mathcal{D}_1 \rangle$ and $\langle \mathcal{B}_2, \mathcal{D}_2 \rangle$ is simply a matter of taste: If we omit the unobservables from $\mathcal{D}_2$ and identify indistinguishable states in $\mathcal{B}_2$, we regain $\langle \mathcal{B}_1, \mathcal{D}_1 \rangle$. Such a "forbidden variable theory" tends to be mathematically more complicated than $\langle \mathcal{B}_1, \mathcal{D}_1 \rangle$ but some might still find it more intuitive. This would be especially true if some intrinsic reason for the unobservability of the effects $F_2 \in \mathcal{D}_2 \setminus T(\mathcal{D}_1)$ could be given[15]. On the other hand, if some of the effects in $\mathcal{D}_2 \setminus T(\mathcal{D}_1)$ are considered merely hidden, but observable in principle, a radical change of the theory $\langle \mathcal{B}_1, \mathcal{D}_1 \rangle$ is proposed. Such proposals have frequently been made dor quantum mechanics, often without due consideration of the far reaching consequences[16]. If (S,T) is an embedding of $\langle \mathcal{B}_1, \mathcal{D}_1 \rangle$ into $\langle \mathcal{B}_2, \mathcal{D}_2 \rangle$ and (s_1, t_1) is an embedding of the frequency function (p,B,D) into $\langle \mathcal{B}_1, \mathcal{D}_1 \rangle$, then $(S \circ s_1, T \circ t_1)$ embeds (p,B,D) into $\langle \mathcal{B}_2, \mathcal{D}_2 \rangle$. However, even if t_1 satisfies the condition $\Sigma_{\delta \in \bar{d}}\, t_1(\delta) = 1$ of definition 7, $\Sigma_{\delta \in \bar{d}}\, T(t_1(\delta))$ may in general differ from 1. Clearly T preserves this condition iff it commutes with sums and $T(1) = 1$, i.e. iff T extends to a linear operator $T: \mathcal{D}_1 \to \mathcal{D}_2$. Similarly S preserves the condition that $\Sigma_{\beta \in b}\, s_1(\beta)$ is independent of b iff it extends to a linear operator $S: \mathcal{B}_1 \to \mathcal{B}_2$. Hence the counterpart on the level of dualities of the independence assumptions used to derive Bell's inequalities is the following concept:

Definition 11: A linear embedding of a directed duality $<\mathcal{B}_1,\mathcal{D}_1>$ into a directed duality $<\mathcal{B}_2,\mathcal{D}_2>$ is a pair of positivity preserving linear maps S: $\mathcal{B}_1 \to \mathcal{B}_2$ and T: $\mathcal{D}_1 \to \mathcal{D}_2$ such that $T(1) = 1$ and $<S(W),T(F)>_2 = <W,F>_1$.

An especially simple kind of linear embeddings is the case where $\mathcal{B}_1 = \mathcal{B}_2$ and $\mathcal{D}_1 = \mathcal{D}_2$ as vector spaces, S and T are the identity maps but $\mathcal{B}_{2+} \supset \mathcal{B}_{1+}$ and $\mathcal{D}_{2+} \supset \mathcal{D}_{1+}$. We call a duality $<\mathcal{B}_1,\mathcal{D}_1>$ saturated if the cones $\mathcal{B}_{1+},\mathcal{D}_{1+}$ are maximal, so that the above inclusions imply $\mathcal{B}_{2+} = \mathcal{B}_{1+}$, $\mathcal{D}_{2+} = \mathcal{D}_{1+}$. The cones $\mathcal{B}_{1+}$, $\mathcal{D}_{1+}$ are then duals of each other. Saturation is equivalent to Ludwig's axiom about the sensitivity increase of one effect, which states that whatever can be measured or prepared can in a certain sense be measured or prepared with optimal efficiency.[17]

For the embedding sketched after definition 10 the map T was linear but S, taking any mixed state into a pure state, was not. The following counterpart of proposition 2 shows that this non-linearity is essential.

Proposition 5: Let $<\mathcal{B},\mathcal{D}>$ be a saturated duality with $\mathcal{B}$ and $\mathcal{D}$ finite dimensional. Suppose that $<\mathcal{B},\mathcal{D}>$ can be linearly embedded into a classical duality. Then $<\mathcal{B},\mathcal{D}>$ is classical.

Proof: Since $\mathcal{B}$ and $\mathcal{D}$ are finite dimensional, $\mathcal{D}$ can be identified with the dual $\mathcal{B}'$ of $\mathcal{B}$ and saturation implies that $\mathcal{B}' = \mathcal{D}$ as ordered vector spaces. Let (S,T) be a linear embedding into $<L^1(\Omega,\mu),L^\infty(\Omega,\mu)>$. Denote the adjoint of S by $S': L^\infty \to \mathcal{B}' = \mathcal{D}$. Then S'T is the identity on $\mathcal{D}$. Now let $F_1,F_2 \in \mathcal{D}$. Then $T(F_1)$ and $T(F_1)$ have a least upper bound $F \in L^\infty$ and it is easy to see that S'(F) is a least upper bound for F_1 and F_2. Consequently $\mathcal{D}$ is a vector lattice, which must be isomorphic to $\ell^\infty(X)$ for a set X with card $(X) = \dim \mathcal{D}$. Saturation implies $\mathcal{B} = \ell^1(X)$. The infinite dimensional generalizations of this theorem require a more careful analysis beyond the scope of this paper. qed.

The most striking feature of Bell's inequalities is that they make possible an experimental test of some fundamental structures of quantum mechanics. It is therefore natural to ask, whether other questions concerning the foundations of quantum mechanics can be decided in a similar way. In particular one might try to justify the use of complex Hilbert spaces in quantum mechanics rather than Hilbert spaces over the reals or the quaternions[18]. If $\mathbb{F}$ is one of the fields $\mathbb{R}$, $\mathbb{C}$, or $\mathbb{H}$ and n is a cardinal, let $\Delta(\mathbb{F},n)$ denote the duality $\langle \mathcal{T}_h(\mathbb{F}^n), \mathcal{B}_h(\mathbb{F}^n) \rangle$ of hermitian trace class and hermitian bounded operators on the n-dimensional Hilbert space over $\mathbb{F}$. We write $\Delta_1 \to \Delta_2$ if the duality Δ_1 can be embedded linearly into Δ_2.[19] Then we have the following scheme of reduction relations:

Proposition 6: (1) For $\mathbb{F} = \mathbb{R}, \mathbb{C}, \mathbb{H}$ and $n \geq m$: $\Delta(\mathbb{F},n) \to \Delta(\mathbb{F},m)$
(2) $\Delta(\mathbb{R},n) \to \Delta(\mathbb{C},n) \to \Delta(\mathbb{H},n)$
(3) $\Delta(\mathbb{H},n) \to \Delta(\mathbb{C},2n)$; $\Delta(\mathbb{C},n) \to \Delta(\mathbb{R},2n)$

Proof: (1) is trivial. For (2) and (3) recall that quaternions can be considered as pairs (z_1,z_2) of complex numbers with the multiplication $(y_1,y_2)(z_1,z_2) = (z_1 y_1 - z_2 \bar{y}_2; z_1 y_2 + z_2 \bar{y}_1)$ and the conjugation $(z_1,z_2)^* = (\bar{z}_1,-z_2)$. $\mathbb{C}$ will be identified with a subfield of $\mathbb{H}$ by $z \to (z,0)$.
(2) Choose a basis in each $\mathbb{F}^n$, so that all operators can be considered as $\mathbb{F}$-valued matrices. Then each $\mathbb{R}$-matrix can be considered as a $\mathbb{C}$-matrix and each $\mathbb{C}$-matrix as an $\mathbb{H}$-matrix, and it is clear that these correspondences preserve algebraic operations, positivity and trace.
(3) We can consider $\mathbb{H}^n$ (resp. $\mathbb{C}^n$) as a normed vector space over $\mathbb{C} \subset \mathbb{H}$ (resp. $\mathbb{R} \subset \mathbb{C}$), which is then a Hilbert space of dimension 2n. An $\mathbb{H}$-linear operator A on $\mathbb{H}^n$ thus becomes a $\mathbb{C}$-linear operator T(A) on $\mathbb{C}^{2n}$. Again this correspondence preserves algebraic operations and positivity and the map $S(W) = \frac{1}{2}T(W)$ satisfies $\mathrm{tr}\, S(W) = \mathrm{tr}\, W$ and hence $\mathrm{tr}\, S(W)T(A) = \mathrm{tr}\, S(WA) = \mathrm{tr}\, WA$. qed.

In the case of infinite n we have 2n = n, so that each of the three dualities $\Delta(\mathbb{F},n)$ can be reduced to each of the other two, although these dualities are not isomorphic. Consequently no finite or infinite set of statistical measurements is capable of discriminating between these theories, and since the embeddings were linear this statement remains true even if the data are supplemented with any number of independence assumptions of the type discussed in sections 3 and 4.[20] If we want to justify the use of complex Hilbert spaces in quantum mechanics we therefore have to introduce assumptions concerning further structures of the theories like, for example, the representation of physical symmetry groups.

This program can be carried out in complete analogy to the above treatment of the principle of directed interaction and its consequences: Suppose that for each g in the symmetry group G and each $b \in B$, $d \in D$ there are transformed procedures

$gb \in B$ and $gd \in D$ with outcome sets $g\bar{b} = \{g\beta \mid \beta \in \bar{b}\}$ and gd, such that $p(g\beta, g\delta) = p(\beta, \delta)$. In the duality $\langle \mathcal{B}, \mathcal{D} \rangle$ constructed from (p, B, D) this induces representations $U: G \to Aut(\mathcal{B})$, $V: G \to Aut(\mathcal{D})$ such that $\langle U(g)W, V(g)F \rangle = \langle W, F \rangle$. The natural constraint for for embeddings (s,t) of a frequency function (p,B,D) into a duality $\langle \mathcal{B}_1, \mathcal{D}_1 \rangle$ then becomes $U_1(g)s(\beta) = s(g\beta)$ and $V_1(g)t(\delta) = t(g\delta)$. Again this assumption cannot be verified experimentally, since it involves universal quantification over the sets of procedures from which $\langle \mathcal{B}_1, \mathcal{D}_1 \rangle$ is assumed to be constructed. Just as the independence assumptions of section 4 suggested the linearity of embedding maps between dualities, this new constraint on embeddings of data into theory suggests the covariance conditions $U_2(g)S(W) = S(U_1(g)(W)$ and $V_2(g)T(F) = T(V_1(g)F)$ for embeddings (S,T).

In the case of spin or polarisation experiments we may take the symmetry group G to be the rotation group. However, it is easy to see that the embeddings $\Delta(\mathbb{F}_1, n_1) \to \Delta(\mathbb{F}_2, n_2)$ constructed in proposition 6 are naturally covariant in the sense that for each representation of the rotation group in $\Delta(\mathbb{F}_1, n_1)$ there is a representation in $\Delta(\mathbb{F}_2, n_2)$ making the embedding covariant.[21] Hence not even the assumption of rotation covariance for the embedding of a frequency function (p,B,D) into the dualities $\Delta(\mathbb{F}, n)$ is strong enough to admit an experimental discrimination between the fields $\mathbb{R}$, $\mathbb{C}$ and $\mathbb{H}$.

Reinhard Werner
Universität Osnabrück, Fachbereich Physik
Postfach 4469, D-4500 Osnabrück

NOTES

1. These inequalities were introduced in Bell (1966). In the sequel I shall use the term "Bell's inequalities" for the whole class of related results[7][13]. Important early generalizations of Bell's original inequalities are to be found in Clausner and Shimony (1969).
2. See Ludwig (1978) for his general reconstruction scheme for physical theories. My presentation of statistical theories in sections 2 and 3 is adapted from Ludwig (1983) and earlier works.
3. Compare Ludwig (1983), p. 8. and axiom APS 7., and Capra (1975), p. 121f.
4. Ludwig's definition of effects (Ludwig (1983), p. 43) is slightly more restrictive. He requires F to be a convex combination of effects of the form $\Sigma_{\delta \in \tilde{\delta}}\, t(\delta)$ with $\tilde{\delta} \subset \bar{d}$,

$d \in D$. Assuming his axiom about the sensitivity increase of one effect (Ludwig (1970), p.275) the two definitions coincide. Note that in the quantum case the effects are precisely the operators $0 \leq F \leq 1$ and not necessarily projections.

5. More general preparators and observables are given by $\mathcal{B}_+$-valued and $\mathcal{D}_+$-valued measures respectively. I use this discrete version in order to minimize measure theoretic technicalities.
6. Cf. Werner (1982), p.52ff.
7. This has been pointed out by Fine (1982b). In Fine (1982a) he demonstrated that "hidden variable" models for any set of data can be constructed, provided the random variables $s(\beta)$, $t(\delta)$ are not required to be defined on a common space (Ω,μ). His "minimal prism models" are precisely of the type discussed in section 2: The "hidden" variables are the observed outcomes. The general form of the "problem of joint distributions" is the following: Given measurable spaces $(\Omega_i,\Sigma_i)_{i\in I}$ and a family M of subsets of I together with a consistent family of measures μ_m $(m\in M)$ on $X_{i\in m}\Omega_i$. When can M be extended to include I? The set of families μ admitting such an extension is completely characterized by a family of "Bell's" inequalities", which are linear in μ. Even in the present case (where M is a bipartite graph) the complete set of such inequalities is not known.
   Cf. Garg and Mermin (1982).
8. See Aspect et al. (1982a).
9. See Aspect et al. (1982b).
10. This argument loses some of its force when one observes that neither quantum field theory nor hidden variable theories like Nelson's stochastic mechanics have yet been given satisfactory relativistic formulations. However, the existing free quantum field theories should suffice to account for the cited experiments. It would be interesting to know whether the vacuum state in such theories already violates Bell's inequalities for suitably chosen spacelike localized observables (S.J. Summers, R.W.: work in progress). For a discussion of Einstein causality in relativistic quantum mechanics see Neumann and Werner (1983).
11. The example of such a system given by Aerts (1982a) is weaker than necessary: the measurement results themselves already violate the independence assumptions of definition 8.

12. This situation is reminiscent of Reichenbach's principle of common cause. A discussion of Bell's inequalities based on Reichenbach's principle was given by van Fraassen (1982). Proposition 4 can be reformulated in those terms.

13. There is a very close connection between Bell's inequalities and the theory of tensor products, i.e. the problem of choosing a positive cone $(\mathcal{B}_1 \otimes \mathcal{B}_2)_+$ for the algebraic tensor product $\mathcal{B}_1 \otimes \mathcal{B}_2$ of ordered vector spaces $(\mathcal{B}_i, \mathcal{B}_{i+})$ $(i = 1,2)$. (See Wittstock (1973)). The cone generated by $\{W_i \otimes W_2 | W_i \in \mathcal{B}_{i+}\}$ is called the projective cone. The set of positive multiples of quasi-classical correlation tables is precisely the projective cone in the tensor product $\tilde{\mathcal{B}} \otimes \tilde{\mathcal{D}}$, where $\tilde{\mathcal{B}}$ is the space of $\bar{p}: B \to \mathbb{R}, \Sigma_{\beta \in \bar{b}}\, \bar{p}(\beta)$ independent of b, with pointwise ordering ($\tilde{\mathcal{D}}$ is defined analogously). Hence the full set of "Bell inequalities" is given by the extreme rays of the dual cone, which is called the injective cone in the tensor product of the dual spaces. The positive cone of $T(H_1 \otimes H_2) \simeq T(H_1) \otimes T(H_2)$ lies properly between the projective and injective cones. A system composed of two subsystems described by dualities $\langle \mathcal{B}_i, \mathcal{D}_i \rangle$ should be described by the duality $\langle \mathcal{B}_1 \otimes \mathcal{B}_2, \mathcal{D}_1 \otimes \mathcal{D}_2 \rangle$ - with a suitable choice of cones. There is no ambiguity in case one of the dualities $\langle \mathcal{B}_i, \mathcal{D}_i \rangle$ is classical. For describing separated experiments, it suffices to take the projective cones in both $\mathcal{B}1 \otimes \mathcal{B}_2$ and $\mathcal{D}_1 \otimes \mathcal{D}_2$. The resulting duality will then fail to be saturated (see section 5) and hence fail to satisfy one of the axioms of quantum mechanics. In a different mathematical framework this has been noted by Aerts (1982b). This can hardly be held against the usual description of composite systems in quantum mechanics, which is designed to describe not only separated systems, but interacting systems as well.

14. Cf. Moulines' article in this volume.

15. For an example how restrictions on the measurability of quantum observables can be motivated by an analysis of measuring processes, see Heinz-Jürgen Schmidt (1980).

16. Mielnik (1974) has pointed out that the non-measurability of non-linear functions on the convex set of quantum states is essential for quantum mechanics. Non-linear Schrödinger equations violate this principle and hence belong to a hidden variable theory in which the hidden parameter space is the set of pure quantum states. For another outrageous example see Helmut Schmidt (1982). Nelson (1983) shows that the measurement of the "hidden"

random variables in his stochastic mechanics requires a measuring apparatus which is not in "quantum equilibrium" with the system and discusses gravitational interactions as a potential candidate.

17. Cf. Werner (1982), p.43f.
18. This problem arises in Ludwig's axiomatic approach to quantum mechanics as well as in lattice theoretic approaches. (See Piron (1976)).
19. This scheme may easily be extended to include all dualities $\langle \mathcal{B},\mathcal{D}\rangle$ with $\mathcal{D}$ a finite dimensional Jordan algebra and $\mathcal{B}$ its dual. (Jordan (1934)). Using Kadison's inequality one can show that any saturated duality, linearly embedded into a Jordan algebra can be equipped with a Jordan product. (Effros and Størmer (1979)). It also follows from the work of Effros and Størmer that the canonical bad apple of the Jordan algebraic approach, the algebra of 3×3 matrices over the Cayley octonians, cannot be properly embedded into another Jordan algebra. (i.e. every linear embedding is an isomorphism with a direct summand.)
20. This seems to contradict Accardi and Fedullo (1982). They work with "transition probabilities" $|\langle\varphi|\psi\rangle|^2$, whose physical interpretation is rather dubious. Their assumptions imply that two-valued observables exist only in two dimensional Hilbert spaces, so that they cannot formulate an analog of the embedding $\Delta(\mathbb{C},2) \to \Delta(\mathbb{R},4)$. Mathematically their work is based on the observation that there are configurations of angles in a ball (i.e. the state space of $\Delta(\mathbb{C},2)$) which cannot be realized in a disc (i.e. the state space of $\Delta(\mathbb{R},2)$).
21. By Wigner's theorem representations of a group in $\mathit{Aut}\Delta(\mathbb{F},n)$ are unitary representations on $\mathbb{F}^n$ up to a factor. If the factor can be chosen to be ±1 as for the rotation group, the embeddings of proposition 6 can easily be made covariant. For an introduction to group representations in quaternion quantum mechanics, see Emch (1965).

REFERENCES

Accardi, L. and Fedullo, A.: 1982, 'On the Statistical Meaning of Complex Numbers in Quantum Mechanics', Lett. Nuovo Cim. 34, 161-172.

Aerts, D.: 1982a, 'Example of a Macroscopical Classical Situation that violates Bell's Inequalities',

Lett. Nuovo Cim. 34, 107-111.
Aerts, D.: 1982b, 'Description of Many Separated Entities Without the Paradoxes Encountered in Quantum Mechanics', Found. Phys. 12, 1131-1170.
Aspect, A., Grangier, P. and Roger, G.: 1982a, 'Experimental Realization of Einstein-Podolsky-Rosen-Bohm Gedankenexperiment: A New Violation of Bell's Inequalities', Phys. Rev. Lett. 49, 91-94.
Aspect, A., Dalibard, J. and Roger, G.: 1982b, 'Experimental Test of Bell's Inequalities Using Time-Varying Analyzers', Phys. Rev. Lett. 49, 1804-1807.
Bell, J.S.: 1966, 'On the problem of hidden variables in quantum mechanics', Rev. Mod. Phys. 38, 447-452.
Capra, F.: 1975, 'The Tao of Physics' (Bantam, New York)
Clausner, J.F. and Shimony, A.: 1978, 'Bell's theorem: experimental tests and implications', Rep. Prog. Phys. 41, 1881-1927.
Effros, E. and Størmer, E.: 1979, 'Positive Projections and Jordan Structure in Operator Algebras', Math. Scand. 45, 127-138.
Emch, G.G.: 1965, 'Representations of the Lorentz group in quaternionic quantum mechanics', In: W.E. Brittin, A.O. Barut (eds.): Lectures in Theoretical Physics 7 A. (Univ. Colo. Press, Boulder).
Fine, A.: 1982a, 'Some Local Models for Correlation-Experiments', Synthese 50, 279-294.
Fine, A.: 1982b, 'Joint distributions, quantum correlations, and commuting observables', J. Math. Phys. 23, 1306-1310.
Fraassen van, B.C.: 1982, 'The Charybdis of Realism: Epostemological Implications of Bell's inequality', Synthese 52, 25-38.
Garg, A. and Mermin, N.D.: 1982, 'Correlation Inequalities and Hidden Variables', Phys. Rev. Lett. 49, 1220-1223.
Ludwig, G.: 1970, 'Deutung des Begriffs "physikalische Theorie" und axiomatische Grundlegung der Hilbertraumstruktur der Quantenmechanik durch Hauptsätze des Messens', (Springer, New York).
Ludwig, G.: 1978, 'Die Grundstrukturen einer physikalischen Theorie', (Springer, New York)
Ludwig, G.: 1983, 'Foundations of Quantum Mechanics I', (Springer, New York)
Mielnik, B.: 1974, 'Generalized Quantum Mechanics', Commun. Math. Phys. 37, 221-256.
Nelson, E.: 1983, 'Quantum Fluctuations', (Mimeographed lecture notes, Geneva).

Neumann, H. and Werner, R.: 1983, 'Causality between Preparation and Registration Processes in Relativistic Quantum Theory', Int. J. Theoret. Phys. 22, 781-802

Piron, C.: 1976, 'Foundations of Quantum Physics', (Benjamin, Reading)

Schmidt, H.-J.: 1980, 'Der Zusammenhang von Wechselwirkung und Meßbarkeit am Beispiel gekoppelter Spinsysteme', In: P. Mittelstaedt, J. Pfarr (eds.): Grundlagen der Quantentheorie, (BI-Verlag, Mannheim)

Schmidt, H.: 1982, 'Collapse of the State Vector and Psychokinetic Effect', Found. Phys. 12, 565-581.

Werner, R.: 1982, 'The Concept of Embeddings in Statistical Mechanics', (Doctoral dissertation, Marburg)

Wittstock, G.: 1974, 'Ordered Normed Tensor Products', In: A. Hartkämper, H. Neumann (eds.): Foundations of Quantum Mechanics and Ordered Linear Spaces, (Springer, New York).

NAME INDEX